International
Telecommunications Handbook

International Telecommunications Handbook

Rob Frieden

Artech House, Inc.
Boston • London

Library of Congress Cataloging-in-Publication Data
Frieden, Rob.
    International telecommunications handbook / Rob Frieden.
        p.   cm.
    Includes bibliographical references and index.
    ISBN 0-89006-568-3 (alk. paper)
        1. Telecommunication policy.   2. Telecommunication.
    3. International trade.   I. Title.
    HE7645.F74   1995
    384—dc20   95-41686
                                                                 CIP

British Library Cataloguing in Publication Data

Frieden, Rob
    International Telecommunications Handbook
    I. Title
    621.382

    ISBN 0-89006-568-3

© **1996 ARTECH HOUSE, INC.**
685 Canton Street
Norwood, MA 02062

International Standard Book Number: 0-89006-568-3
Library of Congress Catalog Card Number: 95-41686

10   9   8   7   6   5

For a complete listing of the *Artech House Telecommunications Library*,
turn to the back of this book.

This book is the result of my family's joint decision to make it possible for me to pursue a career in academics. I am grateful for their support and for the comments, advice, and suggestions I received from Professors Jerry Brock, Wilson Dizard, Eli Noam, Joe Pelton, and Chris Sterling, and from a number of unknown reviewers.

Contents

Introduction: Understanding International Telecommunications

1.1 OVERVIEW

After years of stable obscurity and predictability, international telecommunications grows increasingly complex and volatile. Today we witness an approach more appropriate for *Let's Make a Deal* than for the post office or telephone company. Entrepreneurs offer satellite parking places for sale, nations run clearance sales to privatize their telephone company services, and hundreds of companies search the world for new market opportunities. Just a few years ago, a small club of carriers held what seemed to be permanent and exclusive franchises.

The study of international telecommunications used to involve a relatively straightforward process of learning the "system is the solution" world of public utilities. Telecommunications carriers willingly traded unlimited freedom to earn profits for predictability and insulation from competition and most financial risks. Customers were neither right nor sovereign. They took what carriers offered and had few options with regard to price, service, and choice of carrier.

If a multinational telecommunications firm maintained such an attitude today, it would have a hard time maintaining a lock even on domestic markets. Technological innovation and changes in policy and regulation make it possible for competitors to vie for previously captive customers. Incumbent carriers have less ability to impose restrictive policies, rules, and prices. Fast-changing technology, new market access initiatives, and liberalizing policies have combined to force sleepless nights on incumbent regulators and managers.

Such a shock to the system provides interesting case studies, compared to the long-established old school scenario, which lacked drama and a diverse cast of characters. Although most nations continue to support exclusive local and international telecommunications franchises, many have allowed market entry in niche markets, such as mobile services, like cellular radio, and services that add enhancements to leased lines, like financial services, airline reservation networks, and information processing systems.

Why do most policy makers maintain monopolies, at least for some key markets, even though they recognize that converging and evolving technologies make new market entry possible? In the jargon of the economist, some markets have become *contestable*, that is, ripe for future market entry, and the sleepy, insular world of public utility telecommunications has become but a memory. But other core markets remain natural monopolies, which are most efficiently provided by one enterprise that can operate at the lowest per unit cost. Even when there is no evidence supporting a monopoly, governments typically hesitate to upset the status quo, for fear that doing so would risk financial damage to incumbent operators.

1.2 AN INTERDISCIPLINARY UNDERTAKING

Analysis of international telecommunications and trade policy requires mental nimbleness. One must keep track of numerous, seemingly countervailing factors that quickly change and involve a wide range of disciplines. These characteristics are complicated because few people have much training in several of the fields involved, and most educational programs have yet to pursue an interdisciplinary approach. One may face new situations and work assignments without knowing what to expect and how to cope. Worse yet, one may lack even threshold credentials for employment. Corporate downsizing and pressures to economize place a premium on the ability of employees to perform several functions.

Universities have just begun to address whether and how to support an introductory international telecommunications course. Few undergraduate programs have mastered the technique of blending previously discrete course tracks into an eclectic approach for developing competency and skills in the areas that a career will tap over time. Does such a course belong in political science, because it involves government ownership and policy making? Does it belong in economics, because the driving force for deregulation and market entry involves application of economic principles? If such a course finds its way into communications, how do issues like access to satellite parking places, infrastructure development, spectrum management, and service tariffing mesh with traditional courses in broadcasting and journalism?

This book attempts to identify both the broad subjects that international telecommunications careers will address, as well as some of the more focused substantive issues currently debated in policy-making forums, courts, and the marketplace. The book also provides some insight into the kinds of skills required in telecommunications and information-processing careers. Telecommunications studies involve an integrated set of subjects, not always viewed as complementary or within the same ambit of investigation. For example, one could assess a nation's decision regarding whether or not to embrace procompetitive deregulation in terms of:

- *Political* policies, including a national commitment to make service available throughout a nation regardless of traffic density and the potential to recoup investments;
- *Economic* policies favoring the view that telecommunications is a natural monopoly, and that the incumbent should retain exclusive control of all or most service sectors;
- *Industrial* policies supporting the insulation of domestic manufacturers and service providers from foreign competition;
- *Technology* policies endorsing the view that no single enterprise can provide all available innovations at an optimal scale;
- *Social* policies mandating government intervention to support universal service objectives or reliance on the free interplay of market forces even if some prices rise;
- *Laws* permitting market entry or reserving markets to a regulated monopoly;
- *Treaties* or other multilateral commitments precluding procompetitive initiatives;
- *Foreign relations* concerns about international comity may temper or foreclose efforts to foster competition, particulary if there is a chance that deregulation will force lesser developed nations to incur higher costs;
- *National security* interests precluding market access liberalization;
- *Labor relations* issues jeopardizing any initiative that could result in reduced employment in the telecommunications sector.

Without considering telecommunications issues in terms of their numerous component parts, one may fail to address all the elements that stimulated formation of a particular policy and why particular technologies and companies succeeded where others failed. Strategies for successful national or corporate policies use balanced assessments, made possible when the participants are well-versed in most, if not all, of the components involved.

Here is a partial list of the disciplines typically involved in international telecommunications and trade policy.

1.2.1 Economics

Policy makers increasingly rely on economic analysis to evaluate arguments about efficiency and equity. Telecommunications policy often constitutes one of the many aspects of a nation's overall campaign to spur industrial development and employment. Initiatives such as a decision to sell some or all of the government monopoly telephone company may parallel similar decisions in other industries (e.g., civil aviation, petroleum, and public utilities like electric companies).

Economic analysis provides the key for calculating social welfare enhancements and efficiency gains. The forum for such decision making will often be the legislature (or ministry of posts and telecommunications), where economic arguments about the virtues of private enterprise and competition are pitted against the political clout of incumbent institutions like the post, telegraph, and telephone (PTT) administration and labor unions.

Economic analysis constitutes a key method for assessing whether and how to reform the telecommunications sector. Therefore, anyone seeking full understanding of this field should comprehend fundamental economic concepts. Will a monopoly administration generate economies of scale and scope, putting it in the best position to achieve such social goals as ubiquitous and affordable telephone service? Or will the monopoly exploit its market power to preempt competition made possible by technological innovations?

Can an unregulated marketplace achieve social goals relating to the price and availability of service, or must government mandate subsidies and prescribe rates of return? Is telecommunications a public good that unregulated private enterprises will not produce in socially desirable quantities at affordable prices? Can the marketplace achieve a balance matching supply with demand? Can policy makers rely on competition to diversify services, reduce rates, and stimulate demand? Will short-term price competition result in destructive competition, where the retained earnings of the incumbent enable it to underprice services and survive an unprofitable period after which it can recoup the losses when competitors exit the market? Will market entrants "creamskim" (i.e., serve only the most profitable routes), leaving the incumbent as the only carrier available for service to low-income users or those in remote locations?

1.2.2   Law

Lawyers dominate telecommunications and trade policy making. While often lacking training in economics, engineering, or business, they provide essential advice or make policy and marketing decisions. While their primary product involves legal analysis and documentation, lawyers have become adept at incorporating, but not succumbing to, analytical models of other disciplines. Many have developed a working knowledge of the telecommunications infrastructure and how the component parts operate. Similarly, many attorneys can incorporate economic terms to support a client's legislative or regulatory agenda.

The application of other disciplines and the work of nonlawyers may become subordinate to a legislative, administrative, and policy-making process managed by lawyers. Lawyer-dominated policy making may emphasize procedural fairness and opportunities for interested parties to participate. This means that much time, money, and effort may be devoted to questions of which organization has jurisdiction to decide a matter, what the intent was of a law or treaty, and what procedures will ensure fairness and opportunities for all parties to participate.

1.2.3 Engineering

Technology has a substantial impact on policy making and marketplace prospects for telecommunication equipment and services. Its development forces policy makers to consider whether and how the status quo must or can change to accommodate new service options and prospective market entrants. For example, more sensitive and selective satellite Earth station equipment makes it possible to position orbiting satellites closer to each other. Policy makers must balance the ability to accommodate demand for more satellite capacity with the extra expense most users would incur to upgrade Earth station receiving equipment to distinguish between the target satellite and the now closer satellites that are capable of causing interference.

Legislators and regulators often try to classify technologies with an eye toward developing markets and regulatory programs according to semantic line drawing; for example, they must decide whether a new service is *basic* telecommunications transport and subject to conventional regulation, or *enhanced*, *value-added* services, which are often subject to less regulatory oversight. Telecommunications regulators and providers think of technology in terms of whether it supports journalism, entertainment, broadcasting, closed-circuit transmissions like cable television, information processing, telephony, and other market segments. Even these classifications, and more legalistic ones such as *common carrier* and *private carrier*, increasingly converge. For example, telephone companies, which are typically regulated as common carriers and public utilities, seek to become cable television operators. Cable television operators, which typically are subject to less regulation and are not classified as common carriers or public utilities, have targeted new telephone service markets. Interactive television and multimedia transmission of voice, data, audio, and video over a single medium frustrate a desire to think in terms of discrete and mutually exclusive industries and service classifications.

1.2.4 International Relations

International telecommunications blends public diplomacy and private commerce. Delegations to bilateral or multilateral conferences will address such esoteric but critical issues as parking places for domestic and international satellites, the amount of spectrum allocated for particular services, and the fundamental rules of the road like the number and sequence of digits one must dial to reach a particular telephone. On the other hand, carriers negotiate business alliances and agree to "match circuits" and deliver the traffic generated by "correspondents."

Currently many nations are rethinking the balance between government regulation, management, and ownership in telecommunications on one hand and the permissible scope of marketplace resource allocation on the other hand.

Some nations now rely on marketplace forces to assign access to spectrum through auctions and to establish de facto operating standards. However, no national telecommunications administration has embraced a totally laissez-faire view. Many governments have rejected marketplace resource allocation and decision making in the belief that they are inconsistent with the pursuit of broader national interests. Even as more nations opt to replace government ownership with private enterprise in telecommunications, they do not cede to individual companies decision making on issues affecting nationhood and sovereignty. Therefore, one should understand the role of intergovernmental agreements and the forums where nations meet to establish global rules and policies.

1.2.5 Business

While international telecommunications can "contribute to world peace and understanding" [1], it primarily constitutes an industry with enterprises vying for billions of dollars in annual revenues. Fundamental elements of business administration apply in telecommunications, particularly now that many nations have "corporatized" the PTT to make it more businesslike and less bureaucratic. Both carriers and users benefit by having staff expertise in how costs are allocated among service categories and how the terms and conditions of services are set out in a tariff.

In telecommunications, social objectives often prevent companies from maximizing profits. Regulators may require carriers to subsidize some services, such as domestic local telephone service, from revenues generated by other profitable services, such as international long-distance telephone service. Carriers may have universal service obligations requiring them to extend service to the hinterland at averaged or below-cost rates, regardless of the potential for recouping such capital-intensive investments. Telecommunications regulation can establish government as a gatekeeper that decides who can enter what markets and how much they can charge. Service providers and government officials apply accounting and financial skills to determine rates rather than allow the marketplace to establish an equilibrium between supply and demand.

An increasingly liberalized and competitive telecommunications marketplace requires contract negotiation skills. As the sector becomes more business-oriented, users need not tolerate a take-it-or-leave-it attitude on the part of equipment supplier or service operator. The greatest degree of negotiating clout will lie with users who generate large traffic volumes and can migrate to other suppliers, or who can install their own equipment. Carriers and manufacturers will perceive the competitive necessity to customize deals, and, accordingly, regulators will need to permit nontariffed service arrangements and adjustments in the process by which they evaluate fairness of negotiated terms and conditions.

Management

an technology or governments, have an increasingly domi-
the telecommunications industry. However, technological
ger changes in regulation and particular policies. The pace
ovations force improvements in the ability of public and
s to manage technology and change.

innovations necessitate modifications in the way managers
sider the telecommunications infrastructure. Users no longer
oice and data networking when a single digital network can
munications requirements. Similarly, it is increasingly diffi-
e transmission facilities used to disseminate news and enter-
broadcasting and cable television, from the conduits used to
ess data, such as fiber-optic cables, satellite links, and intelligent terminals.

Telecommunications professionals have begun to employ technological developments in ways that previously would have violated regulations designed to segregate markets and insulate them from some types of competition. For example, managers and users of cable television once viewed their distribution plant as a one-way vehicle for delivering video channels to homes. With minor modification and investment, this plant can provide two-way distribution of voice, data, facsimile, and video traffic. Analysts now consider coaxial cable more advanced and flexible than the telephone company's pair of twisted wires, which has provided "first and last mile" access to telecommunications users. Many cable television operators plan on providing telecommunication services including local and long-distance telephone services. While technologically feasible, interconnection of cable television and incumbent carrier facilities may require legislative and regulatory changes. Such promising market opportunities challenge the status quo both in terms of conventional thinking in each industry and the regulatory structure erected to oversee that industry.

Reference

[1] *Agreement Relating to the International Telecommunications Satellite Organization "Intel-sat" (Intelsat Agreement)*, preamble, 20 August 1971, entered into force 12 February 1973, 23 U.S.T. 3813, T.I.A.S. No. 7532.

The next two chapters will attempt to establish a foundation for understanding the often complex, fast changing, and interdisciplinary world of telecommunications. We will first examine a number of descriptive words that expose the breadth of the subject. Afterward we will consider several explanatory models with an eye toward understanding how nations organized the telecommunications sector and what many have done differently in the last few years.

Until quite recently, the rules of the road in international telecommunications were absolute, straightforward, and unavoidable. To understand how so much has changed in the last few years, as compared to the dozens of preceding ones, one should consider descriptive, key words and underlying operational models. Only after such a foundation can we begin to see how conditions have changed, making it possible for new models and options to evolve after so many years of successful resistance. A number of words beginning with the letter *c* describe many of the still applicable, but no longer absolute, key words in international telecommunications.

2.1 CONSENSUS/COMPROMISE

The terms and conditions under which nations communicate require consensus: a common denominator and standard procedure under which all international carriers operate. The International Telecommunication Union (ITU) and other forums provide a basis for establishing the basic rules of the road for everything from the type of plug, to dialing sequences, to tariff terms and conditions. International telecommunications is a collaborative undertaking between two or more equal *foreign correspondents.* Their objectives must mesh, and if they do not, conflict and the need to negotiate a compromise results. Foreign carriers typically want to migrate traffic to usage-sensitive, switched routings, which generate high toll revenues. Multinational users want to route traffic over private lines or software-defined networks that can be priced without regard to toll revenue sharing arrangements for regular international long-distance calls.

Nations with cutting-edge deregulatory policies might want to see other countries embrace a procompetitive, deregulatory agenda, but no single nation can expect to have its policies and philosophies widely adopted. This means that even when telecommunications initiatives have accrued consumer dividends without financially harming incumbent carriers in one nation, few other nations will simply follow along.

For example, in the late 1970s and early 1980s, the U.S. Federal Communications Commission (FCC) allowed entrepreneurs to acquire bulk long-distance calling capacity and to make available discounted services to users who individually would not generate the necessary demand. The FCC authorized domestic shared use and resale of leased private lines and wide area telephone service (WATS). These initiatives met with universal opposition because they threatened a PTT's ability to:

- Charge for long-distance on a usage- and mileage-sensitive basis (i.e., toll charges by the minute or ten-second pulse and priced as a function of the distance between caller and call recipient);
- Discriminate between user groups so that discounts could be limited to a select group of high-volume users.

Nations typically do not respond favorably to a nation's unilateral initiatives, particularly ones that are not introduced and widely ventilated through international forums like the ITU. Similarly, nations do not welcome campaigns originating in other countries that may adversely affect their own revenue streams.

The ITU provides the primary forum for such consensus building among nations. It establishes technical and operating standards, defines services, allocates spectrum, and promulgates international regulations and recommendations that govern the business of telecommunications. Other bilateral and multilateral forums, often convened on a regional basis, augment the ITU process.

On occasion, nations deviate from the ITU-promulgated consensus. The governing documents of the ITU acknowledge the sovereign right of nations to enter into bilateral "special arrangements" [1] that deviate from ITU rules, or to address issues not yet considered by the ITU. Nations can also state "reservations" about a consensus-reached decision, in effect opting out of the general agreement to implement the ITU solution. Alternatively, nations can seek insertion of a footnote into a spectrum allocation or other regulation stating how they will deviate from the consensus.

Generally, nations recognize that universally agreed-upon terms and conditions serve enlightened self-interest by reducing costs and expanding market access for equipment manufacturers and service providers. However, that recog-

terms and conditions by which they will accept traffic and deliver it to the intended recipient or to another carrier. Correspondents "match" transmission circuitry and coordinate the provision of service between nations.[2] Everything from corporate policies and new services to provisioning time usually takes longer and requires more personal attention. Operating agreements typically are short, vague documents leaving much open to interpretation as you go along.

International carriers pool financial resources in multimillion dollar transmission facility investments.[3] They allocate investment and operational respon-

---

1. "The term 'operating agreement' traditionally has been used to refer to the contract or other arrangement in which an overseas administration agrees to operate with another administration or U.S. carrier and which sets out all the terms and conditions under which service is offered." (International Communications Policies Governing Designation of Recognized Private Operating Agencies, Grants of IRUs in International Facilities and Assignment of DNICs, CC Docket No. 83-1230, Notice of Proposed Rulemaking, 95 FCC 2d 627, 630 (1983), Report and Order, 104 FCC 2d 208 (1986), recon. den., 2 FCC Rcd. 7375 (1987).

2. See also R. Frieden, "International Telecommunications and the Federal Communications Commission," *Columbia J. Transnational Law*, Vol. 21, 1983, p. 423: "Various foreign PTTs are equal partners with U.S. Carriers and decide with which U.S. carrier(s) to connect 'half-circuits.' The financial stakes are extremely high in engineering a 'whole circuit' link between nations." (See also p. 425, note 8.)

3. The U.S. international service carriers participating in the consortium planning the construction and operation of the TAT-9 transatlantic fiber-optic cable estimated the cost for the cable at approximately $400 million in 1987 dollars, with additional land facilities and interest costs of $53 million. The U.S. carriers' ownership share in the facility is approximately 40.2%. (Inquiry into the Policies to Be Followed in the Authorization of Common Carrier Facilities to Meet North Atlantic Telecommunications Needs During the 1991–2000 Period, CC Docket No. 79-184, 3 FCC Rcd. 3979, 3987 (1988).) The cost to construct one INTELSAT VI satellite, (launch, insurance, and other incentive and support expenditures) has been estimated at $232 million. (Cummins et al., "Satellites Versus Fiber Optic Cables," in *Proc. of the Seventh Annual Conference of the Pacific Telecommunications Council (PTC'85)*, D. Wedemeyer, ed., Honolulu: Pacific Telecommunications Council, 1985, pp. 422, 425.) INTELSAT's net investment in international satellite capacity exceeds $2 billion.

sibility on a half-circuit basis, with correspondents theoretically matching half circuits into whole circuits at an international boundary or at the midpoint of a route. One half circuit of a carrier's inventory represents the smallest common unit of capacity corresponding to one analog voice channel. For digital fiber-optic submarine cables, carriers designate capacity in terms of minimum assignable units of ownership (MAUO),[4] a pathway comprising 64,000 bits per second (bps) of throughput and corresponding to approximately four voice-grade channels derived through the use of circuit multiplication technology.

Carriers must coordinate the planning, design, investment, construction, and maintenance of international facilities. Such pooling of resources can lead to a shared view of the need to preserve the status quo through policy inertia and efforts to foreclose competition.

2.3 CONSULTATION

Both governments and carriers, which may still be government-owned, participate in the *consultative process*, in which regular meetings are held and information is shared, ostensibly to estimate demand and schedule future deployment of transmission facilities. Access to international transmission capacity depends largely on how carriers assess demand and when they decide to expand capacity and routing options. Consultation regularly occurs between international carriers and their regulators. In many instances the line blurs between carrier-to-carrier business negotiations versus intergovernmental policy coordination. Some nations merge the function of carrier and regulator within a single ministry, which has the effect of blurring the regulatory/government oversight function versus the management/operational role. Even where a separate regulatory agency exists, most delegations to the ITU and other forums include both carrier representatives and regulators. Likewise, when carriers assemble to address business or policy matters, the group of participants typically will include both operational and regulatory constituencies.

The combination of regulatory and business delegates in a variety of forums has the consequence of merging points of view and institutionalizing a role for government even in matters affecting the strategic business planning of an individual carrier.

FCC delegates in the 1970s actively brokered deals concerning when and where carriers should deploy new international submarine cable transmission facilities. The FCC justified such intervention on grounds that its public interest

---

4. "A MAUO is the Minimum Assignable Unit of Ownership and provides an equivalent digital channel operating at 64,000 bits per second." (American Telephone and Telegraph Co., 5 FCC Rcd. 7331, 7338, note 5, (1990) (authorizing construction of the PacRim East fiber-optic submarine cable).)

view that competition will promote service diversity and downward rate pressure and enhance consumer welfare.

The FCC no longer actively participates in the consultative process. However, it still performs some degree of independent analysis to support its approval of carrier applications to construct or participate in the construction of new facilities. To satisfy its public interest mandate, the FCC requires international carriers under its jurisdiction to submit traffic statistics[5] and provide evidence of increasing demand for service and new transmission facilities. In conjunction with these filings, what it derives from its observer role in carrier consultations and its own analysis, the FCC prepares to consider facility construction applications filed by carriers as required by Section 214 of the Communications Act of 1934.

The FCC used to consider the Section 214 review process an opportunity to closely scrutinize carrier submissions of demand projections and evidence supporting the need for additional transmission capacity. The FCC now refrains from undertaking a thorough demand assessment on the grounds that it should not second guess carriers on capacity requirements, particularly because it now believes that no carrier can afford to tie up capital in unnecessary projects.

Prior to its decision to monitor the consultative process rather than actively participate in it, the FCC predicted when carriers subject to its jurisdiction would need to activate additional transmission facilities and by what ratio they must activate circuits from their inventory of cable and satellite capacity. The FCC made such projections and loading decisions for a specified time period by ocean region. It no longer interferes, because it now believes that carriers should freely activate transmission media on the basis of their assessment of consumer requirements and comparative technological features.

---

5. See Amendment of Sec. 43.61 of the Commission's Rules, CC Docket No. 91-22, Report and Order, 7 FCC Rcd. 1379 (1992) (simplifying and streamlining reporting requirements, but adding reporting requirements for traffic to and from Canada, Mexico, and St. Pierre and Miquelon).

2.4 CULTURE

The culture of international telecommunications blends the diverse folkways of individual nations with shared characteristics of the officials involved in its provision and regulation. Despite significant differences in politics and overall culture, nations and their telecommunication service providers usually agree on shared goals articulated in forums like the ITU and facilities cooperatives and consortia. Until nations like the United States and the United Kingdom undertook an aggressively deregulatory approach, governments and carriers throughout the world agreed on common objectives such as universal service, interconnected networks, price averaging between dense and sparse routes, and single composite rates to users regardless of differences between cable and satellite circuit costs.

While many nations continue to support long-standing goals, deregulation and privatization have spawned a variety of new players who lack the view that their corporate objectives should parallel governmental or social goals. Because market entrants typically have no exclusive service franchise and lack government-mandated insulation from competition, they seek to maximize profits and object to any requirement to provide some services at below-cost rates. New regulatory classifications, like private, non-common carriage, exempt operators from most public service duties in the recognition that they lack the power to control price or availability of service.

Successful market entry by new international telecommunications players challenges the viability of the view that a single PTT monopoly must operate with an exclusive franchise. Incumbents may have met both national and international goals with a public utility culture, and may have operated as a "benign" or "enlightened" monopolist serving both corporate and public objectives. However, recent legislative, regulatory, technological, and marketplace changes have fostered the onset of a corporate, profit-maximizing culture. Corporate players will have less to gain in serving social goals when they no longer enjoy insulation from competition. Likewise, technological innovation makes it easier for outsiders to infiltrate previously closed markets. Lacking a public service culture, private ventures may face mandatory service obligations and pricing limitations to help achieve social policy objectives.

The clash of cultures makes international telecommunications more challenging and confusing:

- How does one do business with a major government-owned or franchised enterprise without knowing its culture and philosophy of business?
- What is the nature of the relationship between the carrier and its government owner or regulator? Does government possess veto power through the option to cast a "golden share" even for privatized carriers?
- Are there major cross-subsidy obligations?

The international transmission facility marketplace continues to have relatively few participants. Despite market entry initiatives in the United States, Canada, the United Kingdom, Japan, Korea, New Zealand, Australia, and a few other nations, typically a single entity provides most international telecommunication services. Even if a nation's policy-making body opts to foster competition, with rare exceptions it has shown no inclination to permit market entry in basic, facilities-based international telecommunications. Competition in international telecommunications primarily occurs in niche markets like mobile, value-added, and enhanced services, including credit card verification, airline reservations, and very-small-aperture terminal satellite networks.

The clubbiness of international transmission facility ownership and operation is promoted by the relative paucity of operators. Even now the predominate model for international service is that of a single entity.

The good news is that the club can no longer blackball every prospective market entrant, and that technological innovations make any network more porous, accessible, and manageable by users. The bad news is that the club still dominates the policy-making and standard-setting process.

2.6 COOPERATIVES/CONSORTIA/CARTELS/COLLUSION

The extent of investment involved in international satellite and submarine cable facilities requires spreading risks and costs over a broad set of carriers. International carriers pool satellite investments and aggregate consumer circuitry requirements through global and regional cooperatives like INTELSAT and the European Telecommunications Satellite Organization (EUTELSAT), whose service mission may mandate cost averaging between dense and sparse routes. A parallel regional investment structure for submarine cables involves the

formation of consortia. Investment shares in such ventures reflect anticipated usage. For example, Comsat Corporation, the sole U.S. representative in
INTELSAT, holds about a 24% ownership share.

Cooperatives structure distribution channels and increase the number of
intermediaries (e.g., Comsat). Consortia favor cost uniformity, since most carriers invest in or acquire Indefeasible Right of User or MAUOs in the same transmission facilities. At some point, however, investment pooling in transmission
facilities can lead to cartel formation. New international satellite and submarine
cable systems, separate from the conventional consortium investment group,
had great difficulty in securing operating agreements with incumbent carriers,
because market entry threatened to move traffic and revenue away from the
incumbent cooperatives and consortia. While economic theories suggest that
technological innovation can support market entry, the international telecommunications marketplace has remained largely closed because carrier incumbents have established and enforced uniform and restrictive rules of the road.

The closed group of incumbent carriers has every incentive to maintain
the comfortable status quo, but technology, consumer demand, and some activist governments, responsive to consumer requirements, make it less sustainable.
Likewise, the cartel has less ability now to discipline members for deviating
from rules that served the collective good but also limited profitability and the
scope of market opportunities. However, while these new incentives challenge
cozy margins, for the time being on many routes, international service rates can
exceed equivalent domestic routes by 300% or more.

2.7 CONSTRAINTS/COMITY

Participants in forming the rules of the road in international telecommunications recognize the need to temper any one nation's policies and philosophies.
The broad goal of international comity means that nations may agree to a least
common denominator solution to achieve consensus. These constraints in the
rules of international telecommunications build in limitations, even at the expense of consumer welfare. International telecommunications users accus-
tomed to the luxury of choice afforded by vigorous competition in domestic
markets find frustrating the high cost and limitations on network control and
flexibility occurring in many domestic markets. Often such constraints represent unnecessary expense and inconvenience built in by the decision of nations
and carriers to refer to a single protocol or standard. This default solution frustrates telecommunications managers who have mastered the art of negotiating
domestic service arrangements customized to meet particular user requirements.

The quest to be all things to all people stems from international carriers'
duty to serve all user requirements, including nonprofitable but socially desirable services. By restricting access to discounted services and perhaps also lim-

...may stall for time to ensure that a new service does not cannibalize existing, more profitable services through customer migration to the new offering.

2.8 COMMINGLED COSTS/CROSS-SUBSIDIES

Government-owned or franchised PTT administrations have traditionally dominated international telecommunications markets. Despite growing interest in privatization and deregulation, many nations still authorize a monopoly for most services. Typically a single player enjoys rate-making flexibility to accrue revenues available for allocation to underprice domestic telecommunications and postal operations. The government PTT or privatized monopoly continues to operate free of competition, but under significant service obligations, including the deliberate underpricing of certain services like domestic telecommunications.

Advocates of cost-based pricing and competition must recognize the threat presented by initiatives that would eliminate high profits from some international services. The typical international telecommunications regulatory regime requires carriers to cross-subsidize basic domestic telephone services using revenues derived from charging excessive rates for services used by international users. Most nations have pricing policies reminiscent of the preaccess charge environment in the U.S, where the Ozark Plan established cross-subsidies from long-distance to local-exchange service. For example, international switched telephone service may generate triple-digit returns so that PTTs can underprice domestic telephone service and undertake major rural construction projects. Such pricing dampens demand for international calling, and the matter grows worse when PTTs cannot use the revenues to improve service, but must contribute some or all to the general treasury. Carriers cannot price all services on the basis of cost, because public policies typically require that pricing help achieve certain socially desired goals, such as widespread availability of basic telephone service even in sparsely populated areas where service is costly to provide. Private lines and software-defined networks alleviate the problem somewhat, but there remain limits on availability and use of such options.

2.9 CLOSED OR CONDITIONAL MARKET ACCESS

Telecommunication service and equipment markets remain less than fully open and competitive. Many nations significantly restrict access to domestic markets, including the opportunity for foreign manufacturers to bid on equipment procurements by the PTT. Sovereignty and national security concerns often serve as a thinly veiled effort to blunt comparative advantages enjoyed by foreign carriers and manufacturers. Nations may have industrial policies designed to favor a "national hero." Even in the most recent initiatives, such as EC'92, which are often hailed as procompetitive, reserved markets tilt the competitive playing field in favor of national or regional players. While the battle for market access may formally take place at the World Trade Organization and other international forums, the front-line player should understand the existing institutional limitations and ones likely to persist. Likewise, nations will continue to insulate carriers from competition that might foreclose below-cost pricing of socially desired services.

Despite the rhetoric about open markets and competition, most nations continue to restrict market access by foreign carriers and manufacturers. These safeguards have become increasingly difficult to enforce as sophisticated users find ways for employing technological innovations to aggregate traffic and route it across political borders without detection. Carriers should expect not to have the option to originate network traffic incountry or to route traffic in the most efficient or cheapest manner. Anticipate the need for incountry representatives to negotiate interconnection arrangements early, often, and perhaps on a circuit-by-circuit basis.

2.10 CORPORATIZATION

Even without the spur of unfettered competition, many international telecommunications carriers have become more businesslike and less bureaucratic—and an increasingly significant number have been privatized outright. This orientation may result from external prodding through legislation, by a revised charter or strategic assessment, or from the recognition that other, more aggressive carriers have begun to serve previously captive customers. Internal changes result from a new strategic assessment that competition and change are inevitable. The proactive incumbent carrier recognizes the need to target and serve

ness-minded government monopoly, perhaps with workers hired outside the civil service, to a shareholder-owned business like any other publicly traded corporation.[6] Unfortunately, PTTs cannot change their corporate culture overnight and may have fewer incentives to act as speedily as, for example, post-divestiture AT&T. Also, the corporatized PTO may exploit artificial competitive advantages when serving new markets like value-added service business.

2.11 CHANGE

The international telecommunications marketplace has begun to change at an increasingly speedy rate, paralleling the changes experienced in the United States over the last 25 years.[7] While it will not match the pace in some domestic

6. See, for example, Ralph H. Kilmann, "Tomorrow's Company Won't Have Walls," *The New York Times*, Sec. 3, p. 3, 18 June 1989; and John Burgess, "'Global Offices' on Rise as Firms Shift Service Jobs Abroad," *The Washington Post*, Vol. 112, No. 136, Sec. E1, p. 1, 20 April 1989.

7. "Today the winds of change are sweeping the telecoms industry worldwide, especially in Europe and particularly in Germany. There are two clear causes. First, [is] the dizzying explosion of technology with its enormous impact in what is possible, the economies of doing it, and the people [both quantity and type] required which has overwhelmed the ability of governments to keep things within the old boundaries. Second, is the rise of the consumer as the master of the economic society—his freedom of choice with increasing complex technology and diverse relationships." (K. Hoyt, *The German Telecommunications Market—Facts, Perspectives and Opportunities*, presented at the Bundespost Telecom Seminar in Washington, D.C., 24 March 1988.) In 1989, legislation in Germany divided the Bundespost into three separate entities: banking, postal service, and telecommunications. The telecommunications enterprise, Deutsche Bundespost Telekom, has embraced a far more flexible and businesslike approach than when it constituted a part of a larger government entity. (See "Germany's TELEKOM: a New Way of Doing Business in a Liberalized Market," *Telecommunications J.*, Vol. 58, No. 10, 1991, pp. 711–715; and also M. Stoetzer, "New Telecommunication Service—Current Situation and Prospects in Germany," *Telecommunications Policy*, Vol. 18 No. 7, 1994, pp. 522–537.)

markets, we can anticipate change and an eclectic blend of initiatives derived from domestic experiments throughout the world.[8] Technological and business imperatives combine to challenge a comfortable status quo and the PTT model. Privatization and deregulation create new businesslike incentives where social engineering and full employment goals previously dominated. The vision of a "global village" has become increasingly possible as nations grow more interdependent through trade, which is facilitated by high-speed, efficient, and ubiquitous international telecommunications.[9] "Telecommunications has become the object of...[a] mad scramble, because it has become so critical to the world's economy" [4]. It constitutes "a key element of any strategy for economic and social revival" [5], but nations have vastly divergent views on how to maximize benefits with the least amount of dislocation, unemployment, and other adverse consequences.

Recent policy liberalization has occurred primarily to spur efficient operations of incumbents and markets. However, it does not necessarily follow that nations will liberalize access to domestic markets by foreign carriers and service providers. Paradoxically, as many nations endeavor to stimulate their economies and promote ease in commercial transactions, they also maintain policies that blunt other nations' comparative advantages and continue to restrict market access. Accordingly, much of the opportunities for strategic alliances and foreign investment are possible only if an individual company's expertise supports evolving governmental policies, or if an incumbent realizes that to serve foreign markets it will have to establish teaming arrangements with other enterprises, often already operating within a targeted region, to tap into needed expertise.

Particularly important and relatively new policies for many nations include:

- Pursuing a conditionally open marketplace;
- Reducing the scope of government ownership of telecommunications carriers and the degree of government oversight;
- Supporting a unified and globally competitive region;
- Mandating that incumbents use a more businesslike, global, and consumer-oriented approach.

---

8. "The increasing trend towards regulatory liberalization has created a new breed of transnational telecommunications enterprise and has opened some doors for them that would have been closed previously." (T. Logue, "Who's Holding the Phone? Policy Issues in an Age of Transnational Communications Enterprises," in *Proc. of the 12th Annual Conf. of the Pacific Telecommunications Council (PTC'90)*, D. Wedemeyer and M. Lofstrom, eds., Honolulu: Pacific Telecommunications Council, 1990, p. 95.)

9. "Money never sleeps nowadays...There are three time zones in the financial world [London, New York, Tokyo]." ("Around the Clock, Around the World with 24-Hour Global Trading Ahead, the Competition Can't Be Far Behind," *Communicationsweek International*, Sec. C2, 6 February 1989.)

...a goal of becoming "a leading worldwide provider of telecommunications and information management services," an enterprise must structure a proposal that "complements the national telecommunications strategies of... [targeted] nations" [7].

New government policies addressing competition, liberalization of policies, and partial deregulation stimulate an expanded and more diversified telecommunication services marketplace. In a growing number of nations, the incumbent PTT has lost the ideological and functional argument that it alone should provide all telecommunication services and equipment as a natural monopoly[10] because it can operate most efficiently and maximize consumer welfare by achieving economies of scale.[11]

---

10. "Traditionally telecommunications has been seen by many people as a classic example of a 'natural monopoly'—said to exist when a single firm is technically able to provide goods or services at a significantly lower cost than two or more competing firms. Such natural monopolies benefit from economies of scale." ("Is There Life After Monopolies?" *Public Network Europe*, Vol. 1, No. 10, October 1991, pp. 27–28.) Professor Faulhaber presents a more comprehensive definition: "A natural monopoly occurs in an industry in which the production technology is such that one producer can supply the entire market more cheaply than two or more producers...Until recently, it was thought that natural monopoly occurs when the average costs of production decline with increasing volume...This allows the resulting monopolist to charge prices substantially above costs and cut back the amount supplied and consumed to the profit-maximizing level. More recent work suggests that the mere presence of cost advantage is not sufficient to enable the monopolist to raise its prices above cost; it appears that only the presence of sunk costs, investments that once made cannot be retrieved for purposes other than original intent, can deter entry and permit above-cost pricing and monopoly profits." (G. D. Faulhaber, *Telecommunications in Turmoil: Technology and Public Policy*, Cambridge, MA: Ballinger, 1987, p. 106.)

11. "Economies of scale refer to the situation where an increase in inputs leads to a proportionally greater increase in outputs. (For example, a doubling of inputs would lead to more than a doubling of outputs)." (K. Wilson, "Deregulating Telecommunications and the Problem of Natural Monopoly: A Critique of Economics in Telecommunications Policy," *Media Culture and Society*, Vol. 14, 1992, pp. 343, 345.)

Deregulation results when policy makers consider telecommunications in terms of a number of submarkets where niche competitors can operate successfully without handicapping the incumbent carrier's ability to provide core services. While preserving a monopoly for basic services, many governments grow to believe that many areas of telecommunications now have low barriers to entry and do not require a substantial degree of up-front investment not easily salvaged (i.e., sunk) in a physical plant. For example, a nation may continue to reserve for the PTT a monopoly of basic switched voice and data services, but authorize competition in mobile communications, such as cellular radio and paging, and for value-added services that enhance lines leased from the PTT. Competition in some market segments can flourish if the regulator can ensure cost-based interconnection with the remaining natural monopoly components of an integrated telecommunications system, such as the public switched telecommunications network (PSTN), which provides the "first and last mile" of a communications link to users and competing service providers.

With some economic support, the incumbent monopolist responds to the campaign for market access by claiming that market entry will foster unfair and inefficient competition, that is, that newcomers will thrive only by "cherry-picking" the most profitable customers for discounted service and by "cream-skimming," that is, serving the most dense routes, leaving the incumbent with fewer resources and revenues to serve remaining users and upgrade the network. Opponents to competition claim that it will result in wasteful duplication of facilities and the bypassing of incumbent facilities, not on efficiency grounds, but simply because regulators have imposed financial burdens on the incumbent, in effect requiring it to raise its rates to underwrite the below-cost provision of other services. "The dilemma for policy-makers is to ensure an adequate contribution to the local network infrastructure without increasing incentives for corporate bypass" [8].

Despite the potential for adverse financial consequences, many governments and most customers demand change [9]:

> Opening the telecomms operating environment to competition is no longer a purely ideological concept: advances in technology now make it practical, policy changes make it possible, and moreover, customers want it.

2.12 THE OLD AND NEW WORLD TELECOMMUNICATIONS ORDERS

The new world telecommunications order does not lend itself to a uniform collection of descriptive key words or to a single definitive regulatory or industrial model. A variety of new descriptive words representing the entire

alphabet have evolved. These include privatization, deregulation, liberalization, and globalization.

Such commonly used words belie the controversy that arose when the first
initiatives were announced. In the early 1980s, the U.K. government decided
first to split postal operations from telecommunications, and then to privatize
the telecommunications enterprise, subject to both independent regulatory
scrutiny and competition. The British initiative stemmed from a largely un-
popular "conviction that private ownership would be more efficient than public
ownership...[and] better able to control the trade unions" [10].

Now, eliminating or limiting the PTT monopoly has expanded support
from economists, policy makers, and the public, despite efforts by incumbents[12]
to generate concerns about lost jobs, reduction in service availability, and rate
increases.[13] Since 1984, major privatizations have included British Telecom,
Teleglobe Canada, and Nippon Telegraph and Telephone. The pace has increased with recent privatizations in Australia, Argentina, Chile, Denmark,
Hungary, Italy, Mexico, New Zealand, Portugal, Puerto Rico, and Venezuela
[10]. As many as 26 countries, representing 95 million telephone lines, may privatize their PTTs in the next few years [10], including such diverse nations as
Germany, Greece, Israel, and Poland.

The study of the international telecommunications revolution requires an
examination of the policies, regulations, procedures forums, and types of business alliances that favor a new telecommunications world order. The old order
supported monopolies, inflexibility, closed markets, nationalism, and pervasive
government involvement. In only a few years, nations that used to own and operate a monopoly PTT now promote competition, deregulation, flexibility, globalization, liberalization, and privatization.

---

12. "For almost a century the key institutional feature of traditional telephony around the world
    has been a ubiquitous network operated by a monopolist...Public telecommunications were not
    merely a technical system, but social, political and economic institutions...The PTTs were sup-
    ported by a broad political coalition, a 'postal-industrial complex.' It included the PTT itself
    and the equipment industry as its supplier, together with residential and rural users, trade un-
    ions, the political left, the newspaper industry (whose postal and telegraph rates were heavily
    subsidized), and affiliated experts. The system worked in no small measure to the benefit of the
    [domestic] equipment industry." (E. Noam, "International Telecommunications in Transition,"
    in *Changing the Rules: Technological Change, International Competition and Regulation in
    Communications*, R. Crandall and K. Flamm, eds., Washington: Brookings Institution, 1989,
    pp. 257, 258.)

13. The support for privatization and deregulation is far from universal. In Poland, increasing con-
    gestion and breakdowns of the PSTN allegedly are the result of "foreign companies coming in
    with their fiber-optic and satellite networks...[and] snatching up the most competent employ-
    ees of the state phone company...In a few years, a new high-tech grid will probably have been
    laid out over the old one, but most people feel certain that they won't be able to afford to tap
    into it, and they will have access only to the old system—and even these rates are rising so rap-
    idly that most people are trying to keep their calls to three minutes or less." (L. Weschler, "A
    Reporter at Large (Poland)," *The New Yorker*, Vol. 68, No. 21, 11 May 1992, pp. 41, 57.)

t is important to understand these frequently used and noninterchange-
terms. Privatization involves the change in legal status of the PTT from
ic to private ownership. However, the newly privatized company may hold
ivate (as opposed to public) monopoly, the government may retain some
ree of stock ownership in the new company, and all preexisting regulations
y remain in force. Typically, but not always, privatization is coupled with
regulation: government streamlines the regulatory requirements of carriers
nd reduces structural safeguards that insulate the PTT from competition in selected markets. Even in a deregulatory climate, government regulators must ensure that newcomers can directly interconnect with the PTT's networks under
fair terms and conditions. Liberalization also may occur around the time the
PTT is privatized. The government allows the incumbent PTT greater flexibility
to identify and serve new profit centers, thereby offsetting market share lost to
new carriers. In many instances, the PTT seeks opportunities in foreign markets.
The changes that have occurred in the last few years contrast with the dozens of
preceding years when change occurred at a glacial pace because a broad "postal-
industrial complex" [11] had a vested interest in maintaining the comfortable
status quo.

2.13 THE CHALLENGE TO SOVEREIGNTY AND CONTROL

Technological innovations in telecommunications coupled with liberalizing
policies threaten the ability of national governments to control the flow of communication and to centrally manage this sector of the economy. Such control
can help bolster a political regime and insulate it from destabilizing, but perhaps also liberalizing, outside influences. Perhaps some nations have shown reluctance to liberalize telecommunications because of concerns for the stability
of the political status quo. Managing access to telecommunications facilities
and limiting or prohibiting market entry may help bolster a regime, particularly
if it can provide employment and below-cost services by enforcing policies that
require cross-subsidization.

While inefficient, the PTT may continue to operate because of the political
power it holds simply by employing large numbers. This large, often unionized
constituency can influence legislators and regulators, encouraging them to establish policies that maintain the status quo, including service pricing requirements that burden large-volume users and international callers with excessive
rates. Technological innovations, market access opportunities, and aggressive
marketing initiatives of some carriers now make it easier for these burdened users to access cheaper alternatives that bypass the PTT entirely, or at least avoid
the most expensive services. Users who previously underwrote cheap domestic
phone and postal service now have more options, and even some of the benefi-
ciaries of cross-subsidies have begun to think that centralized planning and mo-

nopolies may lag in development as compared to the level of progress mandated by a more business-oriented and competitive environment.

Integrated telecommunications networks, composed of geographically dispersed access points called *nodes* and traffic aggregation points called *hubs*, challenge national sovereignty and conventional concepts of control over territory. Broadcast signals and telecommunications networks traverse borders with the consequence that individuals in one nation have increasingly easy and cheap access to databases, news, and entertainment from other nations. While economic theory endorses the view that the utility and value of such networks increase with the number of users, simple politics confirms the potential destabilizing force such access creates.[14]

Mulgan views control in the narrow context of telecommunications law, regulation, and policy, where governments can sponsor control by vesting powers in a PTT monopoly, restrict the monopoly through regulation, and arbitrate on matters of access to frequency spectrum, networks, and markets. For these types of controls, technological innovations present two kinds of pressure: (1) political pressure for government to exit or reduce the scope of its control, and (2) economic pressure for government to avoid sponsoring monopolies and to eliminate restrictions on the scope of activities that former monopolies can pursue after entrants have made the marketplace more competitive.

References

[1]   General secretariat of the ITU, *International Telecommunication Convention* (Nice, 1989), Article 31, Special Arrangements, p. 22, reprinted in ITU, *Final Acts of the Plenipotentiary Conference Constitution and Convention of the International Telecommunication Union, Optional Protocol, Decisions Resolutions, Recommendations and Opinions*, Geneva: ITU, 1990, p. 29.

[2]   American Telephone and Telegraph Co. et al., 73 FCC 2d 248 (1979), on recon., 83 FCC 2d 1 (1980).

[3]   ITT World Communications, Inc., 77 FCC 2d 877 (1980), reversed and remanded sub nom., ITT World Communications, Inc. v. FCC, 699 F.2d 1219 (D.C. Cir. 1983).

[4]   "Telecommunications: The Global Battle," *BusinessWeek*, Special Report, 24 October 1983, p. 128.

[5]   Narjes, K. H., preface to *Telecommunications in Europe: Free Choice for the User in Europe's 1992 Market*, by H. Ungerer and N. Costello, Luxembourg: Office for Official Publications of the European Communities, 1988.

---

14. Geoff J. Mulgan presents a different view of control. The ubiquity, number, and pervasiveness of new telecommunication technologies accrue greater strength to for government to monitor the populace and to determine who should have access and who should not. (G. J. Mulgan, *Communications and Control: Networks and the New Economies of Communication*, New York: Gilford Press, 1991.) But with increasing distribution of network control to end users and with proliferating access points, the ability to monitor and control may in fact decline. The number of successful penetrations of secure computer networks by hackers corroborates this view.

[6]   "The New Boys: Politics on the Line," Telecommunications Survey, *The Economist*, 5 October 1991, p. 1.

[7]   Frost, D., "Pacific Rim Telecom Status Report: Privatization and Competition Strategies in Mexico and the Lands Down Under," in *Proc. 14th Annual Conf. of the Pacific Telecommunications Council (PTC'92)*, M. Lofstrom and D. Wedemeyer, eds., Honolulu: PTC, 1992, p. 176.

[8]   Wilson, K., "Deregulating Telecommunications and the Problem of Natural Monopoly: A Critique of Economics in Telecommunications Policy," *Media Culture and Society*, Vol. 14, 1992, p. 362.

[9]   "Is There Life After Monopolies?" *Public Network Europe*, Vol. 1, No. 10, p. 27.

[10] Dixon, H., "The Sleeping Giants Awaken," *Financial Times Survey*, Sec. III, World Telecommunications, 7 October 1991, p. 1.

[11] Noam, E., "International Telecommunications in Transition," in *Changing the Rules: Technological Change, International Competition and Regulation in Communications*, R. Crandall and K. Flamm, eds., Washington: Brookings Institution, 1989, p. 258.

Current and Developing Models in International Telecommunications

3

3.1 INDUSTRIAL AND REGULATORY MODELS: AN INTRODUCTION

Until the late 1980s and early 1990s, few nations had adopted procompetitive policies in telecommunications. The procompetitive model, defined by law and codified regulations[1] and implemented by an independent expert agency, significantly contrasted with the rest of the world, which vested both policy oversight and operational management within a state bureaucracy. Even now many nations adhere to a model emphasizing public administration, government operation, and centralized planning.

To understand why nations have embraced conditional competition and have changed the industrial structure, we will start with an examination of the traditional PTT model. We will then consider the procompetitive model with an eye toward assessing what must occur to make this model acceptable to governments. The book will examine the technologies, multilateral institutions, services, regulations, business arrangements, laws, players, policies, and trends in international telecommunications and information processing. Throughout the book, we will consider both the old and new rules of the road with an eye toward exploring whether and how competition will occur. Our examination will identify the limitations to the idea that the international telecommunications marketplace can duplicate the deregulatory course implemented in the United States and the United Kingdom.

---

1. See Communications Act of 1934, as amended, 47 U.S.C. Sec. 151 et seq., (1990); Communications Satellite Act of 1962, as amended, Pub. L. 87-624, 76 Stat. 419 (1962), codified as amended at 47 U.S.C. Sec. 701-744 (1990); and 47 C.F.R. Telecommunications, Parts 0–100.

3.2 THE PTT MODEL

The PTT model represents a monopolistic postal and telecommunications enterprise owned, operated, and regulated by government. Advocates for this model believe that only centralized, public management can balance profit-maximizing strategies and the duty to achieve the following social policy goals:

- Universal service, or at least efforts to provide affordable basic telecommunication services to rural areas and low-income users;
- Rate setting that prevents price gouging, except where socially desired (e.g., overpricing outbound international telephone rates to generate a source of revenues for cross-subsidizing local telephone and telegraph services);
- Price averaging that blends high- and low-cost routes (urban versus rural and dense versus sparse) and transmission technologies (cable versus satellite) into a single, composite rate;
- Rate setting that serves social goals (e.g., subsidized "lifeline" rates to low-income users);
- Long-range planning that achieves development objectives (e.g., deploying advanced broadband digital facilities and services to the hinterland even in the absence of demand and the likelihood of fully recouping investment);
- Political brokering functions that balance policies aiming to stimulate efficiency by eliminating government and interservice subsidies on the one hand, and securing political support from the large, often unionized body of telecommunications employees, whose numbers may decline as a result of streamlining initiatives, on the other hand.

Nations unwilling to pursue the difficult and possibly painful migration from monopoly-based service to competition remain convinced that a single network service provider will maximize economies of scale and scope. They believe centralized planning can optimize investment in infrastructure improvements and best serve the national interest. Even if a nation pursues structural modifications aimed at eliminating impediments to efficiency, such as the spinning off of postal operations, most remain unconvinced that the marketplace can support multiple service providers and networks.

PTTs justify monopoly service on philosophical and economic grounds. Despite reevaluation and limited experiments with competition, most foreign governments still consider telecommunications a matter of such social importance that they cannot primarily rely on private enterprise, open markets, and competition. Presumably, a single service provider can operate efficiently (i.e., at the lowest cost) and therefore accrue ample revenues to underwrite unprofitable operations for areas that might not otherwise receive adequate service from private enterprise.

While some nations might create a private monopoly, many still maintain a public monopoly. The PTT model predominates because many governments continue to believe that fundamental economic principles and political pragmatism support the status quo. PTTs, particularly in developing countries, have invested in system improvements with the expectation of having captive users for the long term. PTTs often affiliate with or own the single "national hero" telecommunication equipment manufacturer. The PTT serves as a virtual captive customer of the domestic equipment manufacturer, and users typically have no service alternative.

Monopoly status and the ability to charge high rates provide an attractive source of hard currency. Many PTTs recoup a significant percentage of their total facility investment through a favorable split of international toll revenues and by exploiting the relative inelasticity of demand for outbound international services, that is, charging business users of international message telephone service (IMTS) a steep premium for immediate access to the rest of the world.[2] Multiple facilities-based carriers (i.e., enterprises with their own transmission networks) might compete on price and might willingly reduce the rate they charge foreign carriers for call completion with an eye toward stimulating demand and capturing a larger share of inbound traffic.

Most telephone administrations in Africa, Central America, and the Caribbean still adhere to the PTT model, despite growing consumer dissatisfaction. Reliance on the PTT model persists in many developed and developing countries, notwithstanding a growing body of empirical evidence that competition typically stimulates lower rates, service diversity, efficiency gains, and infrastructure development as measured in such quantitative indexes as number of telephone lines per 100 inhabitants.[3] The status quo typically persists when governments target the telecommunications sector as a source of revenue to subsidize other government services or the general treasury. As long as the PTT operates profitably, government has little incentive to demand streamlining and economizing, or to believe that the market can support competition.

---

2. See *International Accounting Rates and the Balance of Payments Deficit in Telecommunications Services*, Report of the Common Carrier Bureau to the Federal Communications Commission, 12 December 1988.

3. See F. J. Cronin, E. B. Parker, E. K. Colleran, and M. A. Gold, "Telecommunications Infrastructure and Economic Growth: An Analysis of Causality," *Telecommunications Policy*, Vol. 17, August, December 1993, pp. 415–430, 529–535; A. Hardy and H. Hudson, *The Role of the Telephone in Economic Development: An Empirical Analysis*, Washington, D.C.: Keewatin Communications, 1981; and H. Hudson, *When Telephone Reach the Village: The Role of Telecommunications in Rural Development*, Norwood, N.J.: Ablex, 1984.

3.3 MOST NATIONS DO NOT DIVIDE MANAGEMENT AND REGULATORY FUNCTIONS

PTTs can establish long-term policies because most combine the functions of service provider and regulator. The PTT is an important government institution in most foreign nations. It is a major employer and often serves as the financial services provider of last resort in rural locations. The PTT's obligation to provide low-cost postal and plain old telephone service (POTS) creates the real or perceived need to extract ample, monopoly profits.

Many telecommunications markets have yet to experience even incremental deregulation, market entry, or service diversification. PTTs and related government ministries have elevated caution over the requirements of large corporate users (often foreign-owned) who chafe at cross-subsidy burdens and limited access to innovative services.[4] Such reluctance to experiment can be explained by the service mandate of the typical PTT, its institutional heritage, and notions of national sovereignty. Telecommunications in most nations is not merely a technical system, but a social, political, and economic institution [1]. Having merged the PTT's telecommunications functions, which can possibly tolerate competition with postal operations, which more clearly qualify for monopoly status, nations seeking greater efficiency may separate out the *P* part of the PTT acronym. But divestiture of postal operations is unlikely when a government would rather tap a somewhat hidden source of subsidies to keep the post office solvent. Certain telephone services, particularly business lines and international long-distance service, generate substantial revenue surpluses that are used to keep local service and postal rates artificially low.

One should not underestimate the caution that PTTs exercise when considering policy and regulatory changes. The legislature typically must take the lead in fostering change because the PTT has two countervailing preoccupations:

1. How to keep ratepayers captive; that is, how to preserve the telecommunication equipment and service monopoly in the face of technological innovations that provide ways to bypass the PTT or avoid the most expensive services;
2. How to maintain exclusive and lucrative revenue streams; that is, how to prevent market entry.

---

4. "Customers should be free to choose from a variety of telecommunication transmission services …They should be free to choose among competitive suppliers of customer premises equipment and providers of value-added telecommunications and information services." (The International Chamber of Commerce, "Worldwide Information Technology Without Barriers—A Business User's Goal," *ICC Policy Statements on Telecommunications*, Position Paper No. 10, February 1988, p. 1.)

PTTs will avoid any procompetitive initiative that threatens profitable revenue streams used, in part, to subsidize postal operations and domestic telecommunication services. Concern over undesirable competition or bypass creates incentives for PTTs collectively to preserve conservative, if not collusive and anticompetitive, policies.

3.4 THE OLD PTT MODEL

Notwithstanding the resiliency of the PTT organizational model, the pace has quickened internationally for less government ownership and other liberalizing initiatives. The international telecommunications and information marketplace today reflects the tensions that predictably arise when parties pursue different, and in some ways conflicting, goals. On the one hand, this marketplace reflects a substantial impact on policies driven by a blend of technological, economic, and multinational competitive forces. On the other hand, it also reflects persistent efforts on the part of some governments to blunt the level of competition to their benefit [2].

The PTT model no longer solely represents how foreign nations structure their telecommunications industries. A number of factors explain this change:

1. Successful experimentation with private initiatives in former PTT model nations like the United Kingdom;
2. Greater user sophistication and ability to bypass inflexible and expensive carriers;
3. The overlap of telecommunications with information processing technology;
4. A more businesslike approach by the operational side of the PTT.[5]

With increasing frequency, nations have considered ways to foster limited competition that enhances user welfare without threatening the PTT's ability to provide subsidized service and to serve as the carrier of last resort to rural users. Nations typically pursue incremental changes to the PTT model such as:

- Spinning off the postal operation into a separate money-losing enterprise, as in the United Kingdom, Germany, Belgium, New Zealand, and Australia;

---

5. "Against this 'conservative' (PTT) perspective, we have to compare the North American approach, which has generally favoured the use of private enterprise for the provision of telecommunications services...It is true that, in part under the pressure of new technology and capital users, we are now witnessing a 'commercialization' of telecommunications together with a weakening of the 'territorial' perspective of how monopolies see their rights. This phenomena will continue." (R. Butler, "The Changing Telecommunications Environment," *Telecommunications J.*, Vol. 55, No. 2, 1988, pp. 130, 132–133.)

- Separating regulation from operations by creating an independent regulator, or spinning off operations from the Ministry of Posts and Telecommunications, as with the Office of Telecommunications, "Oftel," in the United Kingdom, the Directorate for General Regulation in France, and the Australian Telecommunications Authority, "Austel," in Australia;
- Liberalizing the environment in which the PTT operates by changing its service mandate and charter, as in France, Belgium, the Netherlands, and Australia;
- Privatizing the PTT, for example, through the issuance of publicly held stock as has occurred in the United Kingdom, Mexico, Spain, Argentina, Portugal, Hungary, Italy, Denmark, Chile, the Philippines, Guyana, New Zealand, and Japan;
- Pursuing deregulatory initiatives, such as authorizing market entry of at least one more basic service carrier, as has occurred in the United Kingdom, Canada, Australia, Korea, and Japan, and authorizing market entry of value-adding carriers who either construct new facilities or lease lines, as has occurred in the United Kingdom, Sweden, Australia, Japan, New Zealand, Chile, and Canada.

3.5 NEW PTT MODELS

3.5.1 The "Rehabilitated" PTT

The "rehabilitated" PTT creates the impression that government has ordered revolutionary changes in the PTT. On closer scrutiny, the innovations are more pragmatic and less extreme. Typically, the telecommunications monopoly remains intact, albeit mandated to become more businesslike or perhaps even privatized. Government may create a new, independent regulatory regime.

If government retains ownership, it may nevertheless pursue an aggressive campaign to upgrade the infrastructure. Germany and France have retained government ownership of the PTT, but have mandated high-quality networks and superior service from employees now exempt from civil service rules and limitations. Developing nations like Vietnam, Thailand, and Indonesia have conferred service- or location-specific franchises on foreign operators like Telstra of Australia and AT&T.

Where the PTT franchise has been sold to a private venture, typically government continues to enforce social obligations, such as universal service and providing POTS at less than cost so that the poor, homebound, and elderly have "lifeline" access to essential services. Government may hold a "golden share," that is, one share in a special class of stock that can be cast to veto any corporate decision deemed not in compliance with the national interest. Nations fitting this model include the Netherlands, the U.K., New Zealand, Belgium, Switzerland, and Portugal.

3.5.2 The New and Improved PTT

The new and improved PTT acquires much-needed flexibility to compete against market entrants. This model contemplates some degree of foreign involvement or investment and deregulation to inject niche competition into services not reserved for the PTT. Liberalizing policies free the PTT to become more efficient and less bogged down with noncompensatory social obligations. If the PTT has not been privatized, it certainly has become corporatized (i.e., granted a new charter to become more like a business than a government agency). The PTT typically has no obligation to return profits to the national treasury, and in fact may use some of these funds to acquire international market share individually through acquisitions and jointly through strategic ventures and alliances. Sweden, Singapore, Japan, Korea, Australia, Canada, and Mexico have applied elements of this model.

Recognizing the inevitability of competition, or at least financial threats to captive markets, PTOs have responded with a number of strategies:

- Improving service and responsiveness with an eye toward becoming the carrier of choice;
- Conceding niche markets to competition in exchange for reserved services and new rate-setting flexibility;
- Investing cash reserves in foreign growth markets to become less reliant on domestic markets:
  - Participating in the privatization sweepstakes;
  - Offering stock to the public—U.K., Denmark, Japan, Mexico, Spain, Argentina, Chile, the Philippines, Portugal;
  - Building, owning, and operating—New Zealand, U.K., Japan, the Commonwealth of Independent States (CIS), Eastern Europe;
  - Building, operating, and transfering—Indonesia;
  - Building, transfering, and operating—Vietnam;
  - Securing a franchise/licensed joint venture—China (AsiaSat);
  - Pursuing a strategic alliance.

3.6 THE PROCOMPETITIVE MODEL

With increasing frequency, both developed and developing nations have rejected the status quo and have embraced competition, at least in segments of the telecommunications industry. Many procompetitive nations have also converted the incumbent telephone company from public to private ownership and have liberalized rules and regulations applicable to the incumbent so that it can operate more flexibly. Other nations have gone further by authorizing market entry by one or more competitors of the incumbent PTT.

3.6.1 Objectives for Introducing Competition

Nations embracing the procompetitive model argue against the natural monopoly justification and reject the view that the incumbent has generated more social benefits than what competition could possibly accrue. Many procompetitive nations deem the incumbent carrier's performance inadequate as measured by such statistical indexes as the number of telephone lines per 100 inhabitants, the number of employees per telephone line, the average waiting time for service installation, and the probability of securing a dial tone. Competition theoretically would stimulate efficiency and responsiveness to consumers, and would spur a race to make facilities and services available to users who might otherwise subscribe to the offerings of competitors.

Procompetitive nations reject the belief that telecommunications supports a natural monopoly, that is, a single enterprise that can provide service at the lower cost (economies of scale) while serving social and political policies like universal service and the deliberate underpricing of some services. Nations favoring the status quo believe that centralized planning can optimize investment without wasteful duplication of resources and without abandoning quality and availability of service concerns.

Lower Rates

Advocates for competition point to empirical evidence that competition will lower rates for service, particularly for offerings that have served as a source of cross-subsidies for other underpriced services. Rates do drop, but they also tend to stabilize over time, perhaps as competitors compete more selectively (e.g., for high-volume users), and perhaps also because of homogeneous costs in providing service. For example, carriers typically have roughly the same satellite and submarine cable costs, unless one could generate savings through larger capacity acquisitions.

In addition, rates may stabilize because only a few operators exist, notwithstanding policies favoring market entry. If a market equilibrium exists with only a few operators, they may conspire explicitly through price fixing or implicitly through consciously parallel-pricing; that is, the dominant carrier sets an "umbrella" price, with competitors matching or slightly undercutting that rate. A market with few competitors tends to develop into an oligopoly with limited price competition.

Robust competition can drive rates toward costs, thereby eliminating the financial burden on users who previously had underwritten cross-subsidies by paying rates well in excess of cost. Cost-based competition may lead to pricing based on the carrier's perception of users' opportunities to take service from another carrier or to construct private facilities. Pricing on the basis of users' elasticity of demand and supply typically means that large-volume users will receive service at the lowest per-unit rates. If these users previously paid significantly higher rates, then competition will present them with a major financial benefit at the expense of those with fewer options who secured service at averaged rates.

Service Diversity and Responsiveness

Users in some countries with monopoly PTT service have complained that half of the population has yet to secure telephone service, while the other half cannot get a dial tone from their telephones. Competition can stimulate supplier responsiveness in terms of price, service packages, and speed in resolving service problems and outages. The Canadian Radio-Television and Telecommunications Commission endorsed this view [3]:

> No matter how benignly disposed a single supplier is, or how technologically advanced it strives to be, it cannot differentiate adequately enough or quickly enough to service the particular requirements of a variety of customers.

Competition can also spur improvements in service quality and the speed with which carriers introduce new services and exploit technological innovations.

Global Competitiveness

Nations embrace competition as a vehicle to make the country attractive to foreign investment and to make national carriers attractive candidates to provide service to multinational enterprises operating in many locations. A world-class domestic telecommunications infrastructure, whether the product of competition or public policy initiatives in nations maintaining a monopoly (e.g., France, Germany, and Singapore), provides a comparative advantage in two ways:

1. It distinguishes the nation as an attractive location for conducting business, particularly for enterprises that require instantaneous, reliable, feature-rich, and inexpensive telecommunications;
2. It stimulates the efficiency, nimbleness, and skillfulness necessary to win global, one-stop-shopping tenders for turnkey (ready-to-use) global or regional networks.

A globally known player in telecommunications and information processing may also showcase the home market as a candidate for aggregating regional traffic, that is, serving as a hub for managing traffic originating and terminating from several locations within a region.

Benefits Outweigh Costs

Procompetitive nations conduct a cost-benefit analysis and conclude that the gains from competition outweigh the costs. This outcome differs from previous evaluations where regulators concluded that competition would ruin incumbent carriers by stranding their facility investments, as high-volume and the most profitable users migrate to newcomer services or find ways to bypass the incumbent carrier's most expensive services and facilities. Decision makers in procompetitive nations recognize the potential for traffic and revenue migration from the incumbent, but expect that this adverse outcome will be offset by the increased demand stimulated by competition. A growing market compensates for lost market share, making it possible that incumbent carriers will not suffer a downturn in either revenues or facilities utilization.

Why Support Competition Now?

Nations embracing competition appear to have an increasing sense that they must act quickly and definitively. The failure or delay in seeking to capture the perceived benefits of competition can generate the risk that other nations in the region will do so. Technological innovations have created a greater premium on timely action. New opportunities to aggregate traffic at hubs means that some of the traffic that otherwise would have directly originated and terminated incountry might in the future follow a circuitous routing via another carrier's hub in another country. Since technological innovations have reduced the distance sensitivity in telecommunication service costs, such routing has no price handicap. The lack of technological wherewithal to determine the origination point of traffic means that traffic routed via a hub may be injected into other countries' PSTN without being designated international traffic. Accordingly, nations inclined to support market entry may perceive the need to act so that national carriers can start the process of protecting their vulnerable markets and entering the vulnerable markets of others.

3.6.2   Revamping Regulatory Oversight

Many nations have also revamped the structure and scope of telecommunication regulation by making the overseer independent of the private or public carrier. Separating regulatory and operational roles enhances the possibility that government can establish rules and regulations that promote full and fair competition. This referee function is particularly important where a nation author-izes facilities-based competition and the incumbent carrier seeks freedom from prior regulations to compete even as it provides facilities needed by new com-petitors to originate and terminate service. An independent expert regulatory agency can balance the interests of incumbents versus newcomers while achiev-ing procedural and substantive fairness by:

- Allowing the public an opportunity to participate in the decision-making process by filing comments and data;
- Generating a factual record;
- Explaining how the record supports the final decision or policy;
- Standing ready to defend its decisions before an appellate court.

Regulation often requires carriers to reduce profits and comply with rules designed to serve the "public interest, convenience and necessity" [4].

3.6.3 Regulated Competition

In nations adopting the competitive model, international telecommunications carriers typically operate with somewhat less government oversight than the PTT model does, but the industry remains subject to some degree of government oversight. Market entry eliminates the need for much of the traditional regula-tion designed to safeguard the public from monopoly abuses like price gouging because, theoretically at least, no competitor can affect the price or supply of equipment and service. Advocates for the procompetitive model seek to limit regulation because it artificially segments the marketplace and creates barriers to competition.

The international telecommunications marketplace in the United States and the United Kingdom represents the regulated competition model. Telecom-munications competition does exist in these nations, but the carriers remain sig-nificantly regulated, which has the effect of curbing the scope of competition.[6] Despite over 15 years of progressive deregulatory initiatives, many U.S. and U.K. international telecommunications markets remain concentrated and domi-nated by a few major carriers.

Fostering facilities-based competition constitutes a major step even for na-tions inclined to support competition. At most, a nation will increase the number of carriers to two or three. The U.K., Australia, and Korea have adopted a duopoly and Japan has three international carriers. In the satellite facilities

---

6. See S. Chiron and L. Rehberg, "Fostering Competition in International Communications," *Fed-eral Communications Law J.*, Vol. 38, No. 1, 1986, p. 1; see also R. Frieden, "International Tele-communications and the Federal Communications Commission," *Columbia J. Transnational Law*, Vol. 21, 1983, pp. 423–485.

marketplace, even fewer nations have supported any type of competition. Except for the U.K. and Chile, nations typically confer a monopoly franchise, thereby requiring other carriers to pay a marked-up rate for their indirect access to the space segment. In the United States, facilities-based competition exists, but Comsat Corporation serves as the sole investor in the world's primary international satellite facilities operator, INTELSAT, which is a cooperative of more than 135 nations that handles approximately 80% of the world's international fixed-satellite requirements (i.e., using satellites to transmit voice, data, video, and other services required by users in different stationary locations). As the preeminent operator of international satellites, INTELSAT operates satellites parked in geosynchronous orbits above the world's ocean regions, where their transmission coverage, known as *footprints*, makes it possible to serve almost the entire world.

The Communications Satellite Act of 1962, as amended [5], authorizes Comsat to serve as the sole U.S. signatory to INTELSAT.[7] The FCC recently confirmed that Comsat also constitutes the sole signatory to the International Maritime Satellite Organization (Inmarsat), a growing worldwide cooperative of over 70 nations that provides maritime, aeronautical, and land mobile services.[8]

As the sole U.S. signatory to INTELSAT and Inmarsat, Comsat has authority to operate as a "carrier's carrier": a wholesaler of international satellite capacity to other carriers who in turn serve end users. U.S. international service carriers (USISCs) must acquire capacity through Comsat. These carriers dispute the value of Comsat's intermediary role, arguing that it inordinately marks up capacity relative to the value-adding functions it performs.

The FCC recognizes that Comsat's market position creates the potential for anticompetitive practices and classifies it as a *dominant carrier* subject to more extensive regulatory oversight and tariff scrutiny than required of *nondominant* carriers.[9] The Commission has scrutinized and ordered changes to Comsat's

---

7. The FCC has investigated Comsat's corporate structure to ensure that it does not subsidize unregulated ventures with regulated service revenues. See Comsat Study—Implementation of Sec. 505 of the International Maritime Telecommunications Act, 77 FCC 2d 564 (1980), 90 FCC 2d 1159 (1982) (changes in corporate structure ordered); and Second Report and Order, 97 FCC 2d 145 (1984) (reaffirming that Comsat can participate in ventures outside FCC jurisdiction with proper structural safeguards).

8. See Provision of Aeronautical Service via the Inmarsat System, Application for Authority to Participate in an Inmarsat Program to Provide for Aeronautical Bandwidth of Second Generation Satellites, CC Docket No. 87-75, Report and Order, 4 FCC Rcd. 607 (1989) (permitting interim service) and 4 FCC Rcd. 7176 (1989) (authorizing Comsat to serve as sole Inmarsat signatory for aeronautical services).

9. See International Competitive Carrier, 102 FCC 2d 812, 842 (1985), recon. den., 60 Rad. Reg. 2d (P&F) 1435 (1986).

corporate structure and some rates.[10] On the other hand, the FCC has supported facilities-based competition in international satellite service,[11] and has considered limited deregulation of Comsat.[12]

3.7 MARKET SEGMENTATION

Telecommunication regulation can result in market segmentation and concentrated market shares if it determines who can enter what markets and consciously limits the scope and extent of competition. Legislation or regulation may reserve a monopoly in basic services for the incumbent carrier. Many nations prohibit or limit facilities-based competition (i.e., competition from carriers that own and operate transmission facilities) rather than lease and resell the capacity of others. Even the most progressive nations act with caution in opening up markets to competition, lest they inadvertently handicap the ability of incumbent firms to achieve social policies such as universal service. In seeking to provide ubiquitous service, carriers typically average costs among profitable, dense routes and money-losing, sparse routes. Likewise, they implicitly or explicitly engage in cross-subsidization between classes of service and types of users.

In view of the typical claim by incumbent carriers that competition will handicap or prevent them from achieving the universal service mission, newcomers must convince both policy makers and the incumbent of the competitive benefits. They must also seek authorization to interconnect facilities with the incumbent carrier, who may claim that such access will provide a way for

---

10. See also Communications Satellite Corp., CC Docket No. 90-634, 2 FCC 2d 3306 (Common Carrier Bur. 1987) (tentative conclusion that Comsat must refund $61.7 million in revenues, exceeding a 1987 rate of return prescription), 3 FCC Rcd. 2643 (1988) (reducing refund amount to $3.9 million), 3 FCC Rcd. 1130 (Common Carrier Bur. 1988), and 4 FCC Rcd. 5250 (Common Carrier Bur. 1989) (refund methodology approved); Communications Satellite Corp. (Caribnet), 2 FCC Rcd. 2420 (1987) (investigating a geographically specific discount tariff alleged to be a predatory response to market entry by PanAmSat) and 2 FCC Rcd. 2430 (1987) (discussing tariff withdrawal).

11. See R. Reagan, Presidential Determination No. 85-2, 49 Fed. Reg. 46,987 (28 November 1984), implemented in Establishment of Satellite Systems Providing International Communications, 101 FCC 2d 1046 (1985), on recon., 61 Rad Reg. 2d (P&F) 649 (1986), on further recon., 1 FCC Rcd. 439 (1986), policy liberalized in, Permissible Scope of United States Licensed International Communications Satellite Systems Separate from the International Telecommunications Satellite Organization (INTELSAT), 7 FCC Rcd. 2313 (1992), further liberalization, 9 FCC Rcd. 1282 (1994) (eliminating all restrictions on separate system access to the PSTN in 1997).

12. For example, in 1993 the FCC granted Comsat a waiver to provide, on an unseparated basis, both Inmarsat space segment and value-added services, which include software and can provide one-stop shopping to users. (Communications Satellite Corp. Petition for Declaratory Ruling or, in the Alternative, for Partial Waiver of Structural Separation Requirement, Mem. Op. and Ord., 8 FCC Rcd. 1531 (1993).)

users to leak traffic into the PSTN without proper payment to the incumbent carrier for access.

Procompetitive initiatives first target nonessential, niche markets to demonstrate the merits of competition without the risk of harm to essential, core services. Even when nations permit core market competition, incumbent carriers tend not to lose significant market share.

3.7.1   Service Dichotomies in the United States

Until 1980, the FCC maintained a voice and record services dichotomy that limited AT&T to primarily voice services, while a group of international record carriers (IRCs)—which included ITT World Communications and RCA Global Communications—were authorized to provide text services such as telegrams and telexes.[13] The dichotomy initially was based on the fact that carriers used different transmission media: high-frequency radio for voice and terrestrial or submarine cables for record transmissions. The dichotomy persisted as a way to forestall AT&T from dominating the record market when both wireless and wire-line facilities could handle any mode of communication. Record services have substantially declined in importance because of their slow transmission speeds, high cost, and the ability of users to send facsimile, electronic mail, and other data services at low cost and often via devices attached to regular telephone lines.

3.7.2   Resale as a Competitive Stimulus

Enterprises that resell leased lines[14] or provide one-stop shopping for complete turnkey networking generate much of the competition and innovation in the international telecommunications marketplace. There are two kinds of resale:

> 1. *Simple or pure resale*: "Purchasing discounted bulk...services and re-selling them to smaller users as substitutes for...the services they normally would acquire at higher rates, [thereby] creat[ing] pressure on the

---

13. See Am. Tel. & Tel. Co., 37 FCC 2d 1151 (1964), policy reexamined in, Overseas Communications Services, 84 FCC 2d 622 (1980), modified, 92 FCC 2d 641 (1982). Previously, the FCC allowed AT&T and the IRCs to operate separate and noninterconnected record services.

14. See Regulatory Policies Concerning Resale and Shared Use of Common Carrier International Communications Services, 77 FCC 2d 831 (1980). "The primary impact which resale and sharing should have on international telecommunications services is to reduce price discriminations among services which customers perceive to be substitutable. When close substitutes are offered at widely differing prices, arbitrage activities can be expected to force an eventual equality in interservice price/cost relationships." (Ibid., para. 9.) See also International Relay, Inc., 77 FCC 2d 819, on recon., 82 FCC 2d 41 (1980) (authorizing telex service resale using store-and-forward and batch-processing technology).

underlying carrier to set rates for the discounted services which fully recover the costs of providing that service" [6];

2. *Enhanced resale*: Applying computerized, value-adding enhancements to leased lines for customized services, such as credit card verification, that are not equivalent to basic services.

For about ten years, the U.S. was the only nation to permit international simple resale. In the mid-1980s, New Zealand, Canada, and the U.K. joined the U.S. in viewing simple resale as a brokerage function that helps foster efficiency, reduce discrimination, and promote cost-based rates for all user classes. Most nations continue to view simple resale as a threat to PTT income and the ability to restrict service discounts to high-volume users. This is the very reason the FCC has promoted resale in both domestic and international markets: to confer cost saving to all types of users and to reduce the higher margin generated from services to small-volume users.

An increasing number of nations permit enhanced resale provided by international value-added networks (IVANs). The services provided by IVANs tend to stimulate demand for PTTs' underlying basic transmission services rather than encourage users to migrate from highly profitable services to less profitable ones or to bypass entirely incumbent carrier facilities.

3.7.3 Facilities-Based Competition

Few nations have authorized facilities-based competition (i.e., entry by carriers that have installed their own transmission capacity). Because these new carriers lack market power, the FCC has created classifications that impose few regulatory burdens.[15] The private, non-common-carrier classification allows operators to use contracts rather than tariffs and to avoid a legal obligation to serve all who request service.[16] The FCC considers the capacity of such carriers non-

---

15. The FCC has also created a distinction within the traditional common carrier classification. Dominant carriers (e.g., Comsat in its role as monopoly conveyor of INTELSAT satellite capacity to U.S. carriers) remain subject to conventional common carrier regulation, although even this degree of oversight has been streamlined. Nondominant carriers lacking market power may qualify for regulatory forbearance, meaning that many of the conventional common carrier regulations are waived or substantially mitigated. See International Competitive Carrier, 102 FCC 2d 812 (1985), recon. den., 60 Rad. Reg. 2d (P&F) 1435 (1986).

16. See National Association of Regulatory Utility Commissioners v. FCC, 525 F.2d 630 (D.C. Cir. 1976) [NARUC-I], cert. den. sub nom., National Association of Radiotelephone Systems v. FCC, 425 U.S. 992 (1976). "Factors that indicate non-common carrier operations include the existence of long term contractual relationships, a high level of stability in the customer base, and individually tailored arrangements." (LIGHTNET, Sec. 214, Application to Construct a Fiber Optic System in Florida as Part of an Interstate Network, 58 Rad. Reg. 2d (P&F) 182, 185 (1985), citing NARUC-I at 643.) See also Domestic Fixed Satellite Transponder Sales, 90 2d 1238 (1982), aff'd sub nom., World Communications, Inc., v. FCC, 735 F.2d 1465 (D.C. Cir. 1984).

essential and "a supplement to, and not a substitute for, common carrier capacity" [7]. Examples of international private carriers include satellite systems separate from the INTELSAT cooperative, such as Pan American Satellite, and private submarine fiber-optic cable systems, such as the PTAT-1 joint venture of US Sprint and Britain's Cable & Wireless.

3.8 INTERNATIONAL TELECOMMUNICATIONS POLICY DOES NOT MIRROR DOMESTIC EFFORTS

Try as they might, competition advocates can only encourage other nations to abandon monopoly support and embrace marketplace competition. They would like to see other nations separate (unbundle) telephone equipment, which they believe can be provided competitively, from regulated telephone service. Procompetitive advocates would like other nations to establish network interconnection policies that do not discriminate against new carriers and individuals who want to attach a device not manufactured or bought from the service provider.

If governments do not permit facilities-based competition, advocates for change at least would like to see liberalized policies regarding shared use, resale, and enhancement [8] of leased lines. The procompetitive model limits regulation to core public utility functions, thereby enabling resellers to provide value-adding services that use basic lines as building blocks. So long as enhanced-service providers do not offer a service directly competing with core services provided by the PTT, the advocates believe they present no risk and should receive authorization to operate.

Given what often appears to be a preoccupation with privatization, liberalization, deregulation, and globalization in telecommunications, some predict that the international telecommunications marketplace may soon duplicate what has occurred in progressive nations like the United States, the United Kingdom, Australia, and New Zealand. But at least for the foreseeable future, few nations fully endorse the view that they should refrain from owning telecommunications operators or from creating a private or public monopoly carrier, or at best a duopoly of two facilities-based carriers. Even in progressive nations like Australia, Canada, Korea, Japan, Sweden, and the United Kingdom, the number of facilities-based carriers is officially limited, or the marketplace has supported only two or three carriers. Relatively few nations have completely abandoned the view of marketplace failure in telecommunications, that is, the belief that the marketplace cannot be relied upon to maximize efficiency, enhance consumer welfare, and reduce costs while at the same time furthering social goals like universal affordable, if not subsidized, basic services.

Shifts in political and economic philosophies resulting from transfers of power make it difficult to predict policy outcomes in any particular nation. In some instances, the imperatives of technology, economics, and competition will prevail and foster progress. In other situations, anachronistic policies remain in the face of technological or economic support for change.

However, two conclusions can be reached with certainty:

1. The international telecommunications marketplace will *not* fully track the deregulatory initiatives of the last 20 years that have been experienced in the United States.
2. The United States cannot unilaterally export its procompetitive philosophy, despite ample proof domestically that such policies have enhanced consumer welfare and promoted efficiency.

The international telecommunications environment often appears to favor institutionalized cross-subsidies, centralized management, and full employment at the expense of efficiency. The resilient PTT model incorporates pricing policies reflecting value of service and social engineering as opposed to disciplined cost allocation studies. Only over time and initially only in niche markets will a new international regime more closely approximate the procompetitive model.

3.9 MORE NATIONS ADOPT SOME PROCOMPETITIVE INITIATIVES AND INCUMBENT CARRIERS ADAPT

Consumers and technological innovation are the primary drivers fostering change in telecommunications. Few countries can tolerate the financial and political fallout from failing to establish a mechanism for guaranteeing reliable telephone service. High-volume corporate users can increasingly resort to strategies that make it possible to access services that the incumbent carrier cannot or will not offer. If the incumbent is so unable to finance infrastructure improvements, then large-volume users may resort to self-help by securing the legislated right to install new technologies that are no longer so expensive or complex that only telephone companies use them. If the incumbent is unwilling to innovate, perhaps because of fear that users will migrate to cheaper services, individual corporate users may attempt to bypass the uncooperative or incompetent PTT. Other users, including small businesses and individuals, have access to new illegal or gray market options that exploit technological innovations and the inability of the PTT to detect and prevent such bypassing. Such self-help efforts include the use of call-back services that allow callers in nations with high international call rates to secure a dial tone in nations with lower rates. In fact any satellite Earth station can provide the basis for accessing a cheaper, but not necessarily legal, transmission option.

Slowly, many governments have grown less willing to unconditionally accept PTT claims that large-volume user migration (i.e., bypassing) would jeopardize the ability to improve service. Governments increasingly side with consumers who demand the right to choose whether to operate their own facilities or to select from a number of carriers and services. Users increasingly view telecommunications as both a key cost center and a vehicle to achieve a comparative advantage in the global marketplace. In turn, governments have come to believe that a more efficient telecommunications system can stimulate the national economy. These officials grow impatient with incumbent claims that they are improving the situation with all deliberate speed.

Nations typically adopt change and procompetitive initiatives in an incremental manner. Even with changed marketplace conditions, national governments do not expect to eliminate their regulatory function. It may change in scope and pervasiveness, but [9]:

> [E]xperience suggests...that particularly during the early period of any transition from a non-competitive (or minimally competitive) market to a competitive (or more competitive) one, the government will have to oversee and manage change.

Government must closely oversee the behavior of incumbents during the transition from a closed, often monopolized environment, to one where market entry exists in an increasing number of market sectors. Existing equipment and service providers recognize the likelihood of significant loss of market share and the potential for reduced earnings unless the overall market grows. Incumbents have every incentive to delay, through litigation and regulatory challenges, the onset and impact of competition. In the marketplace, they may stifle competition by anticompetitive practices that include selective price cuts, where the incumbent sets prices below their costs to drive out competition. Incumbents may also try to deny competitors access to their facilities or provide inferior and overpriced interconnection to facilities needed by market entrants who have installed only limited networks or who operate using leased lines.

The history of interconnection disputes in the United States points to the ability and potential for incumbents to abuse their control over essential "bottleneck" facilities. Interconnection between the facilities of different carriers is essential to maximize network access and promote universal service goals. Market entrants typically do not erect networks that are completely parallel with that of the incumbent carrier. To avoid waste and to concentrate on productive deployments of network facilities, newcomers typically rely on existing *local loop* facilities of incumbent carriers to originate and terminate service. Newcomers often rely on incumbents carriers to provide call aggregation functions and to carry traffic over the "first and last miles" closest to the caller and call recipient.

Facilities of incumbent carriers used by newcomers are the choke points of potential competition, because new competitors cannot survive without efficient and cost-based access. Governments must ensure that the incumbent carrier provides equal access to such facilities, as is required of common carriers and public utilities.

The installed base of network equipment and a large subscriber base provide incumbents with the capacity to respond to competition. Governments must ensure a fair competitive response rather than anticompetitive practices that tilt the newly competitive playing field even farther in the direction of the incumbent. The potential for market manipulation, disputes between newcomers and the incumbent carrier, reductions in service, and immediate and substantial rate hikes mean that governments will have to maintain a role as regulator and referee.

There have been several challenges to the traditional PTT model. Concerns about spreading risk, achieving scale economies, and institutionalizing cross-subsidies favored government-owned or franchised "natural" monopolies. The same factors supported cooperation or collusion in international facilities investment, management, and toll revenue division. The only conflict was over which "national hero" would win in a relatively open equipment RFP.

Increasingly, nations have challenged the theoretical underpinning for a single PTT and for government ownership. Telecommunications is viewed as an essential stimulus to the national economy and nations cannot afford to have half the population waiting for a phone and the other half waiting for a dial tone!

More nations have undertaken all of some of the following:

- Creating an independent regulatory agency separate from the telecommunications operator;
- Incremental or partial deregulation while reserving core services for the incumbent carrier;
- Encouraging the PTT to make foreign investments, provided it does not handicap domestic operations;
- Compelling PTTs to provide cost-based access to the basic network.

3.10 THE CUTTING EDGE

Nations like the United States and New Zealand have adopted the cutting-edge model, where the support for competition leaves little, if any, services reserved for the incumbent. Other nations (e.g., the United Kingdom) intend on migrating to this model in time by announcing a future date when open entry will replace a facilities-based duopoly. Typically, carriers in this model generate 25% or more of their revenues from international ventures and strategic alliances with other carriers, service providers, and equipment manufacturers. They intend to operate globally and will exploit opportunities to create traffic routing hubs that dominate a region. Regulators in nations supporting this model permit pricing flexibility, but may impose access requirements to ensure that newcomers can compete on a level playing field (i.e., that they pay the same rate for access to the PSTN as the incumbent carrier charges corporate affiliates).

New Zealand, rather than the United States or the United Kingdom, reflects the cutting edge in telecommunications policy. This nation has substantially deregulated the entire sector and has abandoned virtually all regulatory barriers to market entry. No nation has so endorsed the concept of governmental noninvolvement.

The different carrier models now in place are summarized in Table 3.1.

Table 3.1
International Carrier Models

Old Model	New Old Model	New Model	Cutting Edge
Greece	Germany	Japan	United States
Africa	France	Korea	United Kingdom
Central America	Netherlands	Australia	New Zealand
Caribbean	Belgium	Canada	
	Switzerland	Sweden	
	Malaysia	Mexico	
	Singapore	Argentina	

The carrier models described in Table 3.1 can be placed in the following categories:

- *Old Model:* Like the predivestiture Bell System, this model represents inflexibility and the mindset of a natural monopolist. The status quo favors the view that "the system is the solution," and "one system with end-to-end responsibility." Customers are captive to a single service provider and perhaps to a single "national hero" manufacturer.
- *New Old Model*: Government refrains from near-term privatization but undertakes a campaign to corporatize and streamline the incumbent carrier through a new corporate charter that exempts employees from the civil service, like Amtrak. "Enlightened paternalism" is an option where gov-

ernment commits to infrastructure improvement. Large-volume users are treated like commercial accounts. Government separates telecommunication operations from regulator.

- *New Model*: Liberated policies free the PTT to compete with one or more new private enterprises. Deregulation injects a degree of competition. The PTT may be privatized, but certainly becomes corporatized. PTT earnings are reinvested and become a source for global pursuits, such as strategic alliances, poaching, and hubbing, rather than being merely cash flow to the national treasury. Foreign investment is welcomed to a point.

- *Cutting Edge*: The government permits little, if any, reserved services to the incumbent carrier. Duopolies are phased out leading to virtually open entry. International ventures generate 25% or more of company revenues. Strategic alliances provide market access where individual approach is not possible or is ill-advised. Regulators permit pricing flexibility, but impose access requirements to establish a level competitive playing field.

References

[1] Noam, E., "International Telecommunications in Transition," in *Changing the Rules: Technical Change, International Competition, and Regulation in Communications*, R. Crandall and K. Flamm, eds., Washington, D.C.: Brookings Institution, 1989, p. 258.

[2] Dept. of Commerce, National Telecommunications and Information Administration, "Maintaining America's Strength in Markets for Telecommunications and Information Services," Chap. 6 in *Telecom 2000: Charting the Course for a New Century*, Washington, D.C.: GPO, 1988, p. 118.

[3] Canadian Radio-Television and Telecommunications Commission, "Competition in the Provision of Public Long Distance Telephone Service and Related Resale and Sharing Issues," Telecom Decision CRTC-92-12, 12 June 1992, p. 57.

[4] Communications Act of 1934, as amended, 47 U.S.C. Sec. 302 (1990).

[5] 47 U.S.C. Sec. 701 et seq. (1990).

[6] Regulatory Policies Concerning Resale and Shared Use of Common Carrier Domestic Public Switched Network Service, 83 FCC 2d 167, 169 (1980), on recon., 86 FCC 2d 826 (1981). (The FCC proposed to apply its domestic resale policy internationally in International Resale Policy, 77 FCC 2d 831 (1980).)

[7] Inquiry Into Policies To Be Followed in the Authorization of Common Carrier Facilities To Meet North Atlantic Telecommunications Needs During the 1991–2000 Period, CC Docket No. 79-184, Report and Order, 3 FCC Rcd. 3979, 3987 (1988) (excluding private international carrier capacity from the North Atlantic consultative process assessment of capacity requirements).

[8] Second Computer Inquiry, Final Dec., 77 FCC 2d 384, mod. on recon., 84 FCC 2d 50 (1980), further mod., 88 FCC 2d 512 (1981), aff'd sub nom., Computer & Comms. Ind. Ass'n v. FCC, 693 F.2d 198 (D.C. Cir. 1982), cert. denied, 461 U.S. 938 (1983); Third Computer Inquiry, Report and Order, 104 FCC 2d 958 (1986), mod. on recon., 2 FCC Rcd. 3035 (1987) (Phase I), further recon., 3 FCC Rcd. 1135 (1988); Phase II, CC Docket No. 85-229, Report and Order, 2 FCC Rcd. 3072 (1987), recon. denied, 3 FCC Rcd. 1150 (1988), partially reversed and remanded sub nom., California v. FCC, 905 F.2d 1217 (9th Cir. 1990), on remand, 6 FCC Rcd. 7571 (1991), partially reversed and remanded sub nom., California v. FCC, 4 F.3d 1505 (9th Cir. 1993) and 39 F.3d 919 (9th Cir. 1994).

[9]   Director, M. D., *Restructuring and Expanding National Telecommunications Markets: A Primer on Competition, Regulation and Development for East and Central European Regulators*, Washington, D.C.: Annenberg Washington Program, 1992, p. 31.

The Technologies in Modern International Telecommunications

4

The recent and dramatic changes in international telecommunications in large part have resulted from the creative and entrepreneurial application of techno-logical innovations. Financial barriers to market entry have fallen to the extent that one can seriously question whether a single international telephone company can operate the most efficiently by achieving scale economies (i.e., the lowest cost per unit of capacity for all service types). Just as carriers like MCI demonstrated that competition could flourish in the domestic U.S. long-distance telephone business, market entrants have challenged the notion that international telecommunications can only support a single, "natural" monopoly.

MCI entered the switched long-distance telephone business with a single microwave radio backbone between Chicago and St. Louis. Thanks to regulatory and technological developments, new international telephone companies can begin to compete without any owned transmission facilities. Innovations in telecommunication switches and interfaces with customers make it possible to enter the long-distance business as a reseller of capacity, leased from facilities-based carriers.[1]

Other entrepreneurial companies stretch domestic law and recommendations of the ITU by installing switching devices that provide call-back services: providing dial tone access to cheap-rate domestic or international calling to users physically located in nations with high rates. The call-back service operator[2]

---

1. Gregory C. Staple provides an insightful analysis of how converging technologies and regulatory changes provide consumers with new opportunities for cheaper and diverse services. (G. C. Staple, "International Telecommunications: The Challenge of Convergence," in *TeleGeography 1994*, G. C. Staple, ed., Washington, D.C.: International Institute of Communications, 1994, pp. 11–33.)

2. For an outline of how call-back service operates, see D. Briere and M. Lagner, "Automated Call-Back Services: An Update," in *TeleGeography 1993*, G. C. Staple, ed., Washington, D.C.: International Institute of Communications, 1993, pp. 38–39.)

makes it possible for callers to establish a virtual presence in another country simply by placing an international call and hanging up as the connection is made. The call-back operator's *boomerang box* can identify the caller from stored information, dial out to that caller, and thereby provide dial tone for outbound calls. The circuitous routing of calls via a boomerang box makes it possible for callers in nations with high rates to avoid being captive to the monopoly PTT. The call-back service operator, physically located in a nation with cheap rates and private-line resale opportunities, can provide a profitable service that bypasses the PTT.

Call-back services demonstrate how the combination of technological innovation and deregulation in some nations increasingly puts pressure on the status quo. Previously, the cost and scale of international telecommunication technology supported aggregating traffic and consolidating investment through cooperatives and consortia. Now, the per-unit cost of telecommunication technologies favors market entry, proliferating services, and pressure on incumbent organizations. For example, new private satellite companies challenge the INTELSAT and Inmarsat cooperatives and new submarine cable joint ventures jeopardize the centralized management achieved when a consortium of many incumbent carriers pools financial resources and traffic requirements.

Technological innovations work to limit or eliminate the need to aggregate traffic and limit consumer options. For example, when satellites were first commercialized, nations universally authorized the creation of a monopoly that initially operated a single Earth station with antennas 30m or more in diameter. Decision makers agreed that monopoly status was necessary to ensure that the operator could recoup a sizable and risky investment. Technological innovation has now shrunk the size of international Earth stations to 1m or less, and the number and location of Earth stations have vastly increased. Rooftop antenna and very-small-aperture terminals (VSATs) do not require monopoly control, despite efforts by the incumbent operator to perpetuate the status quo.

4.1 TECHNOLOGY DEVELOPMENTS

4.1.1 Satellites

Satellites have traditionally provided a "bent pipe" in space for receiving signals and retransmitting them back to Earth.[3] In this mode, satellites operate like

---

3. For a further examination of the technology and economics of satellite telecommunications, see M. L. Smith III, "The Orbit/Spectrum Resource and the Technology of Satellite Telecommunications: An Overview," *Rutgers Computer and Technology Law J.*, Vol. 12, 1987, pp. 285–304; H. Hudson, *Communications Satellites: Their Development and Impact*, New York: Free Press, 1990; and M. Giget, "Economics of Satellite Communications in the Context of Intermodal Competition," *Telecommunications Policy*, Vol. 18, No. 6, 1994, pp. 478–492.

the ionosphere for short-wave radio broadcasts: they simply reflect signals transmitted to them. The conventional term used to describe this function and to measure satellite capacity is the *transponder*, typically a unit equivalent to 36 MHz of bandwidth. A transponder receives a signal, amplifies it, and transmits it to Earth stations.

To minimize the potential for geographic coverage of downlinked signals (its footprint), satellite operators position satellites in an orbit where the satellite appears stationary relative to Earth, making it easier to point transmitting antennas called *Earth stations*. Only a small sliver of space, approximately 22,300 miles above the equator, has the physical properties of holding an object in synchronicity with Earth orbit and in an apparently stationary location as well. This orbital arc constitutes a scarce, shared resource, because only in this relatively narrow sliver of space do satellites and the Earth travel at the same speed relative to each other, making the satellite a stable target for transmitting signals upward (also known as *uplinking*).[4]

Recent innovations in satellite technology add intelligence and versatility to the processing capability of the station or change the orbital location of the satellite. New satellites have onboard signal processing, known as *cross-strapping*, which enables operators to transmit on one frequency and receive signals on another frequency. In addition, such processing capability can enable users to change the beam size or location of the signal footprint. For example, a video programming company seeking to distribute a cable television channel might use a large *global, hemispheric,* or *zone* beam to distribute the programming throughout a wide geographical region where cable facility operators have installed large satellite dishes at their Earth stations. The video programmer might want the satellite carrier to cross-strap the programming onto another frequency band or to narrow the footprint to a spotbeam to permit reception of the concentrated signal by smaller dishes, which end users might install at their homes and offices.

Another satellite innovation, known as *intersatellite links* (ISL), enables carriers to transmit signals between satellites. ISLs reduce expense by eliminating the need to transmit down to an Earth station and back up to the another satellite. They also eliminate the approximately 0.5-sec delay a user would incur with such a 44,600-mile routing. A carrier seeking to establish global or multiple-region coverage can link the satellites in orbit, thereby reducing the resulting cost and echo when signals must take a double hop up and down via two Earth stations to link two satellites in different regions of the world.

New satellite service markets support the use of orbits closer to the Earth's surface. For services that require the use of portable, preferably handheld transceivers like those used for terrestrial cellular radio, satellite operators must de-

4. See R. F. G. Hart, "Orbit Spectrum Policy-Evaluating Proposals and Regimes for Outer Space," *Telecommunications Policy*, Vol. 15, No. 1, 1991, pp. 63–74.

ploy their stations closer to Earth. Satellites orbiting closer to Earth need less power to receive and transmit signals. On the other hand, their closer proximity to Earth means that individual footprints will be smaller, and their orbital speed relative to Earth will increase. The combination of smaller footprints and nongeosynchronous satellite operation requires a constellation of satellites whose number grows as the orbits get closer to Earth. Proposed mobile satellite services to handheld transceivers range from 840 refrigerator-sized satellites (Teledesic) operating in *low Earth orbit* to 12 large satellites operating (Odyssey) in *middle Earth orbit*.[5] The Iridium satellite network will employ ISLs to transfer telephone calls from a satellite near the caller to a satellite near the gateway Earth station closest to the intended call recipient.

For some services, satellites have achieved a comparative technological advantage over other transmission media, such as submarine cables and terrestrial microwave facilities. The broad geographical coverage of a satellite footprint favors point-to-multipoint services, such as the delivery of video programming to a number of broadcast or cable television distribution facilities. Additionally, satellites can efficiently and economically deliver traffic to interior locales far from coastal points where submarine cable makes a landfall. Satellites may lack a comparative advantage for high-volume traffic routes, particularly ones with low-cost access to submarine cables.

The cost of launching satellites and the limits on fuel used to keep a satellite in the proper orbit are two major factors affecting the economic viability of satellite technology. On average it costs approximately $75 to $100 million to launch a satellite. Approximately one in four launches fails to place a satellite in the proper orbit. Satellites typically reach the end of their usable life in 10 years, primarily because they have exhausted station-keeping fuel. Expanded launch options and new satellite station-keeping techniques promise improvements in the financial viability of satellites. Until the middle 1980s, only three nations or national alliances had commercial satellite launching capabilities: the United States, the European Space Agency, and the former Soviet Union. In recent years, China has entered the commercial launch marketplace and other nations, which include Israel, Japan, Norway, and Australia, have developed launching capabilities that may have commercial applications. A new station-keeping technology, involving ion thrusters rather than gaseous fuel, promises to help maintain satellite orbits for 15 or more years. As long as the satellite power generation and electronic equipment remain usable, a station-kept satel-

---

5. See R. Frieden, "Satellites in the Wireless Revolution: The Need for Realistic Perspectives," *Telecommunications*, Vol. 18, No. 6, June 1994, pp. 33–36; R. Frieden, "Satellite-Based Personal Communication Services," *Telecommunications*, Vol. 17, No. 12, December 1993, pp. 25–28; and "WARC-92 and Low Earth Orbiting Satellites: A Case Study for the Process for Accommodating Spectrum Requirements for New Technologies," *Proc. of the 15th Annual Conf. of the Pacific Telecommunications Council (PTC'93)*, Honolulu: Pacific Telecommunications Council, 1993, pp. 271–287.

lite can continue to provide service well beyond the current lifetime expectations. Figure 4.1 depicts the Iridium low-Earth-orbiting satellite constellation.

4.1.2 Submarine Cables

Undersea cables provided the first telecommunication links between nations separated by large bodies of water. The first cables supported only a few telegraph channels, but made it possible for social and commercial transactions to occur on a far speedier basis compared to a multiweek transoceanic crossing by ship. After a period of time during which innovations in high-frequency radio made it possible to transmit voice conversations in addition to the previously available telegraphy, submarine cables expanded capacity and again became the lowest-cost routing option for many country pairs. In fact, for many years after the commercial debut of satellites, regulators ordered carriers to activate satellite circuits regardless of whether the medium provided service at a lower cost. Such *balanced loading* of active circuits bolstered the use of satellites to ensure the redundancy of traffic routing and to support the commercial viability of satellite operators.

Most international telecommunications carriers and customers view satellites and submarine cables as complementary media. Satellites provide cost-effective, optimal service for point-to-multipoint applications such as the

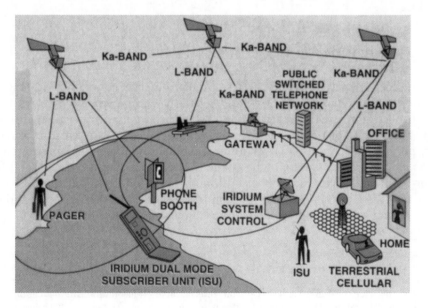

Figure 4.1 Iridium low-Earth-orbiting satellite constellation. (*Source: Telecommunications,* December 1993, p. 27.)

distribution of video programming to a number of broadcast stations and cable television facilities. Cables possess a comparative advantage with their echo-free voice transmissions and are especially suitable for high-volume, point-to-point routes, particularly between countries close to the point where cables make landfalls, thereby reducing the cost of terrestrial *tail circuit* links to the cable. The use of glass fibers in lieu of copper wire, and the associated migration from direct current amplification to low-powered laser transmission, has increased the bandwidth and reliability of submarine cables.

Fiber-optic cables, scheduled for installation in the middle 1990s, will provide transmission capacity equivalent to 600,000 simultaneous voice conversations. New generations of such cables can space signal amplifiers, known as *regenerators*, several hundred miles apart, thereby reducing installation labor and risk of outages. Fiber-optic cables are increasingly suitable for some types of video transmissions, in addition to high-speed data and voice traffic. Recently installed submarine cables provide increased reliability and lower costs because they are configured in a "self-healing" network. Instead of having to resort to satellites for circuit restoration, carriers can achieve a near instantaneous ability to route around a cable outage. An intelligent submarine cable network can sense a break in a cable and route traffic via a parallel cable on the fly. Such functionality means that users will perceive no significant disruption in service, since traffic can continue to the destination via another, perhaps circuitous routing.

4.2 "ENABLING" TECHNOLOGIES

Rarely does a week go by without some mass-market newspaper or magazine heralding the convergence of technologies that will generate a global information infrastructure (GII).[6] A multimedia GII or information superhighway results from the merger of previously discrete technologies and the erosion of regulatory policies that supported such mutual exclusivity. In a digital environment, a bit is a bit is a bit, meaning that a single transmission facility (e.g., a satellite or fiber-optic cable) can support a multimedia environment that bundles voice, data, facsimile, video, text, information processing, financial services, and electronic commerce, including catalog sales, news, and entertainment.

---

6. Vice President Albert Gore envisioned the construction of information superhighways on the basis of five principles: (1) primarily private investment, (2) competition, (3) flexible regulation, (4) open access to the network for all information providers, and (5) universal service. (Remarks of Vice President Albert Gore at International Telecommunications Union, World Telecommunications Development Conference, Buenos Aires, 21 March 1994.) See also U.S. Dept. of Commerce, Information Infrastructure Task Force, "The National Information Infrastructure: Agenda for Action" (Washington, DC: GPO, Sept. 1993); and Progress Report, September 1993–1994 Washington, DC: GPO, Sept. 1994).

The migration from analog to digital transmission technologies, in conjunction with increasing reliance on software-based networks challenges the status quo. Carriers must develop a newfound nimbleness in responding to user requirements, because they can no longer expect any customer base to remain captive. Digitization and software network management makes political boundaries more easily penetrated and accessible to services provided by outsiders. In a software-defined networking environment, service providers can avoid the unwieldiness of hardware and electromechanical switching. Easy network reconfiguration and flexible service arrangements make it possible to aggregate traffic at regional hubs, much like airlines route feeder traffic into and out of airport hubs.

Facilities-based carriers aim to exploit market access opportunities exclusively where possible, but often find it advantageous to negotiate global or regional alliances with other carriers to secure market access opportunities and the ability to provide one-stop-shopping solutions for the requirements of multinational enterprises. Likewise, regulatory initiatives have made it possible for market entry by non-facilities-based resellers of leased lines. Even if some regulatory authorities wish to insulate the national carrier from competition, the porousness of networks makes competition all but unavoidable. Sophisticated users have found ways to migrate some traffic streams away from high-cost, inflexible carriers and into low-cost, software-defined virtual networks created by innovative low-cost carriers or service providers.

Single facilities-based carriers, global strategic alliances of such carriers, and new private-line resellers seek to engineer a patchwork network of lines throughout the world or a region. The software used to configure switches can also manage ad hoc networks that can be reconfigured as a function of user requirements. The lines themselves have grown speedier in terms of bit rate, also known as *throughput,* and are able to handle larger traffic capacity. Digital transmission technologies such as integrated services digital networks (ISDN) and asynchronous transfer mode (ATM) provide a standardized vehicle to carry various kinds of high-volume traffic over the information superhighway. While ISDN involves a standard bit rate of 144 Kbps for the basic rate interface and 1.544 Mbps for the primary rate interface, ATM provides a more flexible, software-defined protocol for switching and routing even larger streams of voice, data, text, and video. Both ISDN and ATM use packet switching, the subdivision of a digital bit stream into uniform packets or cells that can be routed over different physical routes and reassembled in the proper order.

Digitization, software engineering, and new switching and routing technologies work to convert the telecommunications infrastructure from hardwired physical links to ad hoc software-engineered linkages. Visions of a global village seem more plausible when the infrastructure begins to look like the Internet, which has millions of users who can communicate using a protocol that makes millions of nodes and users accessible on a global basis.

4.3 FUTURE SCENARIOS

Marketing hype and blue sky visions aside, there is a bright future for new services and technologies that blend aspects of information processing and telecommunications.[7] The merger of technologies and markets results when creative engineers and entrepreneurs think of new ways to use equipment like the telephone and personal computer and facilities like the twisted wire pair (local loop) and the cable television coaxial cable. The future promises to make it possible for television sets and computer terminals to provide a more diverse array of services and applications than what we now expect from them. An information appliance may integrate the functions of both television sets and computer terminals. Likewise, service providers will identify and serve new markets, with telephone companies entering entertainment markets and cable television companies entering telecommunication markets. Miniaturization of electronic components and users' desire for access to the rest of the world anyplace and anytime favor development of new terrestrial and satellite-delivered mobile services.

No-holds-barred competition contrasts with an international telecommunication industrial structure that until quite recently emphasized monopolies, discrete markets, and rules designed to maintain the status quo. Depending on one's perspective, the status quo supported worthwhile public policies like universal service at affordable and often subsidized rates, or a sleepy, insular, and coddled world of industries administered by government bureaucrats who provided central mismanagement. Even in the United States, less than 20 years ago, FCC policy prevented AT&T from providing textual services and the now defunct IRCs from providing voice services. Well after technological innovations made it feasible for the same line and the same carrier to provide both text and voice services, the FCC sought to render the markets mutually exclusive ostensibly to prevent AT&T from dominating the record service marketplace. Governments throughout the world have attempted to insulate their national carriers from competition, but technological innovations makes such protection less likely. The merger of telecommunications and information processing technologies means that unregulated data processing companies and systems integrators will become involved in the provisioning of leased telecommunication lines, or will at least order them for customers. A volatile clash of market philosophies occurs between such relative newcomers to telecommunications as Electronic Data Systems (EDS), founded by Ross Perot, and government-dominated and monopoly-oriented PTTs.

---

7. See U.S. Congress, Office of Technology Assessment, *Electronic Enterprises: Looking to the Future*, OTA-TCT-600, Washington, D.C: GPO, May 1994.

The stakes have grown in telecommunications, since the sector now constitutes an integral part of information-based economies and is the perceived locomotive for jump-starting national economies.[8] Companies like EDS have entered the telecommunications arena to provide users with one-stop-shopping access to efficient and versatile global networks that are maintained by professionals and enable users to concentrate on their particular line of business, which happens to require cutting-edge telecommunications capabilities. Global markets, operating 24 hours a day, have no tolerance for a nine-to-five service mentality. Additionally, technological innovations make it possible for EDS to apply its systems integration skills to a larger set of customers who rely on telecommunications. Little if any exclusivity remains in information, entertainment, data processing, financial services, message transport, catalog sales, movies, video games, and software delivery. Put another way, few stand-alone markets and product or service distribution channels remain. A business executive can send a hard-copy facsimile via a machine connected to a telephone line, or a soft-copy resident in a computer can be transmitted via the same line. One can view a movie on a pay-per-view basis via satellite, cable television, wireless cable television, or telephone company delivery.

Versatility in communicating means that more options will exist and more players will have an opportunity to seek a market niche. Barriers to market entry will fall as a function of declining facility costs and robust demand generated by multinational enterprises and by small businesses and individuals who for the first time can qualify as potential consumers at home or at work with relatively minor investments. The proliferation of cable television set-top converters, personal computers, modems, wide-screen televisions and satellite dishes attests to consumer interest in new methods for engaging in commerce, communications, and social interaction. The technological innovations that foster merged or newly competitive industries also provide new opportunities for different media to serve different human senses. Multimedia telecommunications blends sight, sound, and human perception into a virtual reality that not only accentuates the enjoyment of video games, but also provides a more realistic simulation environment for testing designs (e.g., in manufacturing) as well as human competency (e.g., jet piloting skills).

Increased interactivity allows people to control their computer-mediated environment. It remains to be seen whether governments can or should try to control the telecommunications environment and achieve a measured and centrally managed introduction of new technologies and services.

---

8. See S. Pitroda, "Development, Democracy, and the Village Telephone," *Harvard Business Review*, November-December 1993, pp. 66–79.

The International Telecommunication Union

5

International telecommunications requires coordination among nations because they must share scarce, collectively owned resources like satellite orbital parking places and radio spectrum. Common rules of the road can promote efficiency, reduce interference, and support ease in interconnection of diverse networks.[1] Governments generally believe that uniformity in terms of technical and operational standards reduces misunderstanding, streamlines equipment production lines, supports universal access to essential services, or fosters market growth. Nations achieve consensus on international telecommunications issues based on a shared view that relinquishing a degree of national sovereignty will accrue ample dividends in terms of enhanced consumer welfare and speedy deployment of equipment and services.

The consensus-building process has grown more difficult. The ITU faces unprecedented challenges from other trade policy and standard-setting organizations to its status as the primary multilateral forum for standard setting in telecommunications and information processing. A rapidly changing telecommunications environment challenges its ability, as a multinational and bureaucratic organization, to operate swiftly and efficiently. As carriers and service providers globalize and become more competitive, individual companies and national governments perceive increasingly attractive incentives to enter markets first and establish rules, regulations, and standards later.

The accelerating pace of technological innovations and the merger of telecommunication and information-processing technologies increase the types of constituencies the ITU must serve. Participants in information-processing markets more frequently operate in a marketplace environment, where enterprises promote incompatible systems operating on different standards, such as Mi-

---

1. For extensive coverage of international satellite communications law, see White and White, Jr., *The Law and Regulation of International Space Communication*, Norwood, MA: Artech House, 1988.

crosoft's Windows versus Apple's System 7 and IBM's OS/2, rather than the ITU model that contemplates cooperation, even among competitors, to establish a single standard.

5.1 WHY NATIONS COOPERATE

The decision to cooperate on telecommunications matters results when nations conclude that they have more to gain by reaching a consensus than what may be lost in terms of advantageous market access and earnings potential. When nations act unilaterally, they expect the innovations and technical standards of national manufacturers and carriers to dominate, at least within the country. When nations go along with an international consensus, they either support widespread sharing of the financial and logistical benefits from single rules and standards, or lack confidence that their domestic companies could dominate in a market-driven process.

Rather than unilaterally establish standards, frequency allocations, and satellite orbital slot assignments, nations typically collaborate bilaterally and multilaterally. They do so out of both enlightened self-interest and the desire to be good global citizens. Self-interest recognizes that a single standard or a reduced number of standards may promote scale economies in production of equipment and reduced transaction costs, such as less inconvenience, delay, and expense in linking national networks. The desire to be a good global citizen stems from the recognition that many of the resources involved, like spectrum and the orbital arc for satellites, are shared, with no ownership right flowing to any single nation.

Nations collaborate on such diverse matters as:

- Reducing actual or potential spectrum interference;
- Ensuring that terrestrial facilities and satellites can operate unimpeded by nearby facilities;
- Establishing standards that, if globally applied, would reduce the number of product lines and improve the potential for manufacturers to achieve production scale economies;
- Agreeing on universal rules of the road that facilitate interconnection of networks and reduce the potential for balkanization, that is, rendering separate networks inaccessible because they operate on incompatible formats, follow different rules, or comply with different standards;
- Promoting the integration of technological innovations and procedures that make telecommunications more efficient and user friendly.

Individual nations risk spoiling the prospects for universality when they act on the temptation to deviate from the consensus. The decision to opt out typically results when a single nation believes its national hero manufacturer or PTT can capture a larger share of financial rewards by establishing a global standard, frequency allocation, or satellite orbital slot claim outside multilateral forums like the ITU. While intellectual property rights such as patents and copyrights were designed to reward innovation, uniform rules of the road allow for multiple manufacturers and carriers while also reducing costs, inconvenience, and misunderstanding.

Other national decisions not to embrace an international standard may result from political or industrial strategies. For example, a nation might enhance the market penetration of its radio stations and satellite carriers by operating at power levels higher than agreed to, on frequencies not allocated for broadcasting, or on unregistered frequencies and orbital slots. The nation may achieve bolstered coverage of a desired geographical area,[2] or attempt to register phantom satellites so that it could resell the orbital slot or delay the onset of competition. It may also improve opportunities to propagandize, or to prevent citizens from receiving signals and programs deemed undesirable.

A nation may also attempt to insulate domestic markets from foreign competition by mandating a standard inconsistent with the international consensus. This strategy may generate short-term benefits by closing procurements or ensuring that national heroes win "public" requests for proposals (RFPs). However, in the long term, such insulation from competition and deviation from international standards may handicap the national hero in competitive tenders outside the home country. Also, such policies may comparatively disadvantage domestic companies, because they may not have access to the latest and most efficient technology, and as a result may not operate the most economically.

In other instances, nations want to support international consensus building, but enterprises chafe at the pace of standard setting and policy making. The ITU process may appear to reach an untimely "consensus by exhaustion," despite speedy technological evolution and marketplace changes. Other concerns include the potential for the process to result in a least common denominator or the involvement of governments and industrial policy, rather than resulting in the marketplace setting rules and standards.

5.2 A BRIEF HISTORY OF THE ITU

Despite growing disincentives and the proliferation of regional bodies, nations still rely on the ITU to establish common rules, regulations, standards, and policies in telecommunications.

---

2. See, for example, United States-Mexico Agreement Concerning Frequency Modulation Broadcasting in the 88 to 108 MHz Band, 9 November 1972, 24 U.S.T. 1815, T.I.A.S. No. 7697.

Founded in 1865, the ITU serves as one of the oldest continually operating international forums.[3] It continues to function because nations recognize the need for conflict management and resolution in telecommunications.[4] Uniform rules of the road make it possible, for example, to dial up any of the more than 600 million telephone lines in the world, for facsimile machines operating over different networks to work, and for transmitting facilities to operate usually without harmful interference.

The ITU evolved when the Austro-German and Western European Telegraph Unions merged to supervise and establish standards for an interconnected regional network in the late 1800s. Operators recognized the need to agree on frequencies and operating rules. Agreement on frequencies meant that international telegrams could be transmitted over radio spectrum in addition to closed-circuit cables. Consensus on operational rules meant, for example, that telegraph companies could designate a particular emergency frequency for 24-hour monitoring and use a single code (i.e., Morse code and a shorthand common sequence of letters to represent frequently used sets of words, such as QRZ to represent the query: "Are you receiving my transmission?"). Such shared rules and recommendations provided measurable benefits, such as a reduction in the number of fatalities resulting from delays in mobilizing for emergency and disaster assistance.

The motivation for collaboration expanded when radio-based services proliferated in the early 1900s. In 1903, a group of nations with maritime interests met to coordinate implementation of radio systems. Twenty-nine nations agreed in 1906 to form the International Radiotelegraph Union (IRU) to coordinate usage, agree on common frequency bands, register station operations, and work to avoid or resolve cases of radio interference. National governments perceived the need to participate because privately operated systems had sought to create equipment and service monopolies by refusing to allow communications with users of other equipment, even in an emergency. It has been alleged that a ship was in the vicinity of the Titanic as it struck an iceberg in 1912, but the radio operator had gone off duty. The IRU created a treaty-level document and began

---

3. For a general description of the ITU prior to its major reorganization in 1993, see G. Codding, Jr., and D. Gallegos, "The ITU's 'Federal' Structure," *Telecommunications Policy*, Vol. 15, No. 5, August 1991, p. 353; M. Rothblatt, "ITU Regulation of Satellite Communications," *Stanford J. International Law*, Vol. 19, 1982, pp. 1–25; G. Codding and A. Rutkowski, *The International Telecommunication Union in a Changing World*, Dedham, MA: Artech House, 1982; R. S. Jakhu, "The Evolution of the ITU's Regulatory Regime Governing Space Radio Communication Services and the Geostationary Satellite Orbit," *Annals of Air and Space Law*, Vol. 8, 1983, p. 380; and D. Gregg, "Capitalizing on National Self-Interest: The Management of International Telecommunication Conflict by the International Telecommunication Union," *Law and Contemporary Problems*, Vol. 45, 1982, pp. 37–52.

4. For an overview of the ITU, see *The International Telecommunication Union: Its Aims, Structure and Functioning*, ITU Press and Information Section, October 1991.

to establish a formal collection of radio regulations, such as the designation of an emergency frequency that would be monitored all the time by radio operators on ships exceeding a certain weight.

The ITU conference held at Atlantic City in 1947 substantially expanded the ITU's mission as nations sought to rebuild and as postwar alliances formed. The ITU voted to become a Specialized Agency in the United Nations System and to adopt its structures and procedures. The ITU's mission grew to:

- Support cooperation among nations for the improvements and rational allocation and use of spectrum;
- Promote development of technical facilities;
- Coordinate efforts to eliminate harmful interference;
- Facilitate worldwide standards;
- Foster international cooperation in the delivery of technical assistance;
- Promote adoption of measures for ensuring safety of life;
- Undertake studies, make regulations, adopt resolutions, formulate recommendations and options, and collect and publish information concerning telecommunications matters.

The ITU emphasizes the right of each member nation to participate in studying issues and expressing views at various world or regional conferences.[5] Nations ratify the final acts of conferences and most subsequently codify in domestic laws and regulations the spectrum allocations, rules, and regulations established at ITU conferences.[6] The future viability of new services and spec-

---

5. For an outline of how the United States prepared for the 1995 World Radiocommunication Conference, see Preparation for International Telecommunication Union World Radiocommunication Conferences, ET Docket No. 93-198, 10 FCC Rcd. 647 (1994), IC Docket No. 94-31, 1995 FCC Lexis 608, 1480 (1995); see also An Inquiry Relating to Preparation for the International Telecommunication Union World Administrative Radio Conference for Dealing with Frequency Allocations in Certain Parts of the Spectrum, GEN Docket No. 89-554, First Notice of Inquiry; 4 FCC Rcd. 8546 (1989), Second Notice of Inquiry, 5 FCC Rcd. 6046 (1989); Supplemental Notice of Inquiry, 6 FCC Rcd. 1914 (1991); Report and Order, 6 FCC Rcd. 3900 (1991); United States Dept. of State, *United States Proposals for the World Administrative Radio Conference Malaga-Torremolinas, Spain 1992*, Washington, D.C. (1991).

6. For an example of how the United States implements international spectrum allocation decisions, see Amendment of Parts 2, 25, 80, and 87 of the Commission's Rules Regarding Implementation of the Final Acts of the World Administrative Radio Conference for the Mobile Services, Geneva, 1987, 4 FCC Rcd. 4173 (1989); 4 FCC Rcd. 7603 (1989); see also Amendment of Sec. 2.106 of the Commission's Rules to Allocate the 1610-1626.5 MHz and the 2483.5-2500 MHz Bands for Use by the Mobile-Satellite Service, Including Non-geostationary Satellites, ET Docket No. 92-28, Notice of Proposed Rule Making and Tentative Decision, 7 FCC Rcd. 6414 (1992) (proposing to implement spectrum allocations for "Big LEO" mobile satellite services decided at the 1992 World Administrative Radio Conference), Report and Order, 9 FCC Rcd. 536 (1994) (implementing spectrum allocation plan); see also spectrum licensing plan in CC Docket No. 92-166, Report and Order, 9 FCC Rcd. 5936 (1994).

trum allocations depends in large part on the willingness of the ITU's community of nations to reach closure. Domestic regulatory agencies may await a global consensus or take unilateral action to expedite availability of new services and technologies and perhaps also to affect the outcome of future deliberations at the ITU.[7]

5.3 ITU STRUCTURE AND FUNCTION

The ITU structure can be divided into permanent, plenary, and ad hoc elements. Supreme authority lies with the over 185 member nations who agree to comply with promulgated rules and regulations and who financially underwrite ITU operations.[8] The Plenipotentiary Conference ("Plenipot")[9] reviews the ITU's basic documents, the Convention, which is subject to revision at each Plenipot, and the Constitution, a permanent document infrequently revised.[10]

At the Plenipot, nations, holding one vote each, establish budgets for future conferences and for the ITU's permanent staff, the general secretariat, based in Geneva. The Plenipot also elects the ITU's secretary-general and other officials. It selects representatives from 46 geographically diverse nations to participate in the ITU Council, which performs executive board functions. The Plenipot also schedules the many specific conferences that modify rules, regulations, standards, protocols, or recommendations. Conferences can address one of the ITU's three major geographical regions of the world or have global appli-

---

7. See, for example, Amendment of the Commission's Rules with Regard to the Establishment and Regulation of New Digital Audio Radio Services, GEN Docket No. 90-357, Notice of Inquiry, 5 FCC Rcd. 5237 (1990), Notice of Proposed Rulemaking and Further Notice of Inquiry, 7 FCC Rcd. 7776 (1992), Report and Order, FCC 95-17, 1995 FCC Lexis 329, 76 Rad. Reg. 2d (Pike & Fischer) 1477 (rel. 12 January 1995).

8. See Constitution of the International Telecommunication Union, CS/Art. 6-34, Execution of the Instruments of the Union in *Final Acts of the Plenipotentiary Conference, Nice, 1989*, Geneva: ITU, 1990, p. 7: "The Members are bound to abide by the provisions of this Constitution, the Convention and the Administrative Regulations in all telecommunication offices and stations established or operated by them." See also Art. 17, Finances of the Union, CS/Art. 16-138, p. 21: "The expenses of the Union shall be met from the contributions of its Members, each Member paying a sum proportional to the number of units in the class of contribution it has chosen from the scale in Article 26 of the Convention."

9. See J. Savage, "The High-Level Committee and the ITU in the 21st Century," *Telecommunications Policy*, Vol. 15, August 1991, p. 365.

10. See *Final Acts of the Plenipotentiary Conference, Constitution and Convention of the International Telecommunication Union, Optional Protocol, Decisions, Resolutions, Recommendations and Opinions, Nice, 1989*, Geneva: ITU, 1990. The Constitution contains the basic provisions and purposes of the ITU. The Convention complements the Constitution and addresses more functional provisions relating to the operation of the ITU and its conferences.

cation.[11] These meetings have addressed such diverse issues as mobile radio, use of the orbital arc, telecommunications development, high- and middle-frequency radio, satellite frequencies, and the terms and conditions for provision of information services. Reforms adopted in 1992 have created a more routine and frequent meeting schedule so that conferences can address and resolve a manageable number of issues.

The council implements ITU policies and regulations, oversees the general secretariat, and establishes questions and issues to be considered at regularly scheduled conferences by the ITU's three sectors:

1. *Development*: "to encourage international cooperation with a view toward harmonizing and enhancing the development of telecommunication services and facilities" [1];
2. *Telecommunications Standardization*: "to study technical, operating and tariff questions and to issue recommendations on them with a view to standardizing telecommunications on a world-wide basis" [2];
3. *Radiocommunication*: to achieve "efficient management of the radio-frequency spectrum in terrestrial and space radio-communications... [including] examining and registering all notices for frequency assignments liable to cause interference outside the territory of the country in which the station is located...[and] all notices for orbital positions of ...satellites" [3].

As part of reforms adopted by the ITU in 1992 to reflect changes in the telecommunications environment, the three sectors have established advisory boards to promote participation by nongovernmental players. The ITU had already authorized participation by *recognized private operating agencies*,[12] nongovernmental carriers and service providers that provide international telecommunications capable of causing harmful interference (e.g., AT&T). The ITU had also authorized participation by *scientific or industrial organizations*,[13] nongovern-

---

11. For an analysis and criticism of U.S. participation in ITU conferences, see United States Congress, Office of Technology Assessment, The 1992 World Administrative Radio Conference Issues for U.S. International Spectrum Policy—Background Paper, OTA-BP-TCT-76, Washington, D.C., 1991, and the 1992 World Administrative Radio Conference: Technology and Policy Implications, OTA-TCT-549 (Washington, DC, 1993).

12. A recognized private operating agency operates a telecommunication or broadcasting service and complies with the obligations imposed by Article 6 of the ITU's Constitution, that is, the treaty-level agreement to be bound by the ITU's Constitution, Convention, and Administrative Regulations. (See ITU Constitution Annex, Definition of Certain Terms Use in This Constitution, the Convention and the Administrative Regulations of the International Telecommunication Union, CS/An 1008 in Final Acts of the Plenipotentiary Conference, Nice, 1989, Geneva: ITU 1990, p. 65.)

13. Scientific or industrial organizations are nongovernmental agencies "engaged in the study of telecommunication problems or in the design or manufacture of equipment intended for telecommunication services." (Ibid., CS/An 1009, p. 65.)

mental organizations that study telecommunication problems or design and manufacture equipment (e.g., IBM). But the creation of advisory boards marked a commitment to solicit participation by former outsiders representing a wide cross section of interests and expertise.

5.3.1 Conflict Prevention in Spectrum Usage

A key function performed by the ITU is serving as a global traffic cop of the airwaves. This role involves defining services and functions that use spectrum, allocating spectrum for particular services and functions on an exclusive or shared basis, and providing a forum to coordinate and register spectrum use. World and regional radio conferences meet on an increasingly frequent and regularly scheduled basis primarily to address issues involving spectrum definitions, allocations, and procedures for coordinating multiple users and services. The Radiocommunication Bureau (BR) performs most ongoing spectrum registration, coordination, and conflict prevention and resolution functions. It maintains a master list theoretically containing all nations' actual and proposed frequency and orbital arc uses, exclusive of military and national security applications.

Before recording a new use, the BR reviews it for compliance with the ITU Constitution, Convention, and applicable rules, regulations, and frequency allocations on a regional or worldwide basis. It also assesses the potential for interference with other registered uses (in operation or planned), and issues an Advance Publication of the proposed new use. The issuance of this document triggers a time period within which nations may report potential interference and express their desire to participate in future meetings convened to resolve such problems. Upon successful conclusion of this coordination process, the BR officially registers the use and notifies all member nations. A reconstituted part-time Radio Regulations Board provides conflict resolution services that used to be provided on a full-time basis by its predecessor, the International Frequency Registration Board (IFRB).[14]

5.3.2 Reforms in the Radiocommunications Sector

Reforms adopted by an Additional Plenipotentiary Conference (APP) in 1992 merged all spectrum planning and management functions into a single radiocommunications sector. These functions include the spectrum management activities of the IFRB and the radio regulation drafting and study functions of the International Radio Consultative Committee and the various world and regional

---

14. As part of the ITU's streamlining efforts, the IFRB was reconstituted to become a part-time forum called the Radion Regulations Board, composed of nine members instead of a full-time operating component of the ITU with five full-time members.

administrative radio conferences. World Radiocommunication Conferences will convene every two years along with a Radiocommunication Assembly that will perform a plenary function, including the provision of technical support for the conferences, and the determination of priorities for study groups.

APP consideration of IFRB reforms proved to be the most controversial, because the board was reformulated to become a part-time body composed of non-ITU employees. In view of the increasing congestion in some frequency bands, the launch of more satellites vying for geostationary orbital arcs and the need to oversee coordination between geostationary and low-Earth-orbiting satellites, some nations expressed concern that a part-time body would not provide adequate and timely consideration of pressing matters.

The new Radio Regulations Board "will approve the Rules of procedure which are used in the application of the Radio Regulations to register frequency assignments, will consider any matter which cannot be resolved through the application of the Rules of procedure and will perform any duties related to the assignment and utilization of frequencies and to the equitable utilization of the geostationary-satellite orbit" [4]. The board will hold up to four meetings a year, meaning that much of the day-to-day operations will be performed by the director of the BR, who will investigate claims of harmful interference and formulate recommendations for their resolution.

5.3.3 Structural Reforms at the ITU

Some nations perceive increasingly significant disincentives to participate in international forums like the ITU. The pace of technological innovation has taxed the ability of the ITU to respond to market developments and user needs with timely spectrum allocations, recommendations on operating procedures using new technologies, and ways to coordinate services. The heightened competitiveness in telecommunications industries means that individual carriers and equipment manufacturers may attempt to introduce products and services in advance of an ITU-promulgated standard, with an eye toward establishing a de facto standard and capturing market share.

Much of the efficiency gains from global rules of the road could be lost without an effort by the ITU to address challenges to its efficiency and relevance. Many of the newfound incentives not to participate stem from fundamental changes in the composition of the telecommunications marketplace. A small number of monopoly carriers and national hero manufacturers dominated the old world order. This model emphasized stability, incremental change, and collaboration among a few players with end-to-end network responsibility and control [5]:

> Like the networks themselves, the old telecommunications world had a relatively simple, hierarchical, star architecture. There were only two tiers: at the national level there were monopoly providers;

and for international coordination there was the ITU around which everyone clustered. It was a relatively small club of constituents, and no one else really mattered.

The ITU's role in the old environment was one of bureaucratically lending "good offices" for the expert participants to share information, develop standards, and adopt rules of the road geared for public facilities and services provided by single national carriers. Most participants in the process shared a common bond as technocratic elites operating a nation's natural monopoly PTT administration and pursuing similar objectives:

- Maximizing revenues;
- Using revenues from overpriced international services to cross-subsidize socially desired services like universal and inexpensive local telephone service;
- Maintaining legislatively conferred or de facto barriers to market entry.

The ITU well served these participants by emphasizing careful deliberation and continuity over speed and flexibility.[15]

The increasingly volatile, complex, and competitive telecommunications environment, closely linked with information service markets, supports a new world telecommunications order. This model supports telecommunications as a key vehicle to stimulate national economies and to promote commerce between nations. High-volume users need and increasingly demand an open, flexible, and diversified telecommunications infrastructure priced on the basis of cost and demand elasticity rather than social policy.[16] Deregulation, privatization, liberalization, globalization, and conditional market entry have become the key elements of change. The new world order emphasizes responsiveness to consumer requirements rather than obedience to the incumbent carrier constituency. Speediness, accessibility, and flexibility are key factors because new

---

15. See A. Rutkowski, "The ITU at the Cusp of Change," *Telecommunications Policy*, Vol. 15, August 1991, p. 292.

16. Beginning in the late 1970s and early 1980s, large-volume users began to consider the consequences of high telecommunication costs and the absence of competition. While few enjoyed a legally possible option to erect a private bypass network, users began to advocate change and recognize technological options in traffic routing, the ability to bargain over terms and conditions, the benefits from reduced regulation and increased competition, and the possibility that resale of value-added telecommunication services could constitute a new profit center. (See P. Cowhey and J. Aronson, "The ITU in Transition," *Telecommunications Policy*, Vol. 15, August 1991, pp. 298, 300.)

carriers, resellers, and equipment manufacturers have secured no reserved market or monopoly franchise. Likewise, they may have little faith in or loyalty to the ITU membership and staff that did not always welcome or accommodate them.[17]

Outsiders to the ITU are more inclined to test their innovations in the marketplace, without resolution of spectrum allocation, standards, service coordination, and interference avoidance issues. As an alternative to a marketplace-driven de facto solution, new players may support bilateral negotiations by their governments or policy making in other forums primarily geared to set standards on a domestic or regional basis.

In an age of divergent yet interconnected networks, a multiplicity of operators, and a more activist user community, the ITU "is no longer on top of the pyramid, but part of a geodesy of organizations all representing different communities that comprise today's complex telecommunications universe. In addition, the organization is compelled to be considerably more effective at what it has traditionally done" [6]. The ITU must now work in a subject area that involves numerous regional standard-setting organizations such as the European Telecommunications Standards Institute (ETSI), international and regional satellite cooperatives such as INTELSAT and EUTELSAT, commercial satellite enterprises such as PanAmSat, submarine cable consortia, numerous trade organizations such as the World Trade Organization, international and regional development bodies, user advocacy groups, and regulatory agencies.

In 1989, the Nice Plenipotentiary Conference passed a resolution calling for the creation of a High-Level Committee (HLC) to carry out an in-depth review of the structure and functioning of the ITU, with an eye toward recommending ways to make it more efficient and effective in a fast-changing environment [7]. In particular, the ITU needed to respond to an overtaxed standard-setting process which could not issue timely recommendations on privatization, liberalization, deregulation, and market entry by private enterprises.

---

17. "It may be argued that the ITU is a specialized agency, and therefore may be immune from changes created by geopolitical shifts. This is a comforting illusion. The ITU may well be narrow in focus but telecommunications is now becoming much too central for competitive global business to be left exclusively to the ITU. Indeed, in many respects the ITU is in danger of suffering sclerosis from its age. In an increasingly borderless global village the ITU is still border bound, reflecting the era of nation states and administering regimes (e.g., accounting rates) which are based on strict respect for national sovereignty with an arm's length type of carrier relationship. At a time when innovation and global networking are essential to maintain competitive edge, intergovernmental multilateral negotiations which cannot keep pace with business practices are in danger of becoming increasingly irrelevant." (J. Solomon, "The ITU in a Time of Change," *Telecommunications Policy*, Vol. 15, August 1991, pp. 372, 373.)

The HLC made 96 recommendations for reform, most of which were considered and adopted, including a more routinized scheduling of conferences.[18] The ITU has operated under a federalist, nonhierarchical structure that confers substantial power and independence to standard setting and radio consultative committees.[19] This diffusion of power will remain in effect, but the secretary general has acquired greater resources, including a small strategic planning and research group and somewhat more power to lead the ITU into a future where its legitimacy and authority are not certain.

5.4   AN EVOLVING THREE-TIERED INDUSTRIAL STRUCTURE

The ITU has available procedures for achieving greater effectiveness. It can modify rules and recommendations to promote the integration of new technologies and it can reallocate spectrum to accommodate new requirements while still deferring to individual nations on the decision of whether to permit such alternatives. However, this "consenting adults policy means that there is no true [single] global framework for commercial competition" [8]. Instead, three elements have evolved:

1. The traditional model of jointly provided service by foreign correspondents, composed primarily of PTTs;
2. Niche market service providers, primarily value-added networks who lease lines from incumbent carriers and whose acceptability arises on a bilateral basis when two nations negotiate "special arrangements," in the ITU lexicon;
3. New, stand-alone facilities-based carriers, such as satellite systems separate from INTELSAT and operators of private cables like PTAT-1 and the North Pacific Cable.

5.5   MANAGING THE SATELLITE ORBITAL ARC

Governments justify their management, regulation, and operation of satellites because such facilities use both scarce frequency spectrum and orbital parking

---

18. The ITU will convene one major conference a year, with a Plenipotentiary Conference for long-term policy planning, followed by Standards, Development, and Radiocommunications Conferences in the subsequent three years. The ITU contemplates convening a Plenipotentiary and Standardization Conference once every four years, with World or Regional Radiocommunication Conferences and Assemblies occurring once every two years.

19. See G. A. Codding, Jr., and D. Gallegos, "The ITU's 'Federal' Structure," *Telecommunications Policy*, Vol. 15, August 1991, pp. 351–363.

places.[20] To maximize the usefulness of these resources, nations must establish policies and rules for interference avoidance in the same manner as they must coordinate radio frequency use.

The geostationary orbital arc nears saturation in some regions, like North America, because of the increasing demand for telecommunication satellite service, growing numbers of nations opting for regional or domestic systems in lieu of global or regional cooperatives, and physical limitations on the number of satellites that can be positioned without causing harmful frequency interference. While the danger of satellite collision is remote, because the closest spacing of 2 degrees still maintains an approximately 900-mile separation, the potential for interference exists because adjacent satellites typically use the same spectrum that is allocated for fixed, mobile, maritime, aeronautical, and other satellite services.

Earth stations using expensive, state-of-the-art technology can communicate interference-free with satellites located as close as 2 degrees apart. This means that for any particular satellite frequency band the theoretical maximum number of operational satellites in the geostationary orbit is 180 (360 degrees divided by 2). The actual number is significantly less because:

- Not all nations will mandate the use of sensitive equipment enabling two degree satellite spacing.
- Operators seeking maximum transoceanic, international coverage will cluster satellites in middle-ocean regions.
- Satellites can be equipped with antennas that concentrate the transmission beam, thereby increasing the probability of interference even from satellites many degrees away.

5.5.1 ITU Conflict Resolution in Satellite Orbital Slots

Notwithstanding shared interests in consensus and conflict prevention, the ITU regularly bears the responsibility of broker compromise and lends its "good offices" to resolve conflicts. Particularly for shared resources like satellite orbital slots, nations vie for a finite number of available positions in the manner of a zero sum game: one nation's orbital slot use can often occur only at the expense of another nation's current or future use.

---

20. The ITU Constitution directs member nations to "bear in mind that radio frequencies and the geostationary-satellite orbit are limited natural resources and that they must be used rationally, efficiently and economically, in conformity with the provisions of the Radio Regulations, so that countries or groups of countries may have equitable access to both, taking into account the special needs of the developing countries and the geographical situation of particular countries." (International Telecommunication Union, Constitution, Art. 33, Para. 175, "Use of the Radio-Frequency Spectrum and of the Geostationary-Satellite Orbit," Geneva: ITU, 1990, p. 30.)

With only recent, limited qualification, the ITU rules favor incumbents and prior registered uses. Typically, developed nations like the United States and global cooperatives like INTELSAT[21] and Inmarsat[22] have established satellite requirements. This means that incumbents will have completed the registration process well before later market entrants in developed nations and lesser developed nations have accumulated the finances to begin procuring a first satellite.

Nations notified of a future satellite deployment have an affirmative duty to avoid causing harmful interference should they seek to use the same orbital location. Developing nations may have to settle for less than optimal orbital slots, because their later launched systems must not interfere with already operating networks and preoperational but registered satellites. The congested orbital arc and limited finances have prompted such nations to lease capacity from cooperatives' satellites, even for domestic applications. Alternatively, they have formed regional coalitions like Palapa, which serves the nations of Southeast Asia.

5.5.2 Inequity in Access to Satellites and Information Resources

The matter of access to orbital slots takes on even greater significance when one considers its impact on nations' collective opportunities to tap information resources and use telecommunications as a catalyst for economic development. Inequitable and inadequate access to orbital slots can exacerbate the gap between nations with a rich and diverse telecommunications infrastructure and ones with second-rate or leased facilities. If one subscribes to the view that the wealth of information resources has a direct and substantial effect on national financial wealth, then access to the orbital arc may have a significant impact on a nation's overall social and economic welfare. Accordingly, the stakes in the orbital slot access sweepstakes involve more than how many television channels a nation can access. It affects the broader issue of whether and how nations share information resources and the technologies for access and distribution.

---

21. INTELSAT is a global cooperative formed by an intergovernmental agreement with a mission of providing ubiquitous satellite communications service. (See *Multinational Communication Satellite System*, opened for signature 20 August 1964, 15 U.S.T. 1705, 514 U.N.T.S. 26 (19-nation agreement establishing interim arrangements for a global satellite cooperative); *Agreement Relating to the International Telecommunications Satellite Organization (INTELSAT)*, opened for signature 20 August 1971, 23 U.S.T. 3813, 1220 U.N.T.S. 21 (INTELSAT Agreement), *Operating Agreement Relating to the International Telecommunications Satellite Organization (INTELSAT)*, opened for signature 20 August 1971, 23 U.S.T. 4091, 1220 U.N.T.S. 149.)

22. Inmarsat is a global cooperative formed by intergovernmental agreement to provide ubiquitous maritime telecommunications to ships in the high seas, with aeronautical and land mobile services available on an ancillary basis. (See *Convention of the International Maritime Satellite Organization*, opened for signature, 16 July 1979, 31 U.S.T. 1, T.I.A.S. No. 9605.)

Likewise, it raises questions on the extent to which developed nations should promote parity of access to resources, both public (orbital slots) and private (programming and databases sent via satellite).[23]

The ITU regularly becomes the forum for fact finding, arbitration, and conflict resolution in the face of growing demands for orbital slots. Developed nations require additional satellites to satisfy demand for more video program options, and developing nations may need a first slot. Congestion grows when nations cannot or will not commit to reduced spacing between satellites,[24] which would expand the number of satellites that can occupy the orbital arc.

The ITU fashions remedies at the macro-level by convening conferences to revise spectrum allocations and to consider changes to the method for reserving orbital arc slots. At the micro-level, the BR publishes prospective uses, coordinates the necessary technical and operational assessment of a nation's interference claims, resolves real interference problems, and formally notifies the ITU membership of newly registered orbital arc uses.

Satellite orbital arc policy raises political questions and juxtaposes equity and efficiency concerns. For its part, the ITU must fashion a compromise that balances financial concerns regarding satellite spacing and efficient frequency use with equity concerns about parity of access to the GSO by developing nations. Under the customary first-come, first-registered process, the later filed registrations of developing nations receive subordinate status.

Representatives from developing countries have advocated the need for equitable access to the orbital arc through a system that guarantees slots, even at the risk of leaving fallow a resource that other nations, singularly or collectively, could put to use sooner.[25] An a priori allotment plan for satellite orbital

---

23. See, for example, B. Harris, "The New Telecommunications Development: Bureau of the International Telecommunication Union," *American University J. International Law and Policy*, Vol. 7, Fall 1991, p. 83; R. Saunders, J. Warford, and B. Wellenius, *Telecommunications and Economic Development*, Washington, D.C.: The World Bank, 1983; and R. Crandall and K. Flamm, *Changing the Rules: Technological Change, International Competition and Regulation in Communications*, Washington, D.C.: Brookings Institution, 1989.

24. In Licensing of Space Stations in the Domestic Fixed-Satellite Service, 54 Rad. Reg. 2d (P&F) 577 (1983), the FCC ordered domestic satellite operators to position satellites within 2 degrees of each other. This reduced orbital spacing accommodates more satellites over the United States and enables the neighboring nations of Canada and Mexico to operate their domestic satellites with wider separation. However, it required higher investment in more sensitive Earth stations.

25. See M. Rothblatt, "ITU Regulation of Satellite Communication," *Stanford J. International Law*, Vol. 18, 1982, pp. 1, 10–11; and H. Levin, "Orbit and Spectrum Resource Strategies: Third World Demand," *Telecommunications Policy*, Vol. 5, No. 2, June 1981, p. 105.

arc deployment[26] reserves slots for developing nations that typically have generated requirements for and ability to finance satellite telecommunications after developed nations.

Professor Harvey J. Levin attempted to quantify the cost handicap incurred by latecomers who may have less attractive orbital slots available to them and who may have to incur added costs to operate on higher frequencies[27] to avoid causing harmful interference to existing satellites.[28] When developing nations have to migrate to higher frequencies to accommodate incumbent users and adhere to the rules of the road articulated by the ITU, they lose much of the financial and operational benefits that would have accrued if they could use older, proven technology. Instead, the orbital slot registration and satellite coordination process may make it difficult for latecomers to find a suitable orbital slot unless they agree to operate on higher frequencies. A developing nation with frequent heavy rain conditions would find that the ITU registration process burdensome by forcing it to procure more powerful and expensive satellites and more sensitive and expensive Earth stations to overcome the signal degradation occurring at higher frequencies during rainfall.

---

26. "An *a priori* system of frequency and orbital position regulation uses administrative conferences to subdivide and allot radio frequencies and orbital positions to countries in advance of need or use. On the other hand, an *a posteriori* [first-come, first-served] system requires subsequent satellite operators to coordinate with pre-existing satellites to avoid harmful interference." (M. S. Straubel, "Telecommunication Satellites and Market Forces: How Should the Geostationary Orbit Be Regulated by the F.C.C.?" *North Carolina J. International Law and Commercial Regulation*, Vol. 17, 1992, pp. 205, 211, note 30.)

27. The first commercial satellites operated exclusively in the C-band, with Earth-to-satellite transmissions occurring in the 6-GHz band and satellite-to-Earth transmission occurring in the 4-GHz band. In the 1980s, satellite carriers also operated in the Ku-band, with Earth-to-satellite transmissions primarily occurring in the 14-GHz band and satellite-to-Earth transmission occurring in the 11-GHz band. Recently, limited commercial and experimental use has occurred in the Ka-band, with Earth-to-satellite transmissions occurring in the 30-GHz band and satellite-to-Earth transmission occurring in the 20-GHz band.

28. "More generally Third World resentment of the practice of awarding rights to build space satellite systems on a first-come, first served basis, is seemingly based on what those nations perceive as the dwindling availability of slots or orbit spectrum assignments. The developing countries (LDCs) also fear the handicaps they suffer due to the higher R&D and engineering costs incurred to open up new bands at higher frequencies." (H.. Levin, "Latecomer Cost Handicap: Importance in a Changing Regulatory Landscape," in D. Demac, ed., *Tracing New Orbits—Cooperation and Competition in Global Satellite Development*, New York: Columbia University Press, 1986, pp. 251, 251–52.) A more extensive and quantitative analysis is available in H. Levin, "Global Claim-Staking and Latecomer Cost in the Orbit Spectrum Resource," *Telecommunications Policy*, Vol. 13, June 1990, pp. 233–248. See also "Regulation of Transnational Communications," 1984 *Michigan Yearbook of International Legal Studies*, Part 1, Regulation of the Geostationary Orbit, pp. 3–70, and Part 2, Regulation of Satellite Communications, pp. 73–82, New York: Clark Boardman Co., 1984 (hereafter cited as *Michigan Yearbook*).

Access to the orbital arc raises questions of equity. While developed nations should not have to handicap or postpone their orbital arc development plans, decision makers will need to consider what, if anything, should be done to support access to satellite technology and orbital slots by developing nations. In view of efforts by INTELSAT and Inmarsat to privatize, developing nations have no certainty that the existing cooperative model will continue to provide access to satellite capacity at averaged rates for any nation, including those unable or unwilling to make a sizable investment in the cooperative.[29]

5.5.3 Space WARCs

The ITU has found itself in the middle of a geopolitical battle of philosophies, particularly at radio conferences with a heavy agenda of orbital slot management issues. Such world radio conferences (WRCs), which used to be called World Administrative Radio Conferences (WARCs), establish rules, regulations, and policies for various types of satellite services, such as fixed (for telecommunications from and to many fixed locations on Earth), broadcast (for satellite broadcasting of video and audio programming directly to dispersed receiver locations), and mobile (for telecommunications between fixed locations and mobile stations, or between mobile stations). They also determine at what frequencies such services should operate, ensuring that these operations do not interfere with other existing satellite networks or other operators whose users are authorized for the same frequencies.

Most importantly, WRCs must anticipate and resolve future bilateral conflicts, possibly involving nations whose representatives might lack the expertise or inclination to help shape a speedy and fair compromise. Accordingly, most member nations of the ITU invest significant resources in extensive preconference preparations and send representatives to marathon international meetings that can run for six weeks or more. Recent Space WARCs have confronted such difficult issues as equitable access to the orbital arc, allocating spectrum for new and expanding mobile satellite services, and developing technical coordination procedures for accommodating satellites in low Earth orbit.

While grandfathering existing registrations, the ITU has established a framework for reserving at least one orbital slot for the fixed-satellite service requirements of each nation and for resolving conflicts on a bilateral, multilateral, and regional basis. The ITU also has expanded the amount of spectrum allocated for mobile satellite services and has begun to establish procedures by which operators of low-Earth-orbiting satellites can coordinate their use with operators of geostationary orbiting satellites and terrestrial applications. WRCs

---

29. See R. Frieden, "Should Intelsat and Inmarsat Privatize?" *Telecommunications Policy*, Vol. 18, No. 9, December 1994, pp. 679–686.

in 1995 and 1997 will emphasize mobile telecommunications and will address a number of satellite coordination and spectrum allocation issues.[30]

Software can compute optimal slotting plans based on the fact that satellites can collocate if they operate in different geographical regions, use different frequencies, and have diverse service parameters, coverage plans, transmission power, and interference levels. But there will always be a human component that requires an honest and impartial brokering like that performed at ITU conferences.

The Space WARCs have also considered the role of global cooperatives like INTELSAT and their status within the ITU. While cooperative satellites situated at midocean can serve the telecommunications needs of many nations, the ITU must acknowledge the interest of individual nations in retaining options for national systems. A proliferation of satellite systems can result in excess capacity and inefficiency, much like what arguably has occurred in international commercial aviation, given the number of national flag carriers. On the other hand, nations with multiple satellite systems cannot seize the diplomatic high ground in orbital slot negotiations, given the number of slots their own satellites occupy.

Cooperatives like INTELSAT have official observer capacity at the ITU and use the services of the nation where the cooperative headquarters are located (the United States in the case of INTELSAT) for satellite registration and official advocacy. Conferring an opportunity for direct participation might bolster such cooperatives' negotiating leverage, but some nations believe this would confer "supersovereign" status. In any event, the ITU and its constituent nations must regularly address how to share the satellite orbital arc to avoid having to consider the tougher task of having to ration slots.

Figures 5.1 and 5.2 depict the spectrum allocation process as executed by the ITU. The footnotes in Figure 5.1 identify exceptions to the consensus adopted by specific nations.

5.5.4 Orbital Slot Reservation Alternatives

Given the prospect for more satellite systems separate from international cooperatives, one can question whether the GSO registration system can remain intact and whether nations like the United States can continue to enjoy the luxury of supporting an open skies policy that currently fills in an excess of 25 geosta-

---

30. See Preparation for International Telecommunication Union World Radiocommunication Conferences, ET Docket No. 93-198, 10 FCC Rcd. 647 (1994); and Preparation for International Telecommunication Union World Radiocommunication Conferences, IC Docket No. 94-31, 1995 FCC LEXIS 608 (rel. 30 January 1995) and 1995 FCC LEXIS 1480 (rel. 3 March 1995).

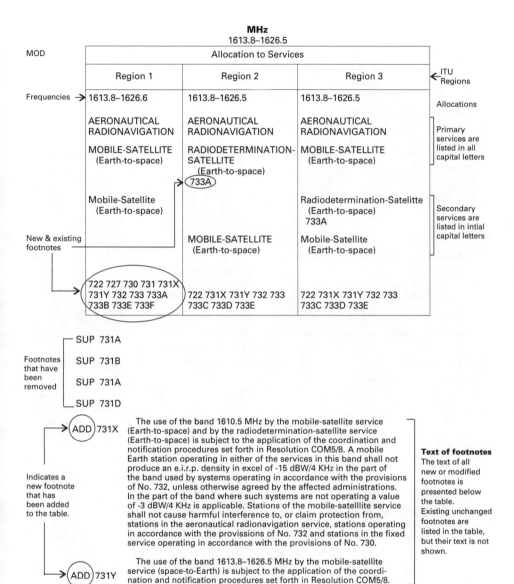

Figure 5.1 Sample page from Final Acts of WARC-92. (*Source:* [9].)

WARC-92 allocated frequencies for a number of new and existing radiocommunication services. The allocations, however, do not tell the whole story of the decisions made at WARC-92. All of the allocations summarized below are subject to limitations and constraints that are described in footnotes to the allocations and in various resolutions and recommendations the conference adopted for each service.

High-Frequency (HF) Broadcasting

A total of 790 kHz was allocated to HF broadcasting with 200 kHz located in frequencies below 10 MHz (the most congested portion of the HF bands) and 590 between 11 and 19 MHz. All the newly allocated bands are allocated on a worldwide basis, all are subject to planning, and all must use single sideband (SSB) modulation for transmission.

5900–5950 kHz	13800–13870 kHz
7300–7350 kHz	13570–13600 kHz
9400–9500 kHz	15600–15800 kHz
11600–11650 kHz	17480–17550 kHz
12050–12100 kHz	18900–19020 kHz

Broadcasting-Satellite Service-Sound (BBS-Sound)

1452–1492 MHz	Worldwide, except the United States.
2310–2360 MHz	Only in the United States and India.
2535–2655 MHz	Various countries in Europe and Asia.

All BSS-sound operations are required to use digital audio broadcasting (DAB) technology.

Broadcasting-Satellite Service-High-Definition Television

17.3–17.8 GHz	ITU Region 2 only.
21.4–22.0 GHz	ITU Region 1 and 3.

Terrestrial Mobile Service

1700–2690 MHz	Upgraded to primary status worldwide.
1885–2025 MHz &	Intended for future public land mobile
2110–2200 MHz	telecommunications systems (FPLMTS) worldwide. Frequencies for a satellite component of FPLMTS were also identified.

Mobile-Satellite Service (MSS)

MSS allocations are made in pairs—one set of frequencies for transmission from the Earth to satellites (uplinks) and one set for transmissions from satellites to Earth (downlinks).

1492–1525 MHz (downlink)	Region 2 only, not allocated in the United States.
1675–1710 MHz (uplink)	Region 2 only.
1525–1559 MHz (downlink)	Portions remain allocated specifically to land, maritime, and aeronautical services, but the United States has allocated almost the entire band to generic MSS.[1]

[1]With the exception of 1545–1555 MHz, which remains allocated to the aeronautical mobile satellite service.

Figure 5.2   Summary of WARC-92 allocations. (*Source:* Office of Technology Assessment, 1992.)

| 1626.6–1660.5 MHz | As above, portions remain allocated to specific (land, maritime, and aeronautical) services, but the United States has allocated the majority of the band to generic MSS.[2] |

1930–1970 MHz (uplink)	On a secondary basis for Region 2 only.
2120–2160 MHz (downlink)	On a secondary basis for Region 2 only.
1980–2010 MHz (uplink)	With an additional 10 MHz allocated at 1970–1980 MHz for Region 2 only.
2170–2200 MHz (downlink)	With an additional 10 MHz allocated at 2160–2170 MHz in Region 2 only.
2500–2520 MHz (downlink)	
2670–2690 MHz (uplink)	

Low-Earth-Orbiting Satellites

Using frequencies above 1 GHz (big LEOs):

1610–1626.5 MHz (uplink)	Upgraded to primary.
2483.5–2500 MHz (downlink)	
1638.8–1626.6 MHz (downlink)	Allocated (secondary status) for downlinks to permit bidirectional use of the band.

Using frequencies below 1 GHz (little LEOs):

137–138 MHz (downlink)	Portions of which are secondary.
148–149.9 MHz (uplink)	Secondary in more than 70 countries.
400.15–401 MHz (downlink)	

Space Services

A number of allocations were made to various space services, including space research, space operations, and Earth exploration satellite services. Frequencies were allocated in the 400- to 420-MHz bands for space communications and research, in the 2-GHz bands for space research, operations, and Earth-exploration satellite service, and in the bands above 20 GHz for intersatellite links, Earth-exploration satellite, and (deep) space research services.

Aeronautical Public Correspondence

1670–1675 MHz was allocated worldwide for ground-to-aircraft communications, to be paired with 1800- to 1805-MHz for aircraft-to-ground transmissions. The United States will maintain its existing system at 849–851 MHz and 894–896 MHz.

Fixed-Satellite Service

A worldwide allocation was made at 13.75–14.0 GHz.

[2]With the exception of 1646.5–1656.5 MHz, which remains allocated exclusively to the aeronautical mobile satellite service.

Figure 5.2 (continued)

tionary orbital slots. Noting the reservation of orbital slots by lesser developed countries (LDC) who might never use them, some policy makers[31] and academics[32] have proposed the use of auctions and other types of market valuation mechanisms that would enable an LDC to transfer a slot for financial compensation.

While the Communications Act precludes United States spectrum licensees to view their authorization as property, the FCC has received legislative authority to auction some spectrum [10]. In the satellite arena, the FCC has approved prelaunch sale of transponder capacity for the lifetime of the satellite [11]. By extension, should an LDC decide against operating its own satellite system separate from INTELSAT, then its reserved slot might have significant resale value if nearby developed nations need additional space.[33] While the GSO has the characteristic of a *res communes*, a shared global resource, consensus-reached decisions to vest certain slots with certain nations could also mean that a secondary market could exist for conversion of a slot into a more desirable resource: hard currency.

5.6 DEVELOPMENT ISSUES

In 1989 the ITU upgraded its commitment to telecommunications development activities by authorizing creation of the Telecommunications Development Bureau (BDT) and by making development assistance a key activity specified in

31. "A market for spectrum licenses or rights, if properly structured, can maximize both 'allocative efficiency' (i.e., prices bid for spectrum reflect the costs to society of spectrum use) and 'distributive efficiency' (i.e., those who value the spectrum most will use it)." (United States Dept. of Commerce, National Telecommunications and Information Administration, *U.S. Spectrum Management Policy: Agenda for the Future*, NTIA Special Pub. 91-23, 98 (1991).)

32. See, for example, H. Levin, "The Political Economy of Orbit Spectrum Leasing," *Michigan Yearbook*, p. 41; H. Levin, "Emergent Markets for Orbit-Spectrum Assignments: An Idea Whose Time has Come," *Telecommunications Policy*, Vol. 12, No. 1, March 1988, p. 68; T. Schroepfer, "Fee-Based Incentives and the Efficient Use of Spectrum," *Fed. Com. L. J.*, Vol. 44, No. 3, May 1992, p. 411; and F. G. Hart, "Orbit Spectrum Policy-Evaluating Proposals and Regimes for Outer Space," *Telecommunications Policy*, Vol. 15, No. 1, February 1991, pp. 63–74. R. H. Coase is credited with initiating in 1959 the debate over whether to use a market-clearing mechanism to allocate frequency spectrum. (See R. H. Coase, "The Federal Communications Commission," *J. Law and Economics*, Vol. 2, 1959, p. 1.) See also De Vany, Eckert, Meyers, O'Hara, and Scott, "A Property System for Market Allocation of the Electromagnetic Spectrum: A Legal-Economic-Engineering Study," *Stanford Law Review*, Vol. 21, 1969, p. 1499.

33. The Kingdom of Tonga advance published with the ITU a total of 31 Tongasat satellites to be located in 26 separate orbital locations. In December 1991, Unicom Satellite, Inc., of Aspen, Colorado, announced a business arrangement whereby it would acquire the rights to two Tongasat orbital registrations. ("Tongasat Authorizes Unicom to Use Orbital Slots Over Asia," *Communications Daily*, Vol. 11, No. 231, 2 December 1991.)

Article 14 of the Constitution. The BDT has been reformulated into the telecommunications development sector, with an eye toward establishing a more structured approach to development in general. The ITU now regularly schedules advisory World Development Conferences.

5.7 CONCLUSION

The ITU has belatedly "adapt[ed] the Union's structure, management practices and working methods to the changes in the world of telecommunications and to the increasing demands placed upon it to keep pace with the ever-accelerating progress in telecommunications" [11]. The HLC recommendations as endorsed by the Additional Plenipotentiary in 1992 created a better foundation for the ITU to become more responsive and effective. Likewise, it placed the ITU in a better position to respond to:

- Other international and regional trade and standard-setting forums;
- The need to respond more speedily to technological change;
- The obligation to find ways to involve the increasingly vocal and active nongovernmental players in telecommunications.

However, the "larger obstacle to the ITU's future may be its 'international civil service' mentality. As the international telecommunications world becomes increasingly commercialized and deregulated, how will the inter-governmental ITU cope?" [12]. The answer lies in responding to changed circumstances in a timely and professional fashion. Likewise, it requires the ITU to accommodate new constituencies such as nongovernmental service providers and users.

References

[1]   International Telecommunication Union, Press and Public Relations Service "The New ITU: Round-up," January 1993, p. 3 (hereafter cited as New ITU press release).
[2]   New ITU press release, p. 4.
[3]   New ITU press release, p. 5.
[4]   New ITU press release, p. 6.
[5]   Rutkowski, A., "The ITU at the Cusp of Change," *Telecommunications Policy*, Vol. 15, August 1991, pp. 286, 291.
[6]   Solomon, J., "The ITU in a Time of Change," *Telecommunications Policy*, Vol. 15, August 1991, p. 297.
[7]   ITU, *Final Acts of the Plenipotentiary Conference (Nice, 1989)*, RES/80, Geneva: ITU, 1989.
[8]   Codding, G. A., Jr., and D. Gallegos, "The ITU's 'Federal' Structure," *Telecommunications Policy*, Vol. 15, August 1991, p. 306.

[9]   Office of Technology Assessment, 1993, with information from ITU, *Final Acts of the World Administrative Radio Conference* (WARC-92), provisional version, Malaga-Torremolinos, March 1992).

[10] 103d Cong., 1st Sess., Omnibus Budget Reconciliation Act of 1993, Title VI, Communications Licensing and Spectrum Allocation Provisions, PL 103-66 (HR 2264), 107 Stat. 312, 379 et seq. (1993).

[11] Domestic Fixed-Satellite Transponder Sales, 90 FCC 2d 1238 (1982), aff'd sub nom., World Comms., Inc., v. FCC, 735 F.2d 1465 (D.C. Cir. 1984).

[12] Savage, J., "The High-Level Committee and the ITU in the 21st Century," *Telecommunications Policy*, Vol. 15, August 1991, pp. 365, 370.

The Standard-Setting Process

6

International telecommunications standard setting typically involves quantitative and qualitative values that may be articulated by engineers, economists, lawyers, business executives, diplomats, government bureaucrats, or a combination forming a national delegation. Standards are developed internationally at the ITU, but increasingly national and regional standard-setting bodies undertake a parallel and sometimes independent process.

Standard-setting bodies provide common specifications running the gamut from the type of plug used for telephones to the format for routing, signaling, and transmitting, to the rate elements in a tariff and the dialing procedure necessary to complete an international telephone call.[1] They provide the basis for uniform, or at least interconnected, operating systems.

There are two primary classifications for standards: (1) end-to-end standards that cover network systems and the facilities necessary to achieve a complete linkage, and (2) interface standards that enable end users to attach customer premises equipment to a network.

Standards provide timely solutions to problems by assigning precise meanings, definitions, and rules. Advocates for standard setting emphasize the need for and the benefits derived from a consistent, systematic, transparent, and comprehensive explanation of how to erect an international telecommunications and information processing network. Common rules of the road make it possible to achieve shared goals such as efficiency, interoperability, ubiquity, portability, ease of use, and a common interface that enables different equipment to access networks, thereby promoting expanded choice and scale economies.

---

1. "Standardization is a common toll of rational, control-oriented societies: it simplifies and brings order, releasing energies for other tasks. To this extent standardization is more than the dull agent of conformity: it is also essential for flexibility and diversity...They become the means for organizing change as technologies change continuously in bursts of innovation and slower periods of modification." (G. J. Mulgan, *Communication and Control: Networks and New Economics of Communication*, New York: Gilford Press, 1991, pp. 184–185 (hereafter cited as *Communication and Control*).)

Successful implementation of a uniform standard can achieve the following benefits for businesses and society:

- Reduced equipment costs by making network elements interchangeable and available from a larger number of competing manufacturers and vendors;
- Enhanced vendor independence, making it less likely for a manufacturer of a particular network element to hold vendors captive to unfair terms and conditions;
- Improved opportunities for manufacturers and vendors to achieve lower per-unit costs by having fewer incompatible product lines to support;
- Network transparency, meaning that services can be provided seamlessly across borders, even where different equipment and carriers are involved;
- A wealth of information on network functionality in the public domain;
- The ability of individuals and multinational enterprises to achieve reliable and efficient networking capabilities.

The failure to reach a single standard, or a consensus-driven standard leached of specificity and beaten down to a least common denominator, may result in reduced efficiency and consumer welfare. Vendors are tempted to use self-help actions when they believe their innovations and proprietary technology may establish a de facto market-based standard in lieu of a mediocre or unavailable ITU-generated standard. They will support open systems and compatibility, but they may worry about having to disclose too much information to competitors and ceding too much power to users who, with ample choice of options, will try to play one vendor against the others in search of a better deal.

Users articulate a need for interoperability and compatibility, but they may be tempted to use products and services with special features. If a core group of trend-setting manufacturers, service providers, and users, including governments, cannot agree on a single standard, then second tier players will not or cannot follow. Breakaway forums and ad-hoc user groups step into the void resulting from the failure of incumbent bodies to reach a timely solution.

The near ubiquity and seamlessness of international direct dialing and facsimile connectivity attest to the potential for success in the standard-setting process. On the other hand, the failure to achieve a single broadcast color television standard, the vast array of incompatible computer languages, and competing formats like Beta and VHS videotapes point to the adverse consequences of failure.

6.1 STANDARD SETTING IN PERSPECTIVE

Standard setting generates public goods in the sense that the product is designed for general application, and one's use does not reduce what others have available. Because international telecommunications involves interconnection of networks, agreement on standards makes it possible to use diverse types of equipment and communicate across national borders. Universal standards promote rationality, conformity, economies of scale, and lower risk in facility and service deployment.

On the other hand, the standard-setting process takes network and equipment design away from the individual, who might be in the best position to customize private applications. When forums like the ITU secure a consensus, they may reduce a standard to a least common denominator designed to serve the typical user. The interest in uniformity may result in standards with less complexity, but also less utility for sophisticated users requiring customized applications.

Once dominated by engineers seeking optimized technological solutions, the standard-setting process has increasingly become politicized and subject to national industrial policies as governments[2] and private enterprises recognize the stakes involved.[3] When individual government or corporate agendas drive or influence standard setting, the process and output reflect a more diverse constituency. Some of this larger set of players may seek to capture the process for monetary or nationalist gain by advocating a home-grown standard developed by domestic manufacturers and service providers. Such advocates infer that regional or global endorsement of a domestic standard will accrue a marketplace advantage. Accordingly, when the stakes are high, as is the case with standard setting for global, integrated telecommunications and information processing, a greater risk exists that ulterior and self-serving motives drive much of the advocacy. While common rules of the road promote network connectivity and seamlessness, participants in the standard-setting process may advocate the solution developed by their domestic manufacturers and service providers.

6.2 INCREASING COMPLEXITY IN STANDARD SETTING

A number of recent factors have increased the stakes and the complexity of the standard-setting process:

---

2. For a recent congressionally commissioned assessment of U.S. interests in the standard-setting process, see Congress of the United States, Office of Technology Assessment, *Global Standards—Building Blocks for the Future*, TCT-512, Washington, D.C.: GPO, March 1992 (hereafter cited as OTA Global Standards Study).

3. "Standards have never been innocent or purely technical. They have been used as a form of discreet (and indeed indiscreet) trade protection…They have also served as stools of corporate strategy and governmental industrial policy. Because of this political dimension orthodox economics has proved ill-suited to describing the dynamics of standards processes; the economists' assumption of optimality offers little insight." (*Communication and Control*, p. 186.) See also G. Wallenstein, *Setting Global Telecommunication Standards: The Stakes, the Players and the Process*, Norwood, MA: Artech House, 1990.

- Increasing numbers of competitors, including newcomers from the computer manufacturing and services industries, have entered the marketplace for telecommunication equipment and services;
- The convergence of telecommunication and information-processing technologies requires standard-setting participation by individuals and enterprises having less experience with and confidence in the process;
- Heightened recognition by governments of the economic rewards in setting standards with an eye toward making output from national heroes (i.e., indigenous manufacturers and service providers) the de facto or ITU-sanctioned global standard;[4]
- Bolstered interest in serving global markets to compensate for lower margins resulting from heightened competition and the need to defray higher development costs;
- The desire for universal interfaces, which permit interconnection of different equipment types, but also the requirement that such interfaces not prevent the use of proprietary equipment designs and novel services interconnected through the interface (e.g., agreeing to standard audio tones for touch-tone telephones, but allowing answering machine manufacturers to design features activated by a sequence of tones).

Standard setting is key to achieving global connectivity and the efficient, worldwide distribution of information, goods, and services at reasonable rates. Pressure has increased to accelerate the process and to specify standards in tandem with preliminary research and development of the underlying technology. Entrepreneurs want to rush products and services to market perhaps even before sufficient time has passed to debug systems and resolve conflicts over how specific the standard should be. Increasingly, it appears that only a narrow window exists when sufficient operating experience meshes with an adequate timetable for international deliberation. Otherwise, operators may concentrate on

---

4. The Office of Technology Assessment reported that critics of the U.S. standards development process have concerns that "other countries are better organized and better able to influence the international standard setting process, to the detriment of U.S. trade. In particular, they fear that the harmonization of European trade law...[and the formation of the ETSI]...will not only make it harder for U.S. companies to trade in Europe, but will also allow the Europeans to take the lead in setting international standards." (OTA Global Standards Study, p. 3–4, citing United States International Trade Commission, "Standards, Testing and Certification," Chap. 6 in *The Effects of Greater Economic Integration Within the European Community on the United States: First Follow-Up Report*, USITC Pub. 2288, Washington, D.C.: USITC, March 1990.) See also Commission of the European Community, *Commission Green Paper on the Development of European Standardization Action for Faster Technological Integration in Europe*, COM(90) 456 final, Brussels, 8 October 1990; and International Trade Commission, "Standards, Testing and Certification," Chap. 5 in *The Effects of Greater Economic Integration Within the European Community on the United States: Fourth Follow-Up Report*, USITC Pub. 2501, Washington, D.C.: USITC, April 1992.

marketplace concerns irrespective of whether it supports or bypasses the process for consensus building in the standard-setting process.

The failure to reach a global standard on a timely basis can result in:

- The development of multiple market-driven standards; such standards may be incompatible with each other, and accordingly may prevent or raise the cost of interoperability, resulting in stifled innovation and decreased options for consumers;
- The establishment of regional standard-setting bodies, perhaps with an implicit mandate to favor indigenous manufacturers and service providers;
- Imposition of higher costs, greater risk, instability, and delay as various enterprises and standard-setting organizations attempt to work out differences that might not have evolved if a centralized standard-setting process had worked;
- Bolstered incentives for enterprises and users to concentrate on the short-term need for specialized networks, which take advantage of liberalized regulatory policies of some nations, rather than address the long-term need for standards and uniform policies that help achieve major changes in the global infrastructure and the available services (e.g., conversion from analog to digital networking).

The kinds of negative consequences in the failure to reach a timely consensus may create incentives to create multiple or provisional standards. In most instances, participants in the process seem to believe that multiple or imperfect standards are better than none at all, particularly if official activity can preempt de facto standard-setting activities elsewhere. If users or manufacturers take matters into their own hands, the outcome may achieve immediate privately beneficial results, with only the possibility that larger public dividends will follow.

International standard setting, whether under the auspices of the ITU or other global forums, succeeds on consensus building and voluntary compliance. A mandatory standard, particularly one created within a single nation or region, may violate national sovereignty and may constitute a nontariff barrier to trade. On the other hand, laissez faire reliance on market-determined, de facto standards can also fail to maximize consumer welfare. Reliance on the marketplace, no matter how consistent with the prevailing philosophy in some nations, can result in stranded investment in equipment by consumers who bet on the wrong standard even though, according to objective technical criteria, it might be superior (e.g., the Beta video standard).

6.3 NATIONAL STANDARD-SETTING STRATEGIES

The standard-setting process forces nations to balance self-interest in promoting the proposals of their national heroes with the global benefits derived from a single optimized standard. Nations generally find that a single standard fosters efficiency and fewer production lines. At least in theory, it also enables manufacturers from various nations the opportunity to compete for procurement tenders in any nation. Issues of sovereignty, national pride, industrial policy, and incentives to press for international application of national standards are usually subordinated to technology optimization and the benefits of having multiple bidders on a project. Similarly, standard setters recognize the incentive to reach consensus through a formal process in advance of market-driven standard setting.

6.4 PREDICTING WHEN CONSENSUS WILL OCCUR

Stanley Besen and Garth Saloner provide an analytical framework[5] for assessing the standard-setting process through four quadrants representing high or low interest in (1) promoting the universal adoption of any standard, and (2) whether individual parties have a vested interest in one particular standard. "Pure coordination" occurs when the per capita rewards to participate in standard setting are large enough to induce participation, even though the parties will compete (e.g., determining where to place the steering wheel in automobiles). Parties have a high interest in a universal standard and no vested interest in any particular company's proposed standard.

Where the per capita gain from standardization is too small for individuals to find participation worthwhile, government may have to intervene to promulgate one. In this scenario, government involvement creates a "public good" and confers a societal benefit, as was the case where government promulgated standards for weights and measures, time, and language. Governmental intervention generates maximum public benefits where the marketplace and prospective manufacturers and service providers are fragmented and have no vested interest in a particular standard.

The standard-setting process generates "pure private goods" when individual participants have a high interest in promoting a particular standard (e.g., a software platform for personal computers), but the industry lacks a single dominant firm who can establish a de facto standard. While society may benefit from a single standard in terms of reduced costs and lower likelihood of stranded investment and incompatible technologies, interested parties may stall proceedings if their particular standard appears unlikely to receive official endorsement.

---

5. The matter of incentives to promote consensus-driven standards is explored by S. M. Besen and G. Saloner, "The Economics of Telecommunications Standards," in *Changing the Rules: Technological Change, International Competition and Regulation in Communications*, R. W. Crandall and K. Flamm, eds., Washington, D.C.: Brookings Institution, 1989, pp. 147–220.

Conflict reflects dueling standards (e.g., Windows versus Macintosh versus OS/2), where individual parties have a keen interest in promoting a particular standard, and society in general has an interest in and would benefit from a universal standard. A dominant firm may try to coerce others into accepting its standard, but typically it cannot dominate standard-setting forums. Without a formalized standard-setting process, manufacturers using divergent standards will vie for consumers in the marketplace.

Single standards enable manufacturers and service providers to reduce costs and more easily achieve scale economies. Standards facilitate access to new markets through product and service compatibility in the absence of artificially created barriers to entry like closed procurements. Single standards promote lower product development costs, speed product rollout, and guard against balkanization (i.e., fractionalizing product and service markets into relatively impenetrable national or regional markets).

6.5 STANDARD-SETTING MODELS

6.5.1 The Traditional Model

Traditionally, standard setting occurred as part of the close collaboration and consultation undertaken by PTTs and the few nongovernmental carriers. While national carriers and manufacturers might promote a standard that would favor national heroes, nations collectively reached consensus at the ITU on operational "international" standards. Nations either fell in line and adopted a single complete standard or agreed to a degree of standardization sufficient to make possible interconnection between incompatible networks. Incumbent carriers typically install international gateways, which comply with global standards, to serve as the point of interconnection with other, possibly incompatible, foreign networks. A nation may only have access to incompatible or obsolete domestic facilities, but access to an international satellite or cable gateway that conforms to contemporary standards will ensure access to and from the rest of the world.

The standard-setting process proceeded at a slow or measured pace and established a least common denominator that "harmonized" existing national standards into an acceptable compromise. The process emphasized voluntary consensus among like-minded players, all of which shared a stake in managing change. The parties also shared an interest in the incremental evolution of a standard over a long term so that the cost of any innovation could be fully amortized.

The traditional model views standard setting as capping off a period of experimentation leading up to widespread diffusion of an innovation. It does not represent a forward-looking process in which innovators propose standards. Instead, it emphasizes a process in which manufacturers and service providers collaborate and then make their offerings comply with the agreed-upon standard.

6.5.2   The New Model

The new model emphasizes speedy and market-influenced standard setting. In this model, some operators and manufacturers will resort to regional bodies, or none at all should the ITU fail to act on a timely basis. This model reflects the more extensive integration of business enterprises and users with governments, while also evidencing increasing tension over shortcomings in centralized standard setting. The process must closely track technological innovation, rather than await a consensus to evolve on which innovation to use as the foundation for a standard.

In this model, standard-setting participants and outsiders, who never achieved access to the process, are less inclined to accept the ITU's centralized standard-setting process as establishing compulsory international rules.[6] More parties want to become proactive standard setters rather than reactive standard takers, particularly because they now recognize the marketplace advantages accruing from aggressive efforts to convert proprietary technology into the industry standard.[7] Manufacturers and service providers want to participate in the process at the outset to ensure that, even if their proposals do not become standards, their equipment can interconnect with other equipment using the established standard.

With increasing frequency, several enterprises involved in the standard-setting process join in strategic alliances even though individually they may have previously developed a product or service to market. Alliances form on the assumption that there is strength in numbers both to achieve economies of scale and scope and to improve the odds for expediting and dominating the standard-setting process. They anticipate improved odds for success by collectively achieving a global marketing presence. Without such collaboration, multiple and incompatible standards fragment markets, making individual countries less likely to be served by multiple foreign vendors, but also rendering each market segment less lucrative than a single, global market. For example, electricity using appliances in the United States and the United Kingdom varies in terms of

---

6. The old model elevated the ITU standard setting to virtual absoluteness. "They were the anchor of a regime that facilitated bilateral monopolistic bargains, reinforced national monopolies, and limited the rights of private firms in the global market. In short, the conventional view of the telecommunications regime as primarily a technocratic exercise in technical collaboration is wrong." (P. F. Cowhey, "The International Telecommunications Regime: The Political Roots of Regimes for High Technology," *International Organization*, Vol. 44, No. 2, Spring 1990, pp. 169, 176.)

7. In telecommunications and information processing, it appears that in increasing numbers manufacturers are willing to adopt a market posture geared to make their technology the de facto standard. Most initial public stock offerings of these companies are touted as "the next IBM," or "the next Microsoft," two companies that succeeded in capturing near monopoly market shares by establishing a de facto industry standard and readily licensing the technology to other manufacturers and operators.

voltage and outlets. For a manufacturer in either country to target and serve the other country, it would have to develop an entirely new product line, or retrofit existing appliances with a new plug and adapter to convert the appliance for operation using a different level and type of electrical power.

6.6 PRODUCTS OF THE STANDARD-SETTING PROCESS

Standardization typically results from a bottom-up development process, beginning with an innovation, leading to experimentation and testing, distilling into building block frames of reference, and finally leading to the formation of uniform standards. Alternatively, standards may be set by adopting an existing technology and conferring upon it official recognition. A provisional standard may refer to building block capabilities at the innovation and experimentation stage of a technology. Enterprises may devise increasingly complex generations of technology that meet a functional least common denominator.

6.7 INTERNATIONAL STANDARD SETTING

The telecommunications standards sector of the ITU has the responsibility to study and issue recommendations on technical, operating, and tariff questions with a view to standardizing telecommunications on a worldwide basis.[8] The ITU can only issue recommendations,[9] which nations and national standard-setting bodies may use in promulgating compulsory or voluntary domestic or regional standards.

The ITU's standard-setting process works in parallel, but not necessarily in total synchronization, with other international standard-setting organizations. These bodies include the International Standards Organization (ISO), whose membership consists of about 90 national standards bodies. The ISO promulgates standards for worldwide use in numerous fields, excluding electrical and electronics matters. The ISO has made a significant impact on information processing, including the creation of the Open Systems Interconnection (OSI) Reference Model, which presents a seven-tiered hierarchy of standards to foster

---

8. The radiocommunications sector has a narrow, nonduplicative standard-setting role relating to spectrum use. It does not include the issuance of recommendations on the interconnection of radio systems to public telecommunications networks or the performance level of such interconnections. These functions are specifically assigned to the telecommunications standardization sector. (ITU, *The New ITU: Round-Up*, Geneva: ITU Press and Public Relations Service, January 1993, p. 4.)

9. The ITU's recommendations are collected in a series of "books" of a color reflecting the year of issuance. The 1989 Blue Book totaled in excess of 5,000 pages.

connectivity between different information processing systems.[10] Other bodies include the International Electrotechnical Commission, the International Maritime Organization, the Universal Postal Union, and the International Civil Aviation Organization.

An ITU-recommended standard typically constitutes the final product of a process involving the generation of questions from the ITU's council and plenary assemblies of the telecommunication standards and radiocommunication sectors. These questions are assigned to one or more study groups, with a particular and sometimes narrowly drawn area of responsibility and expertise. Study group participants include government officials and representatives from individual companies and trade associations. The findings and recommendations of study groups are considered at World Telecommunication Standardization Conferences and Radiocommunications Assemblies convened every four years. The ITU has arrangements for ratification by correspondence if the involved study group unanimously endorses an expedited process.

The ITU standard-setting process involves government and industry representatives. The ITU governance structure permits participation by recognized private operating agencies (e.g., AT&T)[11] and scientific or industrial organizations (e.g., IBM).[12] Other intergovernmental or standard-setting organizations participate as observers.

Both the radiocommunications and telecommunication standards sectors now have advisory groups, whose membership includes trade associations, manufacturers, and other nongovernmental entities. This new component reflects recognition by the ITU's governmental members that the forum must make the standard-setting process more representative of the increasingly diverse and nongovernmental set of players. The ITU advisory groups review priorities, strategies, and work progress. Additionally, they provide guidelines and recommendations on how to coordinate with other standard-setting bodies.

---

10. "The Open Systems Interconnect model is the computer industry's attempt to come to terms with the convergence of computing and telecommunications. A general law of information technologies states that in design, assembly, testing and networking costs rise with complexity and complexity rises with the number of objects to be managed…The OSI project, launched by the ISO in 1978, had the aim of achieving the interworking of any computer system irrespective of its manufacturer, operating system or location." (*Communication and Control*, p. 199.)

11. A recognized private operating agency operates a telecommunication or broadcasting service and complies with the obligations imposed by Article 6 of the ITU's Constitution (i.e., the treaty-level agreement to be bound by the ITU's Constitution, Convention, and Administrative Regulations). (See ITU Constitution Annex, Definition of Certain Terms Used in This Constitution, the Convention and the Administrative Regulations of the International Telecommunication Union, CS/An 1008.

12. Scientific or industrial organizations are nongovernmental agencies "engaged in the study of telecommunication problems or in the design or manufacture of equipment intended for telecommunication services." Ibid., CS/An 1009.

Nations and individual companies commit the resources required by the ITU standard-setting process primarily because it confers a recognized seal of approval. Adopting an ITU-recommended standard generates consumer confidence that they will not get stuck with an incompatible or short-lived system. Nevertheless, in many high-stakes standardization contests, such as broadcast color television and cellular radio, nations have failed to reach closure on a single standard.

Disincentives to an international consensus standard exist when:

- Two or more members from different countries have developed—and heavily invested in—an incompatible means of achieving the same technological goal. For example, if a nation or regional alliance has invested heavily in a color television or mobile radio transmission system, it may attempt to prevent foreign systems from acquiring the certification needed for lawful sales incountry.
- A technology is not perceived as international in nature and hence can be protected against foreign competition through using incompatible standards as nontariff barriers (NTB). For example, nations initially viewed cellular radio as a local, noninterconnected service. Because there was no perceived benefit in coordinating with other nations on a regional basis to support wide-area roaming by visiting business executives, a nation might decide to promulgate a standard incompatible with other nearby nations to prevent those nations from exporting equipment.
- A national telecommunications equipment manufacturing industry is threatened by foreign competition, and incompatibility may serve as an effective NTB. For example, U.S. manufacturers of digital multiplexing equipment must set up a separate product line or software support for an incompatible system in Europe and elsewhere. U.S. T-1 circuits have 1.544 Mbps of throughput while European E-1 circuits represent 2.048 Mbps.
- Basic philosophical differences concerning the need for standardization cannot be reconciled. For example, the FCC could not decide whether to sponsor a standard for AM radio stereo transmissions, or to defer to a market-driven solution. The FCC's failure to promulgate a standard may have constituted a key reason why no single standard and no market developed.
- Developing nations have considered the standard-setting process a means for developed nations to maintain market domination and to prevent them from establishing an indigenous manufacturing capability [1].

Nevertheless, in most cases, the need for uniformity and global connectivity will constitute strong incentives for differing factions to reach closure.

6.8 PROLIFERATING REGIONAL STANDARD-SETTING BODIES

Having recognized the strategic and marketplace impact of standard setting, some nations perceive an incentive to form regional organizations to accelerate the process and to harmonize standards. A fast-paced effort to set a standard for a new technology improves the odds that the ITU and other international forums will subsequently adopt a regional standard.

6.8.1 ETSI

The standard-setting process can force nations to balance self-interest in promoting the proposals of their national heroes with the global benefits derived from a single optimized standard. The European Telecommunications Standards Institute (ETSI)[13] provides a case study of the balancing that regional standard-setting bodies must undertake. As part of the campaign to foster a fully interconnected and interoperable pan-European telecommunications infrastructure, the European Economic Community agreed to form a "jointly financed… institute…draw[ing] flexibly on experts from both the Telecommunications Administrations and industry, in order substantially to accelerate the elaboration of standards and technical specifications" [2].

ETSI was founded in 1988 with three primary missions:

1. To accelerate development of European telecommunications standards;
2. To achieve speedy action by using weighted voting instead of consensus building from the 21 participating nations, not all of which belong to the European Community;
3. To expand the scope of membership beyond PTTs, which previously established standards through the European Conference of Post and Telecommunications Administrations (CEPT), to include manufacturers, users, and private service providers and research bodies.

ETSI has primary responsibility for broadcasting and telecommunications matters, but must share responsibility for computer technology standards with two preexisting organizations: the European Committee for Standardization (CEN), which produces European standards in all fields except electrical and electronic engineering, and the European Committee for Electrotechnical Standardization (CENELEC). Through either ETSI or CEPT, standard setting in Europe has become regionalized, particularly as applied to public procurements by the region's telephone companies.

---

13. For a comprehensive report on ETSI, see S. Temple, *European Telecommunications Standards Institute: A Revolution in European Telecommunications Standard Making*, Sophia Antipolis, France: ETSI, 1991; and S. Besen, "The European Telecommunications Standards Institute," *Telecommunications Policy*, Vol. 14, No. 6, December 1990, pp. 521–530.

6.8.2 Japan

Historically, Japan's incumbent telephone companies Nippon Telephone and Telegraph (NTT) and Kokusai Denshin Denwa (KDD) handled standard-setting matters, including ITU conference preparations. In 1985, passage of the Telecommunications Business Law authorized partial domestic and international competition and created the need for broader participation in the domestic standard-setting process. While the law expressly authorized the Ministry of Posts and Telecommunications (MPT) to specify terminal equipment standards and to ensure conformance, three nonprofit committees have evolved to achieve greater participation in preparing standards for MPT consideration:

- The Telecommunications Technology Committee (TTC) to help manage standard-setting in telecommunications, with an eye toward maintaining network integrity among an increasing number of carriers;
- The Research and Development Center for Radio Systems;
- The Broadcast Technology Association.

The TTC works in tandem with a separate Telecommunications Technology Council, an arm of the MPT. Both bodies draw participants from the same constituency of carriers, manufacturers, and major users, but the government council serves as the official Japanese representative to the ITU. In effect, the TTC reaches private industry consensus on standards issues, and the government council develops and advocates official positions.

Japanese standard setting emphasizes the downstream implementation by the national standard-setting body of internationally developed or developing standards. Only in rare instances, where there is a need for a domestic standard and the ITU has not started work, will the TTC develop a temporary standard. The nature of Japanese governance fosters a degree of ambiguity in terms of authority and responsibility between MPT bureaucrats, NTT and KDD executives, and the rest of the industry. One observer concluded that "so long as NTT (and, where international interconnection is involved, KDD) use their dominance in the standards process not only to maximize the international market opportunities of the major Japanese telecommunications manufacturers, the present system will continue to work" [3].

6.9 U.S. PLAYERS IN THE STANDARD-SETTING PROCESS

Telecommunications standard setting in the United States "reflects American political culture [of due process and pluralism] and the manner in which industrialization took place" [4]. It emphasizes open access to the process by any stake holder, and the balanced participation of representatives from both the public and private sectors, including local and interexchange carriers, manufacturers, consortia, users, consultants, trade associations, professional societies, and general membership organizations. U.S. delegates to ITU conferences may represent a single company or serve on behalf of a broader constituency (e.g., the Electronic Industries Association, American Electronics Association, Telecommunications Industry Association, Institute of Electrical and Electronics Engineers). Government participants typically include representatives from the FCC, National Telecommunications and Information Administration (NTIA), State Department, Department of Commerce's National Institute for Science and Technology, and the defense and intelligence agencies.

The American National Standards Institute (ANSI), founded in 1918, serves as a private, nonprofit umbrella agency for coordinating voluntary consensus-based standard setting in the United States. ANSI operates as the official U.S. representative to the International Organization for Standardization and the International Electrotechnical Commission. It does not set standards, nor can it dissuade major standard-setting bodies from pursuing unilateral standards. Instead, ANSI serves as a clearinghouse and coordinating body that accredits other member organizations to set standards based on their specific areas of expertise. With ANSI's approval, a standard becomes an American National Standard. The primary role of the U.S. government in the standard-setting process occurs in organizing delegations to ITU conferences and study groups and in procurement.

After divestiture of AT&T's Bell Operating Companies and the dismantling of a de facto Bell System standard-setting process for many of the issues affecting local and long-distance telecommunication services, ANSI accredited a replacement: the Accredited Standards Committee for Telecommunication, commonly referred to as the T-1 committee. Infrequently, specific industry coalitions are formed where the standard-setting process has reached a particularly important stage. In computerization, the Corporation for Open Systems fosters information system connectivity and certifies standards.

6.10 A CASE STUDY OF BROADCAST TELEVISION STANDARDS

The history of broadcast television standards and its impact on current standard-setting efforts for high-definition television (HDTV) provides an instructive case study. Because nations were unable to reach consensus on a single broadcast color television standard—whether out of legitimate concerns over quality or more parochial market concerns[14]—three major standards evolved: (1) the

---

14. "So, what should have been a technical decision—the choice of a single worldwide color television standard—became one dominated by international politics and inadequate American responses. What should have been a resounding victory for American interests, with a flagship

National Television Systems Committee (NTSC) standard adopted in the United States and Japan, (2) the phased alternation by line (PAL) standard adopted in Europe and Asia, and (3) the sequential color with memory (SECAM) standard adopted by France, French-speaking counties, Eastern Europe, and the former Soviet Union.[15]

The financial stakes, divergent constituencies, and conflicting standard proposals threaten to create multiple advanced and HDTV standards as well. Such an outcome has broad-ranging consequences that impact the price, availability, quality, and medium of transmission for HDTV, and also affects a wide array of digital interactive video, computers, flat-screen-delivered graphics, and the like. If U.S. standards emissaries fail in their effort to create an HDTV broadcast standard that requires compatibility with existing non-HDTV transmissions,[16] then a terrestrial-based broadcast HDTV system seems less likely than systems delivered via satellite and fiber-optic cable.

The existing broadcast television industry, which uses 6 MHz of frequency spectrum, cannot migrate to current proposed HDTV standards without more spectrum or application of data compression technology. Without a standard that requires compatibility, much like the ability to play stereo records or listen to stereo FM signals on monophonic equipment, an incumbent video distribution industry may lose the opportunity to remain a technology leader, and over time may lose market share.

---

technology, became an exercise in futility as US government interests undermined US corporate interests, and effectively destroyed the broader national interest." (R. J. Crane, "TV Technology and Government Policy," in *The Telecommunications Revolution Past Present and Future*, H. M. Sapolsky, R. J. Crane, W. R. Neuman, and E. M. Noam, eds., New York: Routledge, 1992, pp. 155, 157.) See also R. Crane, *The Politics of International Standards: France and the Color TV War*, Norwood, N.J.: Ablex, 1979.

15. "The French SECAM colour television standard offers one of the clearest examples of standards serving industrial and foreign policy. The French government's strategy was a response to an attempt by the ITU's CCIR (consultative committee for radio) to set a standard for European colour television...The French government seized on SECAM as a means to develop its electronic industry. By offering financial inducements to communists and Third World countries France ensured that the CCIR failed to reach a consensus, despite the apparent technological superiority of PAL. Europe remains divided by two incompatible television standards." (*Communication and Control*, p. 191.)

16. See Advanced Television Systems and their Impact on the Existing Television Broadcast Service, MM Docket No. 87-268, Notice of Inquiry, 2 FCC Rcd. 5125 (1987), Tentative Decision and Further Notice of Inquiry, 3 FCC Rcd. 6520 (1988), First Report and Order, 5 FCC Rcd. 5627 (1990), Notice of Proposed Rulemaking, 6 FCC Rcd. 7024 (1991), Second Report and Order, 7 FCC Rcd. 3340 (1991), Third Report and Order and Third Further Notice of Proposed Rulemaking, 7 FCC Rcd. 6924 (1992). For a report on how the United States, ITU and other standard-setting organizations interact, see U.S. Department of State, Strategic Planning Group, U.S. Organization for the International Telegraph and Telephone Consultative Committee, *CCITT Interactions With Other Standards Organizations*, Spring 1991.

6.11 STANDARD SETTING IN A MULTIMEDIA AND MULTINATIONAL ENVIRONMENT

Technological developments in telecommunications and information processing place a premium on open network architectures, which will enable an increasingly diverse set of services to interconnect with facilities providing basic transport. Multimedia services will integrate the functions of the telephone, television set, computer terminal, and wireless radio into one device. Similarly, providers of services previously separated by consumer perception and delivery method will attempt to widen their market wingspan.

The confluence of technologies and previously discrete markets places a premium on a process that can generate a single standard for open interconnection between different equipment and services operating at increasingly speedy digital transmission rates. This means that standard setting will serve as a forum for establishing rules for a global digital highway composed of high-speed networks operating terrestrially (e.g., cellular radio and personal communication networks) and via satellite (e.g., low-Earth-orbiting satellites providing ubiquitous voice and data services). Likewise, the ITU must work to ensure that users of equipment, including handheld transceivers, can activate the same device while temporarily in another country. Such transborder usage exemplifies the balance needed between respecting national sovereignty and the right to license radio transmitters and the need to support ease of use and interconnection of transceivers used in different nations and over diverse networks.

Nations will have to agree on digital switching, routing, and transmission standards to make more commonplace such innovations as packet switching, digital mobile radio via terrestrial or satellite options,[17] and ISDNs. Also, they will have to set a definitive line of demarcation between telecommunication equipment, which most nations agree can be competitively provided, and the transmission network, which may be monopolized or subject to limited competition.

6.12 THE NEED TO BALANCE NATIONAL SOVEREIGNTY AND MARKET ORIENTATIONS

The regulatory and legislative initiatives that have stimulated market entry have also fragmented markets and created an even greater need for standards to promote connectivity between networks. It has become increasingly rare for a single enterprise to own and operate all telecommunication facilities in a nation. Accordingly, numerous competing enterprises need a mechanism to ensure that their equipment can interconnect with the facilities of other carriers in the same

---

17. See ITU, *Global Mobile Personal Communications Systems (GMPCS)*, Report of the Third Regulatory Colloquium, Geneva, 9–11 November 1994.

manner as customer premises equipment, such as telephones, connect with any telephone network.

Increasingly, nations have accepted the view that market-based resource allocation serves efficiency, consumer welfare, and equity interests. The marketplace can reward an entrepreneurial company by making its innovation a de facto voluntary standard. But in many cases, numerous incompatible standards may arise. In these instances, the standard-setting process responding to consumer demand must follow up and find ways to promulgate common and open architectures so that equipment using different standards and providing different functions can interconnect much like the manner in which Apple and IBM computers can now operate jointly.

Market entry, deregulation, technological innovations, and an increasingly globalized economy combine to place a greater emphasis on service provisioning by multiple carriers accessed through a diverse array of equipment. Users will have more options available, but with choice comes complexity and the need to ensure connectivity. The standard-setting process becomes even more essential when users increasing rely on enhanced networks to provide increasing diversity in transporting and processing information.

References

[1] Savage, J., *The Politics of International Telecommunications Regulation*, Boulder, CO: Westview Press, 1989, pp. 215–216.

[2] Commission of the European Communities, *Green Paper on the Development of the Common Market for Telecommunications Services and Equipment*, COM (87) 290 final, 22, Brussels, 30 June 1987.

[3] Oniki, H., and T. Curtis, "BISDN Standards in the US and Japan," p. 4, unpublished paper presented at the Pacific Telecommunications Council, Honolulu, Hawaii, 1992.

[4] Congress of the United States, Office of Technology Assessment, *Global Standards—Building Blocks for the Future*, TCT-512, Washington, D.C.: GPO, March 1992, p. 39.

Submarine Cables, Facility Loading, and Private Carriers

7

Submarine cables have provided international transmission linkages for over 100 years and for many routes carry significantly more traffic than satellites. The first transoceanic, international cables provided a few telegraph channels. In the 1950s, new submarine cables provided voice transmission capacity, and in the 1980s, fiber-optic submarine cables provided broadband digital capacity capable of handling video signals.

The structure of constructing and maintaining cables involves a straightforward business relationship among international carriers. The carriers typically form a consortium with investment tied to anticipated usage. Submarine cables generally provide high-volume, primarily point-to-point service between nations in close proximity to oceans. They gradually lose cost advantages over satellites for routes requiring long terrestrial circuits to link callers at interior locations with the point where the cable makes its first landfall. Cables typically have no cost advantage for point-to-multipoint applications like video program distribution to numerous broadcast and cable television affiliates.

With the onset of fiber-optic cable technology, carriers have substantially expanded investment in submarine cables. Likewise, users have begun to consider these broadband facilities for a greater percentage of their total traffic requirements. Cost per unit of capacity continues to drop,[1] making cable prices competitive even with new generations of satellites that can operate with circuit multiplication and compression technologies. Carriers have addressed reliability concerns about cable outages from line cuts and equipment malfunctions by installing two or more interconnected cable routes, thereby providing a "self-healing" network that can route around problem sites.

---

1. Appendix A provides estimates of submarine cable capacity and cost by ocean region and cable facility.

7.1 THE IMPORTANCE OF INTERMODAL COMPETITION

For most international telecommunications routes, users now have access to both satellite and submarine cable routings. Such access is important because it provides the basis for intermodal competition (i.e., facilities-based competition between two different transmission media) and the prospect for price differences on the basis of routing, carrier, and restoration options (i.e., the availability of alternative routing and backup in the event of an outage or busy conditions).

The possibility of intermodal competition began in the 1920s with the deployment of international high-frequency radio facilities. While vulnerable to atmospheric changes, radio provided a medium for sending both voice and record (textual) services. However, regulatory policy historically has blunted comparative advantages of one medium over the other ostensibly to ensure that carriers use both technologies and no single carrier dominates both media.

Nations perceive several different types of benefits in having different kinds of transmission media available for telecommunications. National security interests benefit when the military and intelligence community can route tactical communications via different media and alternative routes. Essential messages can route around unavailable or damaged facilities. Commercial interests also benefit by having alternative routes available.

7.2 LIMITED INTERMODAL COMPETITION

Many nations prohibit or limit the scope of intermodal competition through:

- Laws that link the opportunity for a foreign carrier to land a submarine cable with equivalent opportunities to make such a landing in the foreign country (i.e., requiring absolute reciprocity in the conferral of authorization to deploy cables that make their landfalls on foreign soil);
- The refusal to authorize market entry by additional facilities-based carriers;
- Denying carriers that operate both satellite and cable facilities the opportunities to price such services separately on the basis of comparative efficiencies, as opposed to a single composite charge that averages the cost of both media.

7.2.1 Reciprocity in Cable Landing Licenses

National sovereignty and security concerns dictate that an international carrier must secure permission to install a submarine cable that makes a landfall on the soil of another country. Nations, including the United States [1], typically impose a reciprocity requirement before they permit foreign carriers to land a cable. The Submarine Cable License Act [1], enacted in 1921, allows foreign enterprises to land an international cable on U.S. soil if and only if U.S. carriers have such access opportunities in the carrier's home country.

Reciprocity has yet to occur in the grant of submarine cable landing licenses. Nations continue to prohibit the landfall of foreign-owned international cable facilities. While serving national security interests, a ban on foreign ownership of submarine cables reduces both the total number of cables and the extent of direct facilities-based competition.

7.2.2 Limited Facilities-Based Competition

The total number of submarine cables is limited by most nations' decisions to confer an international cable services monopoly franchise. Likewise, most nations refuse to allow other carriers, monopoly or not, to operate cables or spurs from a cable that would provide facility-based competition with the incumbent cable operator. The cable monopoly franchise typically invests in a global inventory of capacity with similarly situated exclusive franchise holders.

Only in rare instances does a nation authorize intramodal competition (i.e., more than one facilities-based carrier). Nations like the United States, Britain, and Japan permit two or more different investor consortiums to invest in separate cables running parallel to each other on the same route. For example, the PTAT-1 cable, owned by US Sprint and Cable & Wireless, a multinational carrier based in Britain, provides additional capacity in the high-volume North Atlantic, where a number of "TAT" cables are owned by the traditional consortium led by AT&T, British Telecom, France Telecom, Deutsche Telekom, and other incumbent carriers. In the Pacific basin, Pacific Telecom, Cable & Wireless, and Japan's International Digital Communications share ownership of North Pacific Cable, which parallels the route of Trans-Pacific Cables, owned by a consortium of incumbent carriers like AT&T and KDD of Japan. Even though there may be two different cables for the same route (e.g., U.S.-Britain and U.S.-Japan), there is no certainty that robust price competition will evolve.

7.2.3 Composite Pricing That Averages Cable and Satellite Costs

Governments blunt comparative advantages that high-capacity submarine cables often have for point-to-point applications by requiring carriers to tariff a single rate that averages cable and satellite costs. Such averaging works in tandem with requirements that carriers maintain an inventory of both cable and satellite circuits. Even at a higher cost and the expense of efficiency, nations require media diversity. They do not permit carriers to activate circuits on the most efficient transmission medium without regard to the usage mix between cable and satellites. Most governments believe that they must ensure the availability of alternative, redundant, and diverse media. Accordingly, they affect or participate in decision making on:

- Where and when carriers should construct and operate a new submarine cable;
- Service rates, often requiring price averaging that denies users the option of selecting the cheaper transmission medium;
- How carriers activate channels from an inventory of cable and satellite circuits.

7.3 THE CONSULTATIVE PROCESS

Governments become immersed in decisions on when and where to construct submarine cables, how to price service, and what ratio of cable to satellite circuits carriers must activate. Such intrusiveness is justified by the view that the marketplace cannot be relied upon to ensure widespread availability of both cable and satellite circuits, particularly when one medium has a significant cost advantage. Governments require carriers to activate circuits in whichever medium is more expensive at the time because they perceive that the failure to do so could result in the elimination of choice between media.

In either their regulatory or operational role, governments collaborate on a regional basis in planning facility deployment. In a manner much like the Organization of Petroleum Exporting Countries (OPEC) management of oil supply and price, governments replace marketplace resource allocation with centralized management. They justify such intervention using the economic theory of market failure: left unregulated, the marketplace for international telecommunication transmission facilities would concentrate on the cheapest medium without regard to the future potential for other media and without regard for countervailing concerns like national security, circuit redundancy, and media diversity.

Critics of this view claim that the marketplace would set a premium price on all cable or satellite routing, near-immediate restoration of channels in the event of an outage, and traffic routing over different facilities to promote service reliability. For example, satellite users with the greatest need for service reliability already pay a premium for dedicated protection transponders that are earmarked for use in the event of a problem with currently used facilities. Users with fewer reliability concerns may acquire unprotected service and even preemptible service, which is the opportunity to use transponders at a discount subject to almost immediate preemption in the event such capacity must be redeployed to restore service to users who have paid for a more reliable grade of service guaranteeing backup capacity.

7.4 INTERNATIONAL FACILITY PLANNING AND DEPLOYMENT

Governments are extensively involved in international transmission facility planning and deployment. Ironically, governments typically do not always consider both cable and satellite capacity when participating in facilities planning. Instead, submarine cables are planned, constructed, and operated by carrier consortia, subject to the consultative process: formal meetings of PTTs, nongovernment carriers, and government regulators to track cable demand projects and determine when and where additional capacity is needed. The traditional satellite planning and construction model involves a cooperative of investors, typically a single satellite services franchisee and often a government-owned enterprise, subject to occasional oversight by national governments. While many nations confer both cable and satellite service licenses to the same carrier, some nations, including the United States, separate the two franchises.

In both cable and satellite facility planning and deployment, national regulatory agencies assert authority to evaluate and approve investments made by individual carriers.[2] In many countries, the regulatory agency participates far more extensively, either because government owns the international carrier, or because involvement in the facility planning process historically has been shared. For example, in Japan the partially privatized international carrier KDD invests in INTELSAT, but virtually all policy decisions require thorough review by the MPT.

Government involvement in facility planning and the broader concept of centralized planning in telecommunications seems anathema given current procompetitive philosophies. Critics of the consultative process allege that in the absence of a carrier's ability to saddle a captive ratepayer population with an unnecessary facilities investment, the carrier and its shareholders should have unfettered opportunities to incur the rewards and penalties commensurate with independent demand forecasting and facilities planning. Advocates for some continuing government oversight claim that a monitoring role is necessary to guard against facilities overinvestment and to prevent carriers from launching short-term price wars with predatory pricing designed to drive out competition. Incumbent carriers may seek to warehouse capacity, thereby rendering it unavailable to market entrants. Conventional rate regulation establishes a virtually guaranteed rate of return and therefore may create incentives for carriers to overinvest in plant, thereby generating a higher revenue requirement.

---

2. For example, the FCC has authority under section 214 of the Communications Act, as amended, to require the filing of an application by any U.S. carrier to invest in the construction of new international facilities. The FCC must determine whether the public interest will be served by such an investment, and until recently this assessment involved a determination of whether current or near-term demand warranted the facility relative to supply. The FCC does not make a demand and supply assessment for private carrier facilities and no longer conducts a serious assessment even for common carrier investments.

Since the mid-1980s, individual carriers, or carrier alliances outside the conventional consortium model, have developed and installed submarine cables on the basis of anticipated market developments and entrepreneurialism. Some carriers, even ones who have worked within the traditional facilities-planning and investment structure, have opted individually to develop projects or to promote an ad hoc alliance of carriers. Rather than develop a single collective view of future supply and demand conditions, both incumbents and new carriers have developed new and independent projects. These include Africa One, a submarine cable loop around the continent of Africa organized by AT&T; PTAT-1 owned by Cable & Wireless and Sprint; North Pacific Cable, owned by Pacific Telecom, Cable & Wireless, and International Digital Communications of Japan; Fiber Optic Link Around the Globe (FLAG), organized by NYNEX; and a trans-Siberian terrestrial cable organized by U S West.

7.5 FACILITIES LOADING

Historically, governments have done more than approve carrier applications to invest in transmission media. Many nations, including the United States until 1988, actively influence carrier decision making in transmission medium investment and circuit activation. Governments justify such intervention on national security grounds and because of concerns for network reliability. They may require carriers to invest in a new and currently more expensive technology to ensure that carriers pursue research and development efforts that may in time improve price performance and media diversity. Industrial policy factors may also infiltrate the decision-making process, particularly if a nation stands to benefit from contract awards to domestic firms for the construction, launch, or management of a new facility.

7.5.1 Facility Loading Policy in the United States

Until 1988, the FCC instituted policies that blunted intermodal competition between submarine cables and satellites. Previously, the FCC directly interposed itself in individual carrier facility planning and investment decisions as well as the consultative process.[3] Upon the introduction of satellites, the FCC directed carriers to activate more expensive satellite circuits, ostensibly to ensure that this newer technology could evolve and over time develop into a cheaper alternative without government support. Until that time, the FCC justified its inter-

---

3. "The Commission has made decisions affecting the distribution of circuits among available international facilities nearly since the advent of communications satellites in 1965." (Policy for the Distribution of United States International Carrier Circuits Among Available Facilities During the Post-1988 Period, CC Docket No. 87-67, Notice of Proposed Rulemaking, 2 FCC Rcd. 2109 (1987), policy abandoned, 3 FCC Rcd. 2156 (1988).)

vention on nonfinancial grounds, which included support for circuit redundancy, facility diversity, and national security.

The FCC's involvement in circuit distribution decisions arose primarily as a consequence of its 1966 "Authorized User" policy [2]. This policy established Comsat's role as the "carrier's carrier," wholesaler of INTELSAT satellite capacity for retailing use USISCs to end users. The policy limited instances in which Comsat could deal directly with end users to "unusual and exceptional circumstances." Accordingly, Comsat had to rely exclusively on unaffiliated carriers to generate satellite circuit demand. To ensure that these carriers made adequate use of the INTELSAT global satellite system, which the U.S. was instrumental in creating, the FCC felt compelled to order the *balanced loading* of cable and satellite circuits.[4] The FCC extensively scrutinized carrier facility investment and circuitry use by ocean region (i.e., separate Atlantic, Caribbean, and Pacific ocean region studies).[5]

The FCC traditionally used its public interest mandate to review Section 214 applications in terms of demand, costs, media and route diversity, restoration, intramodal and intermodal competition, technological innovations, and international comity.[6] But starting in the late 1980s,[7] the FCC concluded that the Communications Act of 1934 did not require it to undertake a facilities planning process before considering an application to construct a new submarine cable.

The FCC justified its involvement in decisions on carrier circuit loading on the following grounds:

- Promotion of both cable and satellite technologies to meet future needs on a timely basis;

---

4. For an example of how the balanced loading policy was applied, see ITT Cable and Radio, Inc.—Puerto Rico, 5 FCC 2d 823 (1966); AT&T, 7 FCC 2d 959 (1967) (FCC determination of the proper mix of submarine cable and satellite facilities for Puerto Rico; Comsat, 29 FCC 2d 252 (1971) ("proportional loading" ordered at the rate of five Atlantic Ocean Region INTELSAT satellites circuits for every one TAT-5 submarine cable circuit activated).

5. See, for example, Policy To Be Followed in Future Licensing of Facilities for Overseas Communications, 30 FCC 2d 571 (1971) (setting forth policies for licensing transmission facilities in the North Atlantic Ocean region during the 1970s); Inquiry To Be Followed in Future Licensing of Facilities for Overseas Communications, 62 FCC 451 (1976) (North Atlantic Ocean region planning policies up to 1985); Policies To Be Followed in the Authorization of Common Carrier Facilities To Meet Pacific Telecommunications Needs During the Period 1981–1985 (POR Planning), 102 FCC 2d 353 (1986). For a detailed history of the FCC's facility loading policy, see Inquiry Into the Policies To Be Followed in the Authorization of Common Carrier Facilities To Meet North Atlantic Telecommunications Needs During the 1985–1995 Period, Third Notice of Inquiry, 98 FCC 2d 1166 (1984).

6. Am. Tel. & Tel. Co., 4 FCC Rcd. 1129, 1131 (1988) (TAT-9 Authorization).

7. Am. Tel. & Tel. Co., 4 FCC Rcd. 8042 (1989) (TPC-4 Authorization).

- Authorization of the most modern and effective facilities available with due regard for efficiency, economy, routing diversity, and redundancy in the event of outages;
- The perceived need to preempt marketplace resource allocation (i.e., user choice of transmission media) if one particular medium experienced such a pricing disadvantage that carriers would refrain from using capacity adequate to stimulate future research and innovation;
- Rate-setting procedures that allowed carriers to earn a rate of return on cable investments, but treated as expenses carrier leases of satellite capacity.

At the onset of satellite service, cable facilities enjoyed a substantial per-unit cost advantage. New, risky, and expensive launch vehicles were needed to deploy low-capacity facilities relative to what submarine cables could offer. The FCC's assumed role of competitive equalizer enabled satellite service to become a viable option as the technology evolved and as capacity costs eventually dropped. The FCC ordered the USISCs to activate more new satellite circuits than submarine cables. At one point, the ratio was seven new satellite circuits for every one new submarine cable circuit. Such mandatory activation of satellite circuits in effect constituted a temporary subsidy to satellite operators until the transmission cost dropped toward parity with submarine cables.

Arguably, any user seeking the benefits of circuit diversity, even though it would involve the use of a more expensive medium, should be willing to absorb the extra cost. The FCC was unwilling to test this economic premise until satellite and cable circuit costs reached near parity. Instead, it ordered carriers to offer composite rates that merged cable and satellite costs into one averaged charge. The FCC's reluctance to allow marketplace resource allocation may have resulted primarily from differences in the regulatory treatment of cable and satellite capacity acquisitions.[8] Because only Comsat has the right to invest directly in INTELSAT satellite capacity, all other U.S. international carriers in effect resell capacity procured through Comsat. Such expenses do not constitute long-term capacity investments that qualify for favorable depreciation and tax treatment. As such, carriers can only recoup satellite expenditures instead of recouping costs plus earning a return on investment. All things being equal, a noninvestor in INTELSAT[9] would favor deploying cable services because it could own and operate such facilities and earn an FCC-prescribed return.[10]

---

8. See Inquiry Into the Policies To Be Followed in the Authorization of Common Carrier Facilities To Meet North Atlantic Telecommunications Needs During the 1985–1995 Period, Second Notice of Proposed Rulemaking, 100 FCC 2d 1405 (1985).

9. Typically, a single PTT participates in both international cable consortia and INTELSAT. Its loading decisions will be based on the needs of users and its inventory of unused but paid-for capacity.

10. Prior to price caps and other forms of streamlined rate regulation, the FCC applied conventional public utility rate setting. Under this type of regulation the carrier determines the total

Over time, as satellite technology matured, the FCC could find fewer justifications for capacity loading and composite pricing policies. Likewise, the proliferation of transmission options from separate satellites and private cables militated against FCC-orchestrated capacity planning, facility construction, and circuit-loading activities. In the late 1980s, the FCC decided against active participation in the consultative process, a forum increasingly suited to carrier-to-carrier discussions without direct involvement by regulators. While most PTTs, as government entities, do not separate business and regulatory functions, the FCC's attendance had become unsettling in that it perceived its public interest mandate as obligating it to leverage its licensing role on submarine cable applications to secure changes in carrier agreements on future deployment of transmission facilities.[11]

The FCC now only monitors the consultative process ostensibly as part of its internal planning process for assessing future facilities applications. The FCC's internal planning process now appears less concerned with statistical justifications, demand projections, and whether a new facility will likely meet short-term future demand. It now refrains from subjecting new facility construction proposals to close scrutiny of whether they are needed to satisfy credible future demand projections [3]. Instead, the FCC uses a loose public interest evaluation and will authorize projects simply because they will promote new technology, increased competition, circuit restoration, national security, and diverse routings irrespective of whether near-term demand projections demonstrate the need for such projects.[12] FCC facility planning assessments are even more generous as a result of the decision to ignore private cable and satellite

---

amount of revenue it should generate in a year, its revenue requirement, with a simple formula that adds operating expense taxes, current year depreciation, and a regulatory agency–prescribed rate of return multiplied by the total value of investment in facilities minus accumulated depreciation.

11. In 1977, the FCC actually held up authorization of the seventh transatlantic submarine cable, having concluded that existing facilities could meet short-term future need. The FCC subsequently acquiesced to the consensus decision in the North Atlantic consultative process, despite its view that the new facility was not economically justified. (See Policy To Be Followed in Future Licensing of Facilities for Overseas Communications, 67 FCC 2d 358 (1977) (rejecting TAT-7) partial reconsideration granted, 70 FCC 2d 1348 (1978), policy modified, 71 FCC 2d 71 (1979), 71 FCC 2d 1178 (1979) (approving USISC participation in construction of TAT-7).) See R. C. Fisher, "Telecommunications in Transition: Private Transatlantic Cable Facilities," *George Washington J. International Law and Economics*, Vol. 19, 1985, pp. 493, 521–531.

12. In authorizing TAT-9 transatlantic submarine cable, the FCC appeared disinclined to consider the amount of digital international satellite capacity that would come online, and it failed to expressly set out quantitative evidence that circuit demand, rather than restoration or other public interest convenience factors, demonstrated the need for the new facility. (See Inquiry Into the Policies To Be Followed in the Authorization of Common Carrier Facilities To Meet North Atlantic Telecommunications Needs During the 1991–2000 Period, CC Docket No. 79-184, Report and Order, 3 FCC Rcd. 3979 (1988) (hereafter cited as TAT-9 Authorization).)

capacity when assessing future needs, deeming such capacity "as alternatives to, and not substitutes for, common carrier transmission systems" [4].

In 1988, the FCC also abandoned its last remaining involvement in circuit-loading decisions.[13] The FCC eliminated all circuit distribution guidelines for AT&T, the only remaining carrier subject to a policy on loading circuits for international message telephone and 800 service, in part because AT&T and Comsat had entered into an agreement extending the term of all existing satellite capacity leases, incrementally freeing AT&T to activate new circuits on the basis of its requirements and not pursuant to an FCC-prescribed formula. The parties also agreed on a formula for determining the minimum amount of future traffic AT&T was obligated to commit to international satellite carriage [5]. In effect, the parties agreed to safeguard Comsat and INTELSAT from traffic migration of existing satellite traffic, and a significant but declining volume of growth traffic.

7.6 MARKET-DRIVEN DECISION MAKING

The FCC's increasingly laissez faire approach to facility planning, cable authorizations, and circuit-loading decisions[14] reflects growing confidence in market-driven, inter- and intramodal facility competition. Although the international telecommunications marketplace has yet to become fully competitive, the FCC believes that private carrier market entry will foster competition. However, the FCC has not aggressively engaged its foreign counterparts in a dialogue aiming to promote competition, particularly that provided by newcomers freed of most conventional common carrier obligations: "private systems...[will] succeed or fail on their own merits and not through Commission action that would guarantee common carrier use of these systems" [6]. Given the limited niche role such new "private" carriers now play, incumbent common carriers, cooperatives, or consortia have little worry of significant economic harm, and in the case of new

---

13. See Policy for Distribution of United States International Carrier Circuits Among Available Facilities During the Post 1988 Period, 3 FCC Rcd. 2156 (1988).

14. In authorizing the Trans-Pacific 4 submarine cable, the FCC refrained from considering the amount of digital international satellite or fiber-optic cable capacity that would come online. It refused to assess quantitative evidence that circuit demand, rather than restoration or other public interest factors, demonstrated the need for the new facility. (Am. Tel. & Tel. Co., 4 FCC Rcd. 8042 (1989).) Recent submarine cable authorizations by the FCC show little scrutiny of the need for new cable facility investment. (See, for example, Am. Tel. & Tel. Co., 5 FCC Rcd. 7331 (1990) (PacRim East), 5 FCC Rcd. 7358 (1988) (PacRim East Cable Landing License), 5 FCC Rcd. 7344 (1990) (Hawaii-5), 5 FCC Rcd. 7360 (1990) (PacRim West), 5 FCC Rcd. 7362 (1990) (PacRim West Cable Landing License).)

private cables, the members of the conventional cable consortia have made spot purchases and leases of capacity primarily for purposes of routing diversity.[15]

Such marketplace accommodation also reflects the growing realization that international fiber-optic cable and digital satellite facilities can coexist. Carriers can exploit comparative advantages: satellites well serve point-to-multipoint, video, and mobile applications. Temporary circuits can be activated quickly to serve news and disaster relief requirements. Despite the cost of launching and insuring satellites, this medium can compete with cable, particularly where long domestic tail circuits are needed to link a cable landing point with interior locales. Cable excels in high-volume point-to-point applications requiring low bit error rates and no transmission delay.

The future international telecommunications marketplace appears ripe for broader bandwidth and increased user demand for information services, HDTV, medical imaging, file transfer, and high-speed computer processing necessary for online financial transactions. Higher speed fiber-optic equipment enhances the cable option, but expanded satellite capacity, the ability to compress channels to conserve capacity, lower cost launch options, and proven reliability promote satellite options. The future international telecommunications facility marketplace appears robust and somewhat less burdened by regulatory intervention, provided the barriers to market entry continue to drop and alternatives to incumbents are afforded a reasonable opportunity to find a market niche.

7.7 CARTEL MANAGEMENT OF INTERNATIONAL FACILITIES GROWS INCREASINGLY DIFFICULT

Market entry initiatives of nations like the United States make centralized control and management less sustainable. Satellites, separate from the INTELSAT cooperative and submarine cables, deployed by new carriers or a faction of the traditional consortium structure, complicate the consultative process. While new operators must convince regulators of the public interest merits of additional transmission capacity, they can enter the market without having first consulted with incumbent carriers. However, they still must acquire operating agreements before the facility can constitute part of a carrier's transmission capacity. The "process of arranging to construct and operate international telecommunications facilities...[has become] more complex and volatile than it was only a few years ago" [7], because the consultative process can no longer control market entry or the amount of capacity construction.

---

15. See Am. Tel. & Tel. Co., Application for Authority To Operate Facilities in the Private Transatlantic Telecommunications Cable System for the Provision of Authorized Common Carrier Services to the United Kingdom and Beyond, File No. I-T-C-91-146, Order and Auth., 6 FCC Rcd. 4875 (1991).

References

[1]　"An Act Relating to the Landing and Operation of Submarine Cable in the United States," codified at 47 U.S.C. Secs. 34–39 (1990). See also Executive Order 10530 (10 May 1954) (delegating to the FCC certain presidential functions relating to submarine cable landing licenses); and Tel-Optik, Ltd., Cable Landing License, File Nos. I-S-C-L 84-002, I-S-C-L-84-003, Mimeo No. 4618 (rel. 17 May 1985).

[2]　Authorized Entities and Authorized Users, 4 FCC 2d 421 (1966), reconsideration granted in part, 6 FCC 2d 593 (1967).

[3]　American Telephone and Telegraph, 98 FCC 2d 440, 467 (1984) (TAT-8 authorization).

[4]　American Telephone and Telegraph, 98 FCC 2d, p. 467.

[5]　Communications Satellite Corp., Transmittal No. 674, 3 FCC Rcd. 1132 (1988).

[6]　Inquiry Into the Policies To Be Followed in the Authorization of Common Carrier Facilities To Meet North Atlantic Telecommunications Needs During the 1991–2000 Period, CC Docket No. 79-184, 3 FCC Rcd. 3979, 3990 (1988).

[7]　Bruce, R.. J. Cunard, and M. Director, *The Telecom Mosaic-Assembling the New International Structure*, London: Butterworths, 1988, p. 339.

Information Services and Intelligent Broadband Networks

8

Telecommunications and computer technologies have merged to create a diverse array of information-processing services. A variety of names are used to describe these services which provide customized solutions for users in multiple locations: enhanced, value-added, and broadband information services. They enhance basic "vanilla" telecommunications capacity by adding value to the building blocks of leased transmission lines. These services often fit within a broad definition of what intelligent networks offer. They typically require large bandwidth to handle the amount of data that must be transmitted and processed. For example, a credit card verification network integrates two key elements:

1. A telecommunication transport function for linking diverse point-of-sale terminals with central databases;
2. An information-processing function for interrogating databases to determine the credit worthiness of a credit card purchaser.

Other examples include electronic messaging, electronic data interchange, airline reservation systems, electronic funds transfer, and videoconferencing.

The convergence of telecommunications and information processing has created new market opportunities that many national governments consider ripe for competition. Even in nations where a basic telephone service monopoly persists, a robust and competitive marketplace can evolve for information services without jeopardizing incumbent carrier revenues. Many governments continue to believe that they must reserve a basic services monopoly for the incumbent PTT or its privatized replacement. Such insulation from competition is less compelling for information-processing markets because value-added service providers:

- Stimulate demand for basic transport capacity from the incumbent carrier rather than trigger traffic migration and stranded investment (i.e., loss of consumer demand sufficient to support already installed facilities);
- Can be prohibited from competing with the basic telecommunications offerings of the incumbent.

8.1 IVANs

IVANs[1] represent a category of information service providers authorized to operate in an increasing number of nations. IVANs represent an important market development, because they often constitute the first licensed competitors to the PTT and a nation's first attempt at fostering a competitive telecommunications market. IVANs also represent the cutting edge in meeting complex networking and information service requirements of multinational businesses, such as:

- *Protocol processing:* Enabling communication between computers and terminals operating with different codes, formats, and protocols;
- *Remote access to databases:* Online retrieval of stored information;
- *Information processing:* Access to databases combined with the ability to manipulate the data (e.g., searching, sorting, collating, calculating, storing, erasing, and printing);
- *Value-added telecommunications:* Employing information-processing features to enhance communications (e.g., electronic mail, voice messaging, temporary computerized storage and forwarding of messages);
- *Electronic data interchange:* High-speed computer-to-computer transmission of data in structured, machine-readable formats, primarily for business applications.[2]

In time, many of the enhancements provided by IVANs will flow to residential and educational applications.

IVANs typically incorporate three service features:

1. They increase the speed at which data traverses public or private networks and the scope of computerized processing features; for example, IVANs have engineered credit card verification systems that take sec-

---

1. For an extended introduction to the IVAN industry, see Dept. of Commerce, National Telecommunications and Information Admin., *International Value-Added Network Services—An Introduction*, NTIA TM-90-258, Washington, D.C.: GPO, March 1990.

2. See R. Austin, *Electronic Data Interchange—The International Environment for Electronic Contracting*, Amer. Bar Assn., Sec. of Int'l Law and Practice, Communications Committee, Monograph Series 1991/1.

onds to process an inquiry even if it requires interrogation of databases in several banks situated in different nations and accessed over diverse networks.

2. They enable networking to take place across vast distances and between numerous points.

3. They promote commerce and the exchange of information by eliminating previous impediments that prevented or slowed transactions and imposed higher costs.

Incumbent carriers and market entrants can develop IVAN services by using basic transmission capacity from private lines or capacity partitioned from public switched services through the use of software. IVANs provide textual, data, or video communication in real time and on a delayed basis through the use of computerized memory to store the information and later forward it in an efficient batch of files. They can package complete one-stop-shopping services that provide the convenience of having one enterprise provide or acquire all necessary transmission capacity linked with the sale, rental, and servicing of telecommunication hardware.

Both developed and developing nations have authorized IVAN competition because it provides customized telecommunications and information processing solutions rather than basic communication services[3] in competition with the incumbent carrier. This means that IVANs by definition provide more than simple voice and data transport services between points. They typically integrate computerization and telecommunications to serve a relatively small set of corporate users with complex telecommunications requirements (e.g., the international airline that needs to process millions of reservations, track flights, and print tickets in thousands of locations).

Nations permit IVANs to operate because they serve niche markets for customized services, stimulate demand for basic transport facilities of incumbent carriers, and do not present a threat to core revenue streams needed by PTTs to cross-subsidize local telephone and postal services. The willingness to authorize a competitive IVAN marketplace juxtaposes with the strong reluctance of all but a very few nations to permit resale of basic services, such as long-distance telephone calling. While IVANs add enhancements to leased lines and subsequently resell them, governments emphasize the nature of the enhancements and assume that most users do not resort to IVANs for the kinds of basic services that remain reserved to the incumbent carrier. Hence, IVANs appear to be less a PTT competitor and more like a system integrator retained by users who decide to rely on experts rather than develop in-house resources.

---

3. This descriptive list is adapted from criteria set out in E. Wittee and M. Dowling, "Value-Added Services—Regulation and Reality in the Federal Republic of Germany," *Telecommunications Policy*, Vol. 15, No. 5, October 1991, pp. 437, 438.

8.2   DISAGREEMENT ON THE SCOPE OF LIBERALIZATION

Telecommunications policy makers in many countries now believe that some degree of market access in information service markets can enhance consumer welfare without burdening incumbent carriers. However, nations have not reached consensus in three major respects:

1. The scope of services that should be reserved for incumbents (i.e., to what extent can PTTs incorporate information processing enhancements into their inventory of services and claim that these additional services fit within a "reserved services" monopoly?);[4]
2. The extent to which users can attach equipment to network services in lieu of carrier payments for such features;
3. The type and nature of competing IVAN services.

Nations like the United States, Canada, the United Kingdom, and Japan support the view that they can use categories to draw a "bright line" distinction between basic and enhanced services for purposes of determining the scope of regulations to apply.[5] Carriers providing basic services (i.e., "plain vanilla" transmission capacity) remain subject to traditional public utility regulation. Enhancements to basic transmission capacity, provided by both incumbent facilities-based carriers and newcomers using leased lines, qualify for less regulation or no regulation (as is the case in the United States).

For example, Japan has three facilities-based Type I international carriers which provide basic services and over 800 Type II carriers which provide enhancements to basic services. Type I carriers are subject to extensive regulation, while Type II carriers need only secure preliminary certification from the MPT, after which they operate largely unregulated.

---

4. For example, policies in the United States attempt to create a "bright line" separation between enhanced service functions, which are unregulated and subject to robust competition and basic transport capacity that is regulated and not robustly competitive. (See Second Computer Inquiry, Final Dec. 77 FCC 2d 384, mod. on recon., 84 FCC 2d 50 (1980), further mod., 88 FCC 2d 512 (1981), aff'd sub nom., Computer & Comms. Ind. Ass'n v. FCC, 693 F.2d 198 (D.C. Cir. 1982), cert. denied, 461 U.S. 938 (1983); Third Computer Inquiry, Report and Order, 104 FCC 2d 958 (1986), mod. on recon., 2 FCC Rcd. 3035 (1987) (Phase I), further recon., 3 FCC Rcd. 1135 (1988); Phase II, CC Docket No. 85-229, Report and Order, 2 FCC Rcd. 3072 (1987), recon. denied, 3 FCC Rcd. 1150 (1988); partially reversed and remanded sub nom., California v. FCC, 905 F.2d 1217 (9th Cir. 1990); on remand, 6 FCC Rcd. 7571 (1991); partially reversed and remanded sub nom., California v. FCC, 39 F.3d 919 (9th Cir. 1994).

5. For an outline and analysis of the *Computer Inquiries*, see R. Frieden, "The Computer Inquiries: Mapping the Communications/Information Processing Terrain," *Federal Communications Law J.*, Vol. 33, 1981, pp. 55–115; and R. Frieden, "The Third Computer Inquiry: A Deregulatory Dilemma," *Federal Communications Law J.*, Vol. 38, 1987, pp. 383–410.

Most nations reject the view that information service providers should operate free of government oversight. Many nations simply extend the PTT monopoly into these markets or impose the kind of Type I common carrier regulation applied in Japan. They do not endorse the view that regulation should apply only to the incumbent facilities-based carrier. Even for nations inclined to favor a deregulated environment, the potential for the incumbent carrier to engage in anticompetitive practices stimulates the need for government oversight when an incumbent carrier provides both basic transport and enhanced services. It may:

- Exploit its ownership and operation of local exchange facilities;
- Impose excessive financial burdens on users of basic services, in effect making them involuntary underwriters of the incumbent carrier's underpriced enhanced services.

8.3 U.S. POLICY ON IVANs AND ENHANCED SERVICES

The FCC's *Second Computer Inquiry* policy required the Bell Operating Companies (BOC) to establish separate subsidiaries when providing unregulated enhanced services. Such arm's length dealing between corporate affiliates providing basic common carrier and enhanced services ensured that the basic services provider could not favor its corporate family member with:

- Cross-subsidies;
- Preferential terms and conditions for interconnection to local exchange facilities;
- Advance disclosure of technical information;
- Shared access to customer proprietary network information (CPNI) (i.e., information on the number and type of basic services used by customers).

The FCC required carriers providing both basic and enhanced services to tariff terms and conditions when leasing basic service facilities to both affiliated and unaffiliated enhanced-service providers. Such arm's length dealing ensured that all enhanced service providers, regardless of affiliation, would achieve equal access to local exchange facilities. Structural separation established a requirement for fair dealing between basic- and enhanced-service providers regardless of corporate lineage. This requirement made preferential dealing between corporate affiliates more easily detected and remedied.

In the late 1980s, the FCC dismantled structural requirements on the grounds that they imposed excessive costs and inefficiency relative to the safeguards. The FCC believed that its auditors and responsiveness to complaints would provide adequate protection. Before the BOCs could reintegrate basic and enhanced functions, they had to file a comparably efficient interconnection (CEI) plan that showed how they would not discriminate against unaffiliated enhanced-service providers.

The FCC also required the BOCs to revamp their local exchange facilities to achieve open network architecture (ONA),[6] a blueprint for specifying and tariffing local exchange functions by their component parts so that enhanced-service providers can select only what they require to engineer customized offerings. Such "a la carte" pricing of access elements enables information service providers to select from a list of network building blocks reduced to their least common denominator. The FCC contemplated ONA as the vehicle for distributing network control to end users and enhanced-service providers so that they could customize services, even though they would need to use the local exchange facilities of a competitor. The FCC envisioned its ONA policy as making the local exchange facilities of incumbent carriers a feature-rich resource for configuring customer-tailored telecommunications and information-processing services.

The ONA goal seeks to maintain a "level competitive playing field" between enhanced service providers, which need to lease transport capacity from incumbent carriers and the incumbent carrier or its affiliates that will compete for enhanced-service customers. Such accommodation of a diverse set of players in the telecommunications and information-processing marketplace requires painstaking coordination and scrutiny on antitrust, regulatory, and technological grounds. The elimination of restrictions on BOC provision of information services[7] makes this task even more essential.

In May 1990, the FCC reaffirmed its ONA model, but also established a mechanism for ongoing oversight of the ONA process, recognizing that it had not completed its deliberations on a number of long-term issues. While generally accepting the initial ONA plans of the BOCs, the FCC emphasized the need for collaboration between users, carriers, and enhanced-service providers to fos-

---

6. See Filing and Review of Open Network Architecture Plans, CC Docket No, 88-2, Phase I, Memorandum Opinion and Order, 6 FCC Rcd. 7646 (1991), partially reversed and remanded sub nom., California v. FCC, 4 F.3d 1505 (9th Cir. 1993).

7. The Modification of Final Judgment in the AT&T divestiture case prohibited the spun-off BOCs from providing broadly defined information services. (United States v. AT&T, 552 F. Supp. 131 (D.D.C. 1982), aff'd sub nom., Maryland v. United States, 460 U.S. 1001 (1983).) As part of its triennial review and determination of whether line-of-business restrictions should continue, the District Court narrowed the information restrictions. (United States v. Western Electric Co., 714 F. Supp. 1 (D.D.C. 1988); 767 F. Supp. 308 (D.D.C. 1991).) On appeal the Circuit Court of Appeals determined that the narrowing was inadequate and used improper evaluative criteria. (United States v. Western Electric Co., 900 F.2d 283 (D.C. Cir.), cert. den. sub nom., MCI Telecommunications Corp. v. United States, 111 S.Ct. 283 (1990).) The Circuit Court of Appeals subsequently lifted the ban on BOC provision of information services. (United States v. Western Electric Co., 1991-1 Trade Cases (CCH) ¶ 69,610 (D.C. Cir. 1991).)

ter greater uniformity in how the component parts of ONA are tariffed and provided.

The FCC envisions a feature-rich broadband digital highway available under fair terms and conditions subject to minor regulation. Notwithstanding its recognition that the ONA blueprint requires refinement,[8] the FCC contemplates that the United States will develop an upgraded telecommunications and information-processing infrastructure that favors competition, equal access, diverse features, and the ability of individual users and commercial enterprises to customize network building blocks with other enhancements.

8.4 MOST NATIONS LACK ENHANCED-SERVICE SAFEGUARDS

IVANs present challenges to current policies on telecommunications and information processing. While policy liberalization made it possible for market entry, IVAN market share depends in large part on whether PTTs conscientiously respond to requests for cost-based and feature-rich access to local exchange facilities.

Few nations have formulated regulatory safeguards to achieve parity of access to incumbent carrier facilities by affiliated and unaffiliated enterprises. These nations also lack an ONA plan for ensuring cost-based access to the basic telecommunications infrastructure needed by IVANs to originate and terminate services. An ONA would make it easier for enhanced-service providers to customize services to meet the growing diversity in user requirements.

Many nations have just begun to permit market entry IVANs unaffiliated with the PTT. The terms and conditions for accessing PTT facilities are typically dictated by the PTT, often without published tariffs and extensive regulatory scrutiny. Most nations still prohibit the shared use or resale of basic services on the grounds that such competition would adversely affect the PTTs' financial wherewithal to cross-subsidize socially desirable services. IVANs qualify for an exception because they generally stimulate demand for basic PTT services and promote the nation's campaign to attract new information-intensive industries.

On one hand, the PTT needs to set the rate of compensation for access to basic transport services high enough to recoup costs, including infrastructure upgrades to handle multiple-carrier access. On the other hand, the regulatory agency should ensure that competitors of the PTT have full and cost-based access to foster competition where permitted. The regulatory agency must guard

---

8. The FCC acknowledges that "ONA is an evolutionary process." Filing and Review of Open Network Architecture Plans, Docket No. 88-2, Phase I, Mem. Op. & Order, 5FCCRed. 3013, ¶11 (1990).

against anticompetitive practices by incumbents, but it must also require new-comers to pay their fair share for access to incumbent carrier facilities.

8.5 EUROPEAN EFFORTS TO SUPPORT INFORMATION SERVICES

The European Union (EU) has initiated a program to revamp the national tele-communications infrastructures[9] as part of the Single Europe Act of 1987 [1]. The EU's open network provision (ONP),[10] plan appears to parallel the FCC's ONA scheme. In application they have significantly different focus. The FCC envisions ONA as a blueprint for supporting a competitive enhanced-services marketplace with operators able to access the local exchange infrastructure un-der fair terms and conditions. The FCC's focus lies in promoting full and fair competition between enhanced-service operators regardless of whether they are affiliated with the local exchange carrier that provides access to essential facili-ties used to originate and terminate service. The EU's ONP plan emphasizes the need for harmonizing national regulatory policies to promote intra-EU com-merce and a bolstered trade posture vis-a-vis other nations whose telecommuni-cations infrastructure may be more advanced and better coordinated. Nevertheless, the ONP/ONA planning process shows the growing similarity be-tween two major trading blocs, each seeking to promote commerce through tele-communications and information processing.

The FCC's concern with semantic line drawing between basic and en-hanced services stems from an interest in promoting full and fair competition. The EU has little interest in using regulatory classifications to segment markets because most incumbent carriers will provide both basic and enhanced serv-ices, subject to limited, if any, IVAN competition. Both the FCC and its foreign counterparts must tread a fine line between foisting unnecessary costs, ineffi-ciency, and duplication of effort on carriers and safeguarding ratepayers and competitors from anticompetitive practices (e.g., abusing control of local ex-change facilities by denying interconnection, providing inferior interconnec-tion, or charging discriminatory rates to competitors relative to the rate charged corporate affiliates).

---

9. For a report on the scope of progress achieved, see Commission of the European Communities, *Communication to the Council and European Parliament on the Consultation on the Review of the Situation in the Telecommunications Services Sector*, COM(93) 159 final, Brussels, 28 April 1993.

10. See The Council of the European Communities, *Common Position Adopted by the Council of February 5, 1990 With a View to Adopting a Directive on the Establishment of the Internal Mar-ket for Telecommunications Services Through Implementation of Open Network Provision*, Council Directive No. 4078/90.

8.6 INTEGRATED SERVICES DIGITAL NETWORKS

The proliferation of IVANs promotes diversifying telecommunications and information networks primarily through leased private lines. To foster a ubiquitous, intelligent network, facilities-based carriers will need to transform analog transmission networks into a digital format. This transition to ISDNs[11] will make available wider bandwidth and greater information processing capability, even to residential users.

ISDN means different things to different people. It constitutes a highly visible subject in the ITU standard-setting process that aims to ensure a trouble-free conversion from analog to digital network transmission formats with higher capacity already packaged for use by typical users. Part of the delay in ISDN deployment stems from the inability of nations to reach consensus decisions within the context of standard setting at the ITU. The technological characteristics of the innovations provide clear evidence of service and efficiency gains. However, nations have not agreed on how to package the technology, who can provide the services, how and where customer-supplied equipment can access ISDNs, and what regulatory classification is appropriate. Disagreements remain over such fundamental tasks as specifying the number and types of permissible ISDNs and drawing the line between network services and the equipment users will attach to the ISDN.

ISDN also has suffered perhaps from overly aggressive, confusing, and conflicting marketing campaigns. Whether the ISDN acronym stands for "I see dollars now," or "innovations subscribers don't need," or even "I still don't know," depends on one's perspective. In theory, telecommunications planners envision ISDN as "innovations serving diversified (user) needs." Rather than pursue technology for technology's sake, ISDN planners anticipate the need to satisfy diverse broadband consumer requirements through a more efficient and faster digital network system that can be all things to all people. Questions remain whether greater efficiency and more throughput from such applications as packet switching are necessary to serve the typical requirements of residential and even small-business users.[12] On the other hand, multinational business enterprises require customized and complex features that a public ISDN might not provide.

If residential telephone consumers continue to require only POTS, then ISDN to them stands for "innovations subscribers don't need." However, technological innovations sometimes take time to win consumer acceptance.

---

11. See A. Rutkowski, "Integrated Services Digital Networks and Options for the World's Future Communications Systems," 1984 *Michigan Yearbook of International Legal Studies,* New York: Clark Boardman Co., 1984, p. 243; and A. Siff, "ISDNs: Shaping the New Networks That Might Reshape FCC Policies," *Federal Communications Law Journal,* Vol. 37, 1985.

12. See Integrated Services Digital Network (ISDN), First Report, 98 FCC2d 249 (1984).

The basic rate interface (BRI) for ISDN comprises two bearer channels, which provide transport services, and one data channel, which provides call setup and network management functions. The BRI will deliver 144 Kbps of throughput compared to the throughput of about 30 Kbps currently available. Over time, the larger capacity standard may represent the minimum bandwidth necessary to access a feature-rich telecommunications and information-processing environment.

The primary rate interface for large-volume ISDN users is composed of 23 bearer channels and one data channel for a total of 1.544 Mbps of capacity. Transmission lines of 1.544 Mbps used to qualify as high-volume facilities required only by governments and large corporate users like airlines, banks, and oil exploration companies. Over time, such line capacity has become more routine and 45-Mbps networks have evolved.

Should the ISDN scenario play out, then the telecommunications infrastructure will have evolved from a narrowband analog system providing POTS, with high-capacity intercity links, to robust and nearly ubiquitous broadband networks.

8.7 WHAT DOES ISDN STAND FOR?

To appreciate the international policy and coordination requirements in ISDN, one should understand what is represented by the ISDN acronym: integrated services digital networks.

8.7.1 Integrated

The ISDN vision anticipates global connectivity of feature-rich telecommunications and information-processing networks. Such connectivity contemplates ubiquitousness, which means that ISDN functionality over time will become as commonplace as POTS. The integration of ISDN into the telecommunications and information-processing infrastructure requires full deployment of the kind of technology currently used in high-capacity lines that link telephone company facilities.

The BRI can be delivered via the twisted copper wires that already extend to residences. This means that ISDN can arrive without the installation of fiber-optic lines needed for ISDN via the primary rate interface. However, carriers may delay ISDN deployment until widespread installation of fiber-optic cable, a medium that will enable them to compete in the video programming marketplace.

8.7.2 Services

ISDN will succeed or fail in large part depending on the services it delivers. For PTTs, the investment decision to upgrade networks to provide ISDN depends on their assessment of whether such capability will preserve or extend market dominance in the face of expanding consumer choices of facilities and services. Incumbent carriers will make the sizable ISDN investment if they believe consumers will demand such services or if regulators confer favorable tax and other financial benefits.

Regulators can substantially affect the pace of ISDN deployment with policies that promote or delay the deployment of new technologies. For example, depreciation schedules (i.e., the stated time period within which a carrier can recoup its plant investment) can be shortened to support faster turnover of equipment. Likewise, an incumbent carrier may be more willing to risk deploying new technologies and costly new equipment if it has greater confidence that it can recoup its investment. While such a reservation of a service monopoly might reduce perceived risk to the incumbent carrier, it might also reduce the efficiency and competitiveness of the carrier, and encourage it to avoid having to work harder and to innovate.

ISDN availability may become enmeshed in the ongoing debate over how best to promote access to expanded video programming options. In the short term, video programming constitutes the most likely occupant of increased bandwidth and processing capability via ISDNs and other broadband digital networks. There exists a proven market for movies delivered on a pay-per-view basis. While incumbent carrier expansion into video services typically faces no regulatory barriers, some nations have concerns about potentially adverse cross-ownership, cost allocation. and anticompetitive consequences. Legislators and policy makers in the United States have debated whether telephone companies should be permitted to install cable television distribution plant and distribute content, particularly in the same geographical region that they provide telephone service.[13]

Opponents fear that telephone companies would use revenues derived from captive POTS ratepayers to predatorily price their video offerings with an eye toward driving competing cable television ventures out of business. They also worry that over time a single telephone company wire into the home would vest too much control in one enterprise, particularly if that enterprise controlled both the conduit and the content carried.[14]

The telephone companies have countered that POTS, as a heavily regulated service, does not accrue excessive revenues or provide subsidies for underpriced video services. The FCC has proposed a compromise in allowing

---

13. See Telephone Co.-Cable Television Cross-Ownership Rules, Secs. 63.54–63.58, CC Docket No. 87-266, Second Report and Order, Recommendation to Congress and Second Further Notice of Proposed Rulemaking, 7 FCC Rcd. 5781 (1992).

14. For an assessment of the future broadband environment, see R. Pepper, *Through The Looking Glass: Integrated Broadband Networks, Regulatory Policies and Institutional Change*, FCC, Office of Plans and Policy, Working Paper No. 24, November 1988.

telephone companies to provide "video dial tone" (i.e., the regulated provision of video transmission capability), but not the content. The telephone companies reject this proposal and have pursued legislative and judicial remedies to permit their investment in both conduit and content.[15]

The issue of telephone company–provided cable television services possibly has an impact on the time period for deployment of the broadband digital transmission lines necessary for both video and ISDN services. While the twisted wire pair already installed to the home can support the BRI, many of the advanced, higher speed ISDN applications will require significant upgrades to the infrastructure.

8.7.3   Single or Multiple ISDNs?

The advanced ISDN version contemplates a network with intelligence distributed throughout, instead of previous information architectures where most of the processing occurred at a centralized computer center accessed by relatively "dumb" terminals. If users have intelligent computers on the premises, they acquire a greater degree of network control, management, and reconfiguration possibilities. In the United States, this flexibility will be reflected in opportunities for users to install customer premises equipment (CPE) to manipulate network features. It also means that the most users and policy makers in the U.S. favor flexibility in the deployment, interconnection, and number of ISDN services.

Few foreign nations embrace the concept of multiple ISDNs, each with a variety of different services and the capability to access each other. Instead, the predominant, PTT-driven ISDN vision contemplates a single "one size fits all" network that in time would become ubiquitous. Should the PTTs' version of ISDN fail to incorporate what had already been achieved in private networks, then corporate network operators will see little value in migrating to the public switched ISDN.

Furthermore, it is quite likely that PTTs will seek to recoup substantial ISDN investments through metered, usage-sensitive charges. Any user who has engineered a private line digital network is apt to face significantly higher rates if forced to move to a public usage-sensitive ISDN network. Without mandatory usage requirements, PTTs may have to develop a diverse array of ISDN features to attract users. Users who can "vote with their feet" may locate new facilities in nations with more accommodating telecommunications policies.

In the short term, the most attractive services and features of ISDN will reflect the ability to consolidate previously separate networks, such as voice and data, onto a single network. The network management and signaling capability

---

15. See Telephone Co.-Cable Television Cross-Ownership Rules, Secs. 63.54–63.58, CC Docket No. 87-266, Second Report and Order, Recommendation to Congress and Second Further Notice of Proposed Rulemaking, 7 FCC Rcd. 5781 (1992).

achieved through ISDN streamlines and accelerates call setup, automatic number identification, network reconfiguration, bandwidth assignment, and switching. The likely increase in per-unit costs in an ISDN will likely be recouped through greater efficiency and the ability to take advantage of routing flexibility and time zone, peak/off peak transmission pricing.

8.7.4 Digital

Digital technology represents one characteristic that all planners agree ISDN will deliver. Given the fact that much of the existing telephone network is primarily analog, the conversion to ISDN involves substantial investment in network upgrades. The ubiquity of ISDN cannot occur until this conversion takes place, with digital technology delivered to the home and office. However, the initial conversion can begin using existing copper wire, thereby obviating the need to await widespread installation of new fiber-optic cables and facilities.

The near-term availability of ISDN makes it an attractive option for users of online information services who now secure access through a modem attached to a conventional telephone line. Even at transmission rates of 28,800 bps, downloading files containing sound, graphics, or video can prove frustrating. A 144-Kbps or higher delivery rate provides relatively blazing throughput.

8.7.5 Networks

The concept of networking in ISDN achieves ubiquitousness through digital pipelines. This feature-rich system attempts to become a single network solution, whereby the ISDN provider achieves both scale and scope economies, and the user can consolidate multiple networks. However, the domestic policies of some nations, designed to promote competition, will impede efforts to create a single, all purpose network.

Although moderated to eliminate structural separation between *basic* and *enhanced services*, the FCC's *Computer Inquiries* maintain a semantic distinction between types of services. While U.S. BOCs may operate a single ISDN network configured for both types of services, regulatory requirements may increase initial investment costs and delay deployment until the BOC demonstrates equal access arrangements for both affiliated and unaffiliated enhanced-service providers.

8.7.6 Customer Equipment

The *Computer Inquiries* also create a dichotomy between the United States and most nations regarding the line of demarcation between the telephone company network and customer premises. The FCC favors line drawing that extends the range of end-user decision making and expands user options involving equipment located on customer premises in lieu of telephone company facilities. This policy applies the long-standing view of the United States that network services and CPE constitute separate markets, with the latter competitively provided in a mostly unregulated environment.

The FCC's support for a competitive CPE market includes network channel terminating equipment,[16] that is, devices that provide supervision, diagnostic, and clocking functions in the circuit interconnection process. Many national regulatory agencies consider the functions performed by such equipment as properly within the exclusive province of the PTT, even if the devices are physically situated on customers' premises. The FCC, on the other hand, believes that such equipment should not extend the scope of telephone company responsibility and presence on user premises.

PTTs seek to extend the wingspan of their networks and the extent of their market domination. Their ISDN visions contemplate fewer user options to attach CPE to the network and more compulsory features available only from their facilities. This position stems from their reluctance to part with the revenues generated by being a nation's sole or primary CPE lessor and apprehension that end users and unaffiliated enhanced-service providers might reduce PTT revenues through remote network control and equipment sales.

8.7.7 ISDN Compatibility

Complete ISDN integration requires an uninterrupted, seamless interconnection of national systems. Even if not all services are supported globally, nations must reach a consensus on core technical and operational standards. Network incompatibility will result in system balkanization (i.e., segmentation of the global ISDN marketplace into incompatible regional or single-nation networks).

The standard-setting process should support the option for nations to support multiple ISDNs, even though most will have single systems. In nations with cutting-edge policies like the United States, the United Kingdom, and New Zealand, several facilities-based interexchange carriers will upgrade their transmission networks to support ISDN. Local exchange carrier provision of ISDN access will constitute a reasonable extension of existing exchange access requirements. Accordingly, it can be expected that some nations will support multiple ISDN systems in both local and long-distance markets. PTTs, on the other hand, will typically develop single unfragmented ISDNs, perhaps with IVANs adding further enhancements to leased ISDN channels.

PTTs' goal of a single ISDN merges with their interest in remaining the exclusive network and service provider for a variety of user telecommunications and information-processing requirements. The ISDN investment will be worth-

---

16. See Furnishing of Customer Premises Equipment and Enhanced Services by AT&T, CC Docket No. 85-26, 102 FCC 2d 655 (1985), on partial recon., 104 FCC 2d 739 (1986).

while if it solidifies a PTT's monopoly and also enables it to expand into adjacent markets. Such an outcome does not jibe with some nations' recent policy initiatives promoting competition, service diversity, and user sovereignty[17]—which in the context of ISDN favors multiple ISDNs, extensive CPE options and distinctions between enhanced and information services.

8.7.8 Line Drawing in the ISDN Schematic

Much of the compatibility problems in ISDN stems from conflicting regulatory philosophies that have delayed progress in establishing global standards. Without consensus, nations proceed with policies and regulations on such issues as where to draw the line between carrier-provided networking and customer provided equipment. ISDN interface standards, configured by the ITU's telecommunication standards sector, provide several alternative demarcation points.

The demarcation point ultimately chosen will divide areas of responsibility between carrier and customer. The United States, which has long permitted users to attach any device to networks, provided it does not cause network harm,[18] favors a demarcation point that expands the reach and versatility of equipment connected to ISDNs by users. Other nations favor using a demarcation point that incorporates more equipment and functionality into the ISDN. Such line drawing has a direct impact on the size and scope of markets reserved exclusively for carriers and markets available to both carriers and equipment manufacturers.

The United States favors the competitive provision of most ISDN equipment and services. This means that customers can attach devices of their own choice, rather than pay for carrier-defined ISDN service that includes unneeded or undesired services, features, and equipment. With a demarcation point physically closer to carrier-switching facilities, users can elect to lease equipment and secure additional services from the carrier providing the underlying facilities to transport ISDN. Alternatively, they can supply and maintain their own equipment.

A demarcation point closer to, or even on, customer premises restricts user options by expanding carriers' wingspan. This market expansion may be necessary to create sufficient profit centers for carriers to finance network upgrades to support ISDN functionality. But it runs counter to goals articulated by the FCC to unbundle equipment from network services and to confer upon users maximum flexibility.

---

17. See ISDN, Notice of Inquiry, 94 FCC 2d 1289 (1983), First Report, 98 FCC 2d 249 (1984).

18. See North Carolina Utilities Commission v. FCC, 537 F.2d 787 (4th Cir.), cert. den., 429 U.S. 1027 (1976) (affirming FCC's registration program for CPE meeting technical standards).

8.8 A MORE DEMOCRATIC AND OPEN INFORMATION SERVICES MARKETPLACE?

To support a more open, accessible, and diversified information services marketplace, the ITU has begun to revise current regulations and recommendations based principally on the PTT model. Heretofore, governments, incumbent carriers ("administrations"),[19] and a small number of entities that provide telecommunications transport services called *recognized private operating agencies* (RPOAs)[20] have made up the complete ITU constituency. These ITU participants and recent market entrants may need to negotiate special arrangements on a bilateral basis rather than await closure at the ITU on new recommendations and standards for ISDN and other information services. Long-standing participants at the ITU have confidence in the ability of the organization to respond to new requirements. On the other hand, recent market entrants like IVANs often lack such confidence, particularly in the timeliness of the ITU standard setting. These operators express greater willingness to pursue solutions in advance of institutionalized changes in recommendations and standards.

For example, entrepreneurs who resell leased private lines to the general public have provided both value-added and basic services that may violate both domestic and ITU regulations. Rather than advocate changed policies at the ITU on private line resale to accommodate new services, these enterprises for the most part have concentrated on acquiring market share. Some of their services fall outside the realm of the ITU's definition of "public correspondence" and are accordingly outside the scope of services the ITU recognizes as within the ambit of what an administration or RPOA[21] provides.

"We interpreted the ITU definition of public correspondence to refer to a 'transport' service (as distinguished from data processing or data-base access) of the type traditionally offered by administrations, in which the service provider conveys information from a sender to a non-related recipient. We also held that providers of enhanced services, such as remote-access data processing that do

---

19. Administrations are "[a]ny governmental department or service responsible for discharging the obligations undertaken in the International Telecommunication Rules and Regulations." (Constitution of the International Telecommunication Union, Annex, No. 1002, Definitions of Certain Terms Used in the Convention and in the Regulations of the International Telecommunications Union, in *Final Acts of the Plenipotentiary Conference*, Nice 1989, ITU: Geneva, 1990.)

20. RPOA is defined as any "individual or company or corporation, other than a government establishment or agency, which operates a public correspondence or broadcasting service...[who agrees to accept] the obligations provided for in Article 6 of this Constitution." (Ibid., Nos. 1007, 1008.)

21. The ITU defines public correspondence as "[a]ny telecommunication which the offices and stations must, by reason of their being at the disposal of the public, accept for transmission." (Ibid., No. 1004.)

not offer a transport service, are not engaged in public correspondence and thus are not eligible for RPOA status."[22] But most resellers, particularly ones offering basic long-distance services would rather acquire a satisfied user base than convince ITU delegates of the need to legalize their services.

In advance of definitive and favorable modifications to the ITU Recommendations, the United States, Japan, Hong Kong, and the United Kingdom have taken advantage of a provision in the ITU Convention that permits nations to establish "special arrangements." Article 31 of the ITU Convention allows "special arrangements on telecommunications matters which do not concern [ITU] Members in general" [2]. Much controversy has arisen over what class of service providers this article addresses and whether nations should resort to special arrangements instead of pursuing policy changes within the ITU to recognize value-added resale performed by IVANs.

The lack of an ITU classification has handicapped IVANs, particularly regarding their opportunity to lease basic transport lines and resell them combined with telecommunications and information-processing enhancements. Recommendations of the ITU's Telecommunications Standards sector contemplate that private leased circuits will customarily be used by a single customer or at most shared by a closed user group, such as a consortium of banks or airlines whose "communications must be concerned exclusively with the activity for which the circuit is leased" [3]. These are the very circuits that IVANs need to lease from underlying facilities providers like PTTs for use in provisioning third parties with value-added/enhanced network services.

The ITU's recommendation on private line resale, which many nations codify or deem mandatory, has the effect of "greatly restricting the growth of IVANs" [4].[23] It suggests limits on who can qualify to lease a private line, the scope of access to the public switched network, and whether a leased line in one country can be interconnected with lines in another country.

Liberalized, flexible, and more widespread IVAN options will require modification of ITU Recommendations, particularly to legitimize IVANs as a category separate from basic service resellers who are less likely to qualify for the right to lease private lines. The recommendations exist primarily to safeguard PTT revenue streams, particularly from basic transport services. A cam-

---

22. International Communications Policies Governing Designation of Recognized Private Operating Agencies, Grants of IRUs in International Facilities and Assignment of Data Network Identification Codes, CC Docket No. 83-1230, Report and Order 104 FCC 2d 208 (1986), on recon., 2 FCC Rcd. 7375 (1987), FCC 87-354, p. 6, note 5 (rel. 10 December 1982).

23. However, the D series does recognize two exceptions to the general prohibition on shared use and resale of leased lines: (1) where the lines are used in conjunction with remote assess to data processing centers (Recommendation D.1, Sec. 8.1, p. 15), and (2) where a closed user group, sharing "common interests and exercis[ing] the same activities in areas other than telecommunications" forms to satisfy specialized international communications needs. (Recommendation D.6, Sec. 2.1, p. 27.)

paign pressing for too liberal a modification of the Recommendations might appear as seeking widespread basic resale of leased lines and legalization of arbitrage: subdivision of bulk-capacity, private lines into smaller capacity individual circuits for use by unaffiliated users who otherwise would not have qualified for bulk-capacity discounts. Typically, private lines are usage-insensitive, meaning that a bulk volume of telecommunications transport capacity is made available for use on a 24-hour-a-day basis at a fixed monthly rate.

PTTs worry about the loss of revenue resulting when users can migrate from metered service, typically measured on a per-minute basis, to unmetered, usage-insensitive services that have flat monthly rates. Such migration would also prompt the carrier to rethink the marketing value of having *any* private line service options [4]:

> The U.S. faces a critical balancing act between favoring liberalized use of private leased circuits for value-added services and maintaining the present lease circuit service [option] for intra-corporate and intra-government use. The U.S. Government and U.S. industry are interested in ensuring that no adverse ripple effect occur because of changes in the D. Series Recommendation, such as imposition of some volume-sensitive pricing scheme in place of flat-rated pricing ...or high access fees when private leased circuits are interconnected to the public switched network.

8.9 ITU CONSIDERATION OF INFORMATION SERVICES

In 1988 the ITU convened a World Administrative Telegraph and Telephone Conference (WATTC) to address IVAN issues and the wider matter of how to define the scope of entities and services subject to ITU Regulations and Recommendations. If international telecommunications and information-processing regulation becomes too broad in terms of scope and coverage, then nations seeking liberalization bear growing incentives to opt for special arrangements or to ignore some Recommendations altogether.[24]

Bilateral governmental agreements executed by such nations as the United States, Japan, Hong Kong, and the United Kingdom have conferred official authorization for private line leasing and resale on a commercial basis to com-

---

24. However, the International Telegraph and Telephone Consultative Committee (CCITT) D series of Recommendations does recognize two exceptions to the general prohibition on shared use and resale of leased lines: (1) where the lines are used in conjunction with remote access to data-processing centers, (Recommendation D.1, Sec. 8.1, p. 15), and (2) where a closed user group sharing "common interests and exercis[ing] the same activities in areas other than telecommunications" need private lines to satisfy specialized international communication needs. (Recommendation D.6, Sec. 2.1, p. 27.)

pletely unaffiliated third parties. Notwithstanding the ITU Recommendations, the nations involved opted to permit a more robust and expansive IVAN marketplace, which would not be limited by the ITU Recommendations on the terms, conditions, and scope of private line leasing and enhancements. The nations also agreed to permit IVANs to use proprietary, nonstandardized access arrangements in recognition that IVANs often customize services and a single ITU-promulgated model would present an inflexible, least common denominator.

These nations felt the time had arrived for the ITU to establish definitive arrangements to legitimize IVANs and to normalize their relationship with other carriers, despite the following risks:

- Revenue and traffic diversion to underlying facility-providing carriers;
- More frequent assertion of national sovereignty to evade ITU Rules, Regulations, and Recommendations which in the aggregate could threaten the legitimacy of the ITU;
- Increased pressure on the ITU to reduce regulatory and policy coverage loopholes by expanding the set of services and service providers subject to mandatory regulations;
- Heightened sensitivity and hostility to any bilateral pursuit of special arrangements, despite their legitimacy.

The controversy in IVAN service and what should or should not be done in the ITU underscore increasing tension on what should be done to accommodate the policy initiatives of some nations. The ITU generate secretariat succinctly articulated the macro-level problem when reporting on preparations for the 1988 WATTC [5]:

PC/WATTC-88 was confronted with two opposing views, one voiced by those who favour the greatest possible liberalization of telecommunications and the development of a competitive environment (including the U.S.), and the other by those who, while favouring some degree of liberalization, hold that international telecommunications should nevertheless be subject to a number of universally applicable regulations.

The IVAN controversy pits many LDCs and their benefactors against more developed and newly developed nations. The philosophical disagreement sometimes extends to a broader referendum on what international forums should do to preserve service universality and standard operating procedures. If the least common denominator becomes too burdensome and takes too long to reach, then nations with investments and policies favoring liberalization may not wait.

References

[1]   Commission of the European Communities, *Towards a Competitive Community-Wide Tele-communications Market in 1992, Implementing the Green Paper on the Development of the Common Market for Telecommunications Services and Equipment*, COM (88) (9 February 1988).

[2]   ITU, *Constitution*, Nice, 1989, Article 31, Geneva: ITU, 1990.

[3]   ITU, Consultative Committee on International Telegraph and Telephone, Blue Book, Volume II-Fascicle II.1, *General Tariff Principles—Charging and Accounting in International Telecommunications Services*, Series D Recommendations, Sec. 1.7, Geneva: ITU, 1989, p. 10.

[4]   Dept. of Commerce, National Telecommunications and Information Admin., *International Value-Added Network Services—An Introduction*, NTIA TM-90-258, Washington, D.C.: GPO, March 1990, p. 9.

[5]   CCITT Circular No. 67, Article 3.4.2, Geneva, 9 November 1987.

The International Telecommunications Toll Revenue Division Process

9

International telecommunication service providers negotiate operating agreements to arrange for the interconnection of lines and the division of toll revenues for calls between nations. This contractual arrangement provides for complete routing of calls, including the domestic tail circuits running to and from international gateway facilities. Operating agreements include an accounting rate to be used as the amount of compensation shared by the carriers.

Carriers theoretically specify accounting rates to represent the approximate costs they incur.[1] However, for most routes, correspondents have failed to reduce the accounting rate to reflect lower costs in new generations of transmission facilities. If two carriers have equal traffic volumes, then a high accounting rate presents no problem because the carriers do not transfer funds from the toll revenues they collect. But many telephone service routes, including most originating in the United States, generate an imbalance in traffic flows; that is, more traffic originates from the United States than terminates into the United States. An imbalance in traffic flows requires a transfer of funds when carriers settle accounts: the carrier generating more outbound traffic must compensate the correspondent for accepting more traffic that traverses the correspondent's international and domestic facilities.

---

1. For a complete history of accounting rate regulation by the FCC, see R. Frieden, "International Toll Revenue Division: Tackling the Inequities and Inefficiencies," *Telecommunications Policy*, Vol. 17, No. 3, April 1993, pp. 221–233; L. Johnson, "Dealing With Monopoly in International Telephone Service: A U.S. Perspective," *Information Economics and Policy*, Vol. 4, 1989/91, pp. 225–247; K. Cheong and M. Mullins, "International Telephone Service Imbalances Accounting Rates and Regulatory Policy," *Telecommunications Policy*, Vol. 15, No. 3, April 1991, pp. 107–118 (hereafter cited as International Telephone Service Imbalances); K. Stanley, "Balance of Payments Deficits, and Subsidies in International Communications Services: A New Challenge to Regulation," *Administrative Law Review*, Vol. 43, Summer 1991, pp. 411–438; and R. Frieden, "Accounting Rates: The Business of International Telecommunications and the Incentive to Cheat," *Federal Communications Law J.*, Vol. 43, No. 2, April 1991, pp. 111–139.

In 1991, U.S. carrier settlements with foreign counterparts resulted in a $3.4 billion deficit (see Figure 9.1).[2] Application of excessive accounting rates has a direct and adverse impact on the rates charged users. While technological innovations like digitization and compression have dramatically reduced the cost of transmission facilities, rates have not dropped commensurately, because the frame of reference for tariffing service is not actual transmission facility costs, but what the carriers agree should be the per-minute level of compensation they will share. The actual cost of providing a minute of telephone service between the United States and Europe is roughly $0.25 a minute, but the accounting rate can exceed that by 400% or more.

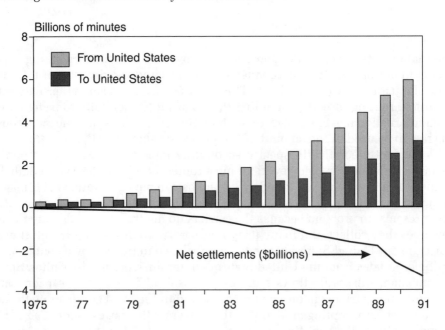

Figure 9.1 Telecommunications traffic balance. (*Source:* Office of Technology Assessment and FCC, 1992.)

---

2. In Regulation of International Accounting Rates, CC Docket No. 90-337, Phase II, 6 FCC Rcd. 3434, 3435 (1991), the FCC reported a $3 billion deficit. The Commission reported the $3.4 billion figure in Regulation of International Accounting Rates, CC Docket No. 90-337, Phase I, Order on Reconsideration, 7 FCC Rcd. 8049, para. 11 (1992) (denying reconsideration petition of US Sprint to expand the time period for considering a wavier of the International Settlements Policy, which inter alia requires uniformity in the accounting rate and division of toll revenues, for telex and packet services, and a request by MCI that the Commission approve only accounting rate reductions that take place at the same time, at the same accounting rate level and with the same terms for all U.S. carriers).

Six major factors explain this substantial transfer of funds:

1. U.S. international service carrier (USISC) outbound traffic volumes greatly exceed the return, inbound flow because of a larger population, widespread multinational business activity, often significantly lower end-user charges that stimulate outbound calling, higher per capita income, and a large immigrant population that seeks to maintain ties with family members abroad.[3]
2. Foreign international telecommunications rates greatly exceed what USISCs charge, thereby retarding demand for inbound U.S. calling.
3. Most current accounting rates substantially exceed service cost because they have failed to reflect the technological innovations that have reduced the per-unit cost of providing service.
4. USISCs, without apparent financial penalty, have burdened customers with the consequences of artificially high accounting rates.
5. With increasing frequency, high-volume international telecommunications users, particularly in nations with high toll charges, have devised traffic routing arrangements exempt from the accounting rate settlement process (e.g., use of private lines and call-back services that deliver dial tone for services in a country with low rates to users in a nation with high rates, thereby reducing the amount of inbound U.S. traffic that would have offset some of the accounting rate surplus generated).[4]
6. The ITU has only begun to address the issue of whether international telecommunication services should be priced on the basis of actual costs, or on other factors that would build in a subsidy (e.g., keep accounting rates high), or on a policy of flowing a higher percentage of the division for lesser developed countries to support infrastructure construction.

A 1988 report of the FCC's Common Carrier Bureau acknowledged that higher outbound U.S. traffic volumes account for half of the growing accounting rate deficit.[5] Low charges to end users (collection rates) in the United States

---

3. A recent study by the FCC's Common Carrier Bureau revealed that charges for calls to the United States can exceed by as much as 80% the outbound rate. (FCC, Common Carrier Bureau, *Calling Prices for International Message Telephone Service Between the United States and Other Countries*, August 1992.)

4. See G. Staple, "Winning the Global Telecoms Market: The Old Service Paradigm and the Next One," in *TeleGeography 1992 Global Telecommunications Traffic Statistics and Commentary*, G. Staple, ed., London: International Institute of Communications, 1992, pp. 32–53).

5. "[I]t is clear that the price of calling the United States has not declined as rapidly as the price of calling from the United States. Because American prices were lower to begin with, the differential in prices has increased. This has undoubtedly contributed to the continued and increasing imbalance in traffic flows and in turn contributed to the rapidly growing balance of payments

stimulate traffic growth, while much higher rates in foreign countries retard demand and reduce return traffic flows [1]. The FCC objects to this "direct underwriting by U.S. consumers of foreign telecommunications administrations" [2], but a solution remains elusive.

On one hand, the FCC has considered using an activist posture to force both domestic and foreign carriers to reduce accounting rates. The FCC has considered mandating lower rates if the carriers fail to act. On the other hand, the FCC recognizes that accounting rates result from business negotiations and that it should support international comity, which it defines as "the mutual recognition and accommodation by nations of their differing philosophies, policies and laws" [3].

The FCC appears ambivalent in how to remedy the accounting rate issue. An activist posture may overestimate what the FCC can order unilaterally, particularly given other nations' concerns about sovereignty and the lack of any substantial consideration of the issue by the ITU until 1992.[6] A passive posture may naively rely on the carriers to remedy the problem by negotiating accounting rates downward without regulatory prodding. Only recently has the FCC undertaken a balanced campaign to create incentives for carriers to act. AT&T in particular has recognized the value in having lower accounting rates, because in a price-cap ratemaking environment, it does not have to flow through all savings and efficiency gains to ratepayers.[7]

The FCC hopes to avoid direct confrontation with foreign governments on the accounting rate issue by linking access to U.S. markets with an assessment of whether equivalent opportunities exist for U.S. companies, particularly resellers of private line capacity.

---

deficit." (FCC, International, Common Carrier Bureau, *Accounting Rates and the Balance of Payments Deficit in Telecommunications Services*, 12 December 1988, p. 29 (hereafter cited as Common Carrier Bureau Report).)

6. Article 6.1.1 of the *International Telecommunication Regulations* recognizes that the level of toll charges is a "national matter," but recommends that carriers avoid "too great a dissymmetry between the charges applicable." (*The International Telecommunication Regulations, Final Acts of the World Administrative Telegraph and Telephone Conference*, Art. 6.1.1, Melbourne, Australia 1988.) Study groups of the ITU's CCITT have frequently considered accounting rate reforms. However, 1992 marks the first time any CCITT study group has suggested on the record that administrations consider actual costs when setting accounting rates. (See Regulation of International Accounting Rates, Docket No. 90-337, Phase II, Second Report and Order and Second Further Notice of Proposed Rulemaking, 7 FCC Rcd. 8040, para. 36 (rel. 27 November 1992) (hereafter cited as Accounting Rate Second Report and Proposed Rulemaking).)

7. See Strategic Policy Research, *The U.S. Stake in Competitive Global Telecommunications Services: The Economic Case for Tough Bargaining*, 16 December 1993 (study conducted for AT&T).

9.1 ACCOUNTING RATE FUNDAMENTALS

Routing international telecommunications traffic involves a contract negotiated between carriers for each type of service between each pair of nations. Such bilateral arrangements promote multilateral collaboration in transmission facility investment. International carriers jointly own, operate, and maintain international submarine cables through consortia and international satellites through cooperatives like INTELSAT.[8] Typically, ownership interests are allocated as a function of anticipated use. Joint ownership means that most carriers incur roughly the same cost per unit of international capacity. It follows that most parallel routes to the same region of the world would have roughly the same total costs, even though expenses for the domestic haul portion of a complete route can vary as a function of equipment vintage and traffic volumes.

A key component to international traffic routing is the financial terms and conditions under which one carrier compensates the other carrier for agreeing to match international circuits and to provide the switching and domestic routing necessary to deliver calls to the intended recipient. An accounting rate serves as the basis for dividing the toll charges[9] collected for the joint provision of an international service. Accounting rates are negotiated by the participating carriers and have two components:

1. The unit of currency used and the applicable rate per unit of traffic carried. For example, in 1991, USISCs and France Telecom divided 1.0 International Monetary Fund Special Drawing Rights ($1.42) per minute of full rate international message telephone service traffic between the U.S. and France.
2. The settlement process—how the accounting rate amount will be divided between correspondents, usually 50:50 when two carriers participate.

Figure 9.2 depicts the settlement process.

---

8. INTELSAT is a global cooperative, formed by an intergovernmental agreement, with a mission of providing ubiquitous satellite communications service. (See *Multinational Communication Satellite System*, opened for signature 20 August 1994, 15 U.S.T. 1705, 514 U.N.T.S. 26 (19-nation agreement establishing interim arrangements for a global satellite cooperative); *Agreement Relating to the International Telecommunications Satellite Organization ("INTELSAT")*, opened for signature 20 August 1971, 23 U.S.T. 3813, 1220 U.N.T.S. 21 (INTELSAT Agreement); and *Operating Agreement Relating to the International Telecommunications Satellite Organization ("INTELSAT")*, opened for signature 20 August 1971, 23 U.S.T. 4091, 1220 U.N.T.S. 149.)

9. The accounting rate does not necessarily equal the rate charged end users. This "collection rate" may be lower, as is the case for some outbound U.S. calls, but typically exceeds the accounting rate, particularly in foreign nations.

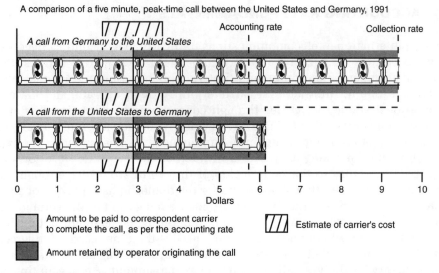

A comparison of a five minute, peak-time call between the United States and Germany, 1991

Figure 9.2 Accounting and collection rates for international telecommunications traffic. (*Source:* Office of Technology Assessment.)

NOTES: The accounting rate with Germany in 1992 was 0.8 special drawing rights or $1.14 (FCC, Statistics of Communications Common Carriers, 1991/1992 Ed.). The collecton rate (i.e., what the caller is charged) for the U.S.-to-Germany call is calculated as $1.77 [for the initial minute] + 4 × $1.09 = $6.13 (FCC). The collection rate for the Germany-to-U.S. call is derived from 5 × $1.88 (TeleGeography 1992, International Institute of Communications). The cost to the carriers are estimated at $0.15 per minute at both the U.S. and German end; this number is conservative.

Carriers set accounting rates to represent the total cost generally incurred to establish a complete international circuit (i.e., both international half circuits and both domestic tail circuits to and from the international gateway). However, the FCC has concluded that "accounting rates for international telephone service (IMTS) are significantly greater than the current costs of providing service" [4].

Disparity between accounting rates and actual routing costs would have little significance if there were parity in traffic volumes from and to the United States. Accounting rates have great significance when an imbalance of traffic streams exists and substantial funds must be transferred, as is the case for IMTS between the United States and many foreign nations, both developed and developing.

Comparatively low IMTS rates in the United States have stimulated outbound IMTS calling, resulting in traffic volumes far in excess of inbound flows. Typically, the IMTS rates in foreign nations vastly exceed carrier cost and the equivalent outbound U.S. rate to callers, thereby dampening demand for inbound U.S. calling.[10] Foreign carriers have viewed accounting rate surpluses as

10. See Regulation of International Accounting Rates, CC Docket No. 90-337, Phase II, Further Notice of Proposed Rulemaking, 6 FCC Rcd, 3434, 3436 (1992).

a painless way to subsidize below-cost provisioning of postal and certain telecommunication services.

Artificially high accounting rates, which have not dropped commensurately with reduced costs per unit of capacity, require net debtor carriers to make even higher settlement payments to compensate for the disparity in traffic volumes. It appears that despite increasing outpayments, USISCs have been less than vigorous in their advocacy of accounting rate reductions. Such high accounting rates do not necessarily reflect foreign carrier "whipsawing"[11] of USISCs by leveraging inbound U.S. traffic to secure concessions. However, they do represent a surcharge to U.S. ratepayers, because USISCs incur financial burdens preventing them from offering further rate reductions.

USISC reluctance to press for lower accounting rates stems in part from concern that such advocacy might prompt PTT retaliation in ways that could further hurt profitability. But the FCC's Common Carrier Bureau implies two other reasons:

1. International calling volumes are rapidly growing, and despite the massive outpayment in accounting rates, IMTS typically "generates at least four times as much revenue as the average domestic interstate call."
2. A higher percentage of international telephone calls are metered and charged on a usage-sensitive basis, contrary to usage-insensitive private lines that are more prevalent for intercity links in domestic markets [5].

With robust traffic growth and high margins, USISCs may not have perceived a need to push accounting rates downward. In fact an aggressive campaign might trigger a foreign carrier backlash, particularly because the FCC requires all USISCs to apply a uniform accounting rate. Higher than necessary accounting rates also may help preserve the appearance of IMTS as an expensive and unattractive business to enter, thereby reducing the scope of competition. If accounting rates were to drop toward cost, market entry would become more attractive and incumbent USISCs would have to reduce international rates significantly, thereby reducing or eliminating a source of revenues to offset losses or low margins in the more competitive domestic marketplace.

---

11. Whipsawing refers to the potential for a monopoly PTT to have a superior bargaining position vis-a-vis multiple USISCs. As long as the PTT has one operating agreement with a USISC, it has national access. Because more than one USISC seeks access to the foreign locale, the PTT may play one carrier against the others by auctioning off access to the carrier willing to accept the lowest settlement rate. "Generally, this has involved a PTT seeking a modification of an accounting rate in a manner more favorable to the PTT—that is, a reduction in the rate paid by the PTT to the U.S. carrier for delivery of traffic in the U.S., or an increase in the rate paid by U.S. carriers to the PTT for delivery of traffic in that foreign country." (Implementation and Scope of the International Settlements Policy for Parallel International Communications Routes, CC Docket No. 85–204, 2 FCC Rcd. 1118 (1986).

9.2 FCC APPROACHES TO THE PROBLEM

After years of neglecting the issue, the FCC in 1985 began to consider high accounting rates a trade issue and a major irritant. The FCC's International Settlements Policy (ISP) sought to prevent foreign carriers from playing one USISC against the others[12] to extract financial concessions [6]. The ISP, like its predecessor, the Uniform Settlements Policy,[13] "requires all [U.S. International Service] carriers providing the same service to the same foreign point to have the same accounting, settlement and division of tolls arrangement with the foreign administration" [7], including an expectation that carriers will negotiate an operating agreement requiring the proportionate routing of return traffic.

The FCC mandated uniform accounting rates ostensibly to prevent monopoly PTTs from whipsawing multiple USISCs. This mandatory uniformity may in fact exacerbate the accounting rate settlement problem because:

- A PTT is less likely to agree to any accounting rate reduction if instead of rewarding one high-volume carrier, the lower rate has to apply to settlements with all USISCs;
- A single USISC cannot opt for a strategy of stimulating traffic volume growth through an accounting rate differential;
- Any single USISC considering hardball negotiations with PTTs to lower the accounting rate has to balance the potential for increased outbound traffic and higher revenues against the prospect for the PTT to retaliate by routing less return traffic, regardless of FCC-articulated concerns that foreign carriers should not discriminate against USISCs.

Despite FCC concerns and policy initiatives, foreign carriers have continued to exploit artificially high accounting rates, while avoiding them where such rates reduce their settlement windfalls. The U.S. net settlement deficit has increased in 4 years from approximately $2 billion in 1988 to 3.3 billion in 1992 (the accounting rate deficit did not increase significantly between 1993 and 1995 [8].

---

12. The international telecommunications marketplace in 1985 could be characterized as "becoming increasingly competitive on the U.S. side, in contrast to the foreign side which continues to be dominated by monopoly post, telephone and telegraph (PTT) carriers." (L. Johnson, "Dealing with Monopoly in International Telephone Service: A U.S. Perspective," *Information Economics and Policy*, Vol. 4, 1991, p. 225.)

13. See Uniform Settlement Rates on Parallel International Communications Routes, 84 FCC 2d 121 (1980). "The policy...was first developed in the 1930's and had its roots in antitrust law." (International Settlements Policy Reconsideration, 2 FCC Rcd. at 1118.) See also Mackay Radio and Telegraph Co., 2 FCC 592 (Telegraph Committee 1936), aff'd, 97 F.2d 641 (D.C. Cir. 1938).

9.2.1   The Whipsawing Concern

Notwithstanding a preference for avoiding marketplace intervention and oversight, the FCC believes that it must intervene and respond to superior foreign carrier negotiating leverage, particularly for routes where only one foreign carrier exists for accepting traffic and delivering it to the intended call recipient. Monopoly PTTs may enjoy bargaining leverage, because they have a number of USISCs available to handle traffic to all U.S. destinations, but all USISCs have to secure an operating agreement with a single carrier. PTTs have few immediate incentives to confer additional operating agreements unless a USISC can generate additional traffic volumes to compensate the PTT for the additional transaction costs incurred to provide interconnection. This disincentive grows more pronounced where new USISCs seeking access cannot propose temporary "deal sweeteners" (i.e., an offer to pay a premium over the applicable accounting rate until a threshold volume of traffic is generated), because the ISP requires uniformity in the terms and conditions for settling accounts.

Without a regulatory mandate of accounting rate uniformity, the FCC theorizes that foreign carriers could auction off inbound U.S. traffic flows to the highest bidder (i.e., the USISC willing to part with the greatest share of the toll revenues). In application, the FCC's policy appears to have mandated accounting rate uniformity stuck at artificially high levels, far exceeding actual service costs.

New USISCs, as potential whipsawing victims, have no way to pressure foreign correspondents for lower accounting rates. Established USISCs have little incentive to force the issue as well. The initiating carrier would risk good corporate relations with the foreign carrier whose managers typically have great latitude in determining how to route return, inbound U.S. traffic, despite FCC concerns about the potential for discrimination. Even if one USISC cared to take the risk, with an eye toward generating even higher returns, the FCC's accounting rate uniformity requirement would confer the benefit on all USISCs.

9.2.2   Avoiding Accounting Rate Liability

Because accounting rates remain at artificially high levels for many routes, carriers and their customers strategize on how to route traffic exempt from the settlement process. The vehicles for avoiding high accounting rates include the use of call-back services, which provide dial tone to end users physically situated in another country, and linking international private lines with a switch that secure access to the PSTN. These options may violate ITU Recommendations[14]

---

14. Recommendation D.1, Sec. 7.1.1 of the ITU's International Telegraph and Telephone Consultative Committee Blue Book, Vol. II, Fascicle II.1, General Tariff Principles, Charging and Accounting in International Telecommunications Services, suggested that administrations can consult and agree to the scope of access to public networks provided to users of international

and carrier tariffs, because they enable end users to secure services in a manner that the carrier did not intend on providing. While such bypassing may expedite reforms, it flouts uniform rules of the road. For example, the ITU Recommendations on leased international private lines contemplate the consultation and agreement on the scope of service. Private lines, by definition, provide closed, intracorporate networking capabilities, not the functional equivalent to switched public long-distance services.

What is occurring in international telecommunications parallels the gray market in international commercial aviation, where carriers look the other way or clandestinely collaborate with ticket resellers, consolidators, and brokers who offer seats at rates well below the published tariff.[15] In international telecommunications, sophisticated users and system integrators design private line networks that avoid accounting rate liability. Carriers originally offered unmetered private lines as a way to fill up excess capacity and satisfy large-volume user requirements for closed, internal networks. Private branch exchanges and other customer-controlled equipment have enabled users to interconnect unmetered international private lines with local public switched telephone networks. Such "leakiness" enables the private line subscriber to access users outside the internal network. Expanded access to a private line "network" means that users, who otherwise would have to use IMTS circuits, can opt for specially configured private line access for functionally equivalent service.

Resellers can expand the reach of leaky private lines with higher capacity switches. Some carriers and their regulatory overseers do not object to this type of "pure resale," which does not enhance leased lines. Resale stimulates overall capacity demand, and it can reduce outbound IMTS accounting rate liability, particularly where regulatory policies block or limit inbound resale. Some carriers, intent on capturing larger market shares by aggregating and routing regional traffic through a hub, may engineer a complex array of private lines and acquire both half circuits on routes to handle accounting rate–exempt traffic. Transiting, the routing of traffic destined for another country across domestic facilities, pre-

---

private leased circuits. To the extent that a private line reseller or end user does not engage in such consultation and erects a system for accounting rate evasion, then the host country may deny access to the PSTN. However, in many instances accounting rate avoidance schemes may go undetected by the carrier providing interconnection.

15. International carriers do provide discounted rates to high-volume users (e.g., as an incentive to migrate from unmetered private lines to metered "virtual" (software-defined) private lines using the public switched network). The carriers avoid application of artificially high accounting rates by creating a new service category and applying a different, and lower, accounting rate. Foreign carriers typically have no obligation to justify how the new rate does not discriminate against users paying higher charges for existing offerings subject to accounting rates.

sents another opportunity for carriers and new international telephone entrepreneurs alike to engineer innovative new arrangements for users.[16]

9.2.3 FCC Strategies for Remedying the Problem

In a 1990 Notice of Proposed Rulemaking, the FCC announced its intent to take a more proactive role in accounting rate oversight, going so far as to conclude tentatively that it had authority "to establish international accounting rates, to determine and prescribe just and reasonable accounting rates, and if necessary, [to] condition Section 214 grants [of authority to operate international facilities] on the implementation of lower, more cost-based accounting rates" [9]. The FCC intended to "jawbone" U.S. and foreign carriers and to use its more direct regulatory oversight on USISCs to achieve cost-based accounting rates.

The FCC proposed to link its accounting rate advocacy with broad objectives like universal service, prevention of uneconomic bypassing, network efficiency, and elimination of discrimination [10].The FCC also sought to impose an expectation of market access parity for users and resellers of leased international private lines. Such a get-tough policy, including prescribed accounting rates,[17] would have generated the potential for serious foreign backlash and retaliation. While the FCC can properly condition grants of regulatory authorizations and prescribe rates for the carriers it regulates, attempts to affect the behavior and the financial performance of other carriers would be viewed as overly intrusive and a failure to appreciate international comity and national sovereignty.[18]

---

16. Even companies with limited budgets can get into the international telecommunications business and exploit high accounting rate and end-user charge differentials. A "boomerang box" enables callers in high-cost foreign locations to place a call to the United States, hang up, and soon *receive* a call from the United States with the intended call recipient on the line. At the micro-level, the foreign caller avoids having to pay the significantly higher charge for originating an international call, the foreign carrier loses some toll revenues, and the USISC handling the international call accrues some additional toll revenues. At the macro-level, the transaction contributes to the expanding U.S. accounting rate deficit, thereby blunting the foreign carrier's revenue losses and the USISC's revenue gains.

17. The FCC proposed to "establish,...determine and prescribe just and reasonable accounting rates" if USISCs and their foreign counterparts failed to negotiate rates downward to an FCC-determined benchmark range. (Regulation of International Accounting Rates, Notice of Proposed Rulemaking, 5 FCC Rcd. at 4950.)

18. When the FCC attempted to influence the timetable for construction and activation of the TAT-7 overseas cable through direct negotiations with foreign governments, foreign carriers deemed such activism as an intrusion on national sovereignty, and the U.S. Court of Appeals for the District of Columbia deemed it a violation of the Government in the Sunshine Act. (ITT World Communications, Inc., 77 FCC 2d 877 (1980) (order denying petition for rulemaking on permissible scope of FCC contacts with foreign administrations to negotiate delayed deployment of a transatlantic submarine communications cable), reversed, ITT World Communications v. FCC, 699 F.2d 1219 (D.C. Cir. 1983), reversed on other grounds, 466 U.S. 463 (1984).)

Similarly, an FCC proposal to impose reporting requirements and other means for overseeing the extent of participation in the U.S. telecommunications market by foreign-owned firms[19] would have violated the commitment to "national treatment" of foreign enterprises (i.e., applying the same regulatory rights, responsibilities, and opportunities for foreign-owned carriers as for domestic carriers). The FCC subsequently decided to calibrate the scope of regulatory oversight of foreign carriers to the degree of market access accorded U.S. carriers, particularly the extent to which U.S. service providers may use leased international private lines to access foreign locales. This mechanism provides strong leverage for achieving market access parity by linking the scope of inbound U.S. market access with reciprocal opportunities for outbound traffic.[20]

Before making adjustments, the FCC's policy triggered significant opposition. Both U.S. and foreign carriers, equipment manufacturers, and government agencies objected to FCC formulation and implementation of trade policy, particularly when such blunt tactics could trigger foreign retaliation. The FCC refrained from instituting a get-tough proposal, instead emphasizing reciprocity of reseller market access, carrier negotiations, and the prospect of revisiting the matter if necessary.

9.2.4   Resale "Solutions" to the Accounting Rate Dilemma

In 1980, the FCC proposed to authorize unlimited international resale and shared use of leased lines.[21] The FCC's initiative languished, primarily because foreign governments feared that liberalized resale policies would siphon IMTS traffic and jeopardize PTTs' ability to subsidize domestic services.

---

19. See Regulatory Policies and International Telecommunications, CC Docket No. 86-494, Notice of Inquiry, 2 FCC Rcd. 1022 (1987), Report and Order and Supplemental Notice of Inquiry, 4 FCC Rcd. 7387 (1988), order on recon., 4 FCC Rcd. 323 (1989). The FCC has modified its policies that impose more extensive oversight of foreign-owned carriers providing international services from the United States. (See Regulation of International Common Carrier Services, CC Docket No. 91-360, Notice of Proposed Rulemaking, 7 FCC Rcd. 577 (1992), Report and Order, FCC 92-463 (rel. 6 November 1992) (retaining more burdensome "dominant carrier" oversight only where the foreign affiliate of a USISC has the ability to discriminate against unaffiliated carriers through control of bottleneck services and facilities in the foreign market).)

20. See Cable & Wireless, Inc., DA-1344, Tele. Div. (rel. 8 December 1994); Cable & Wireless, Inc., 8 FCC Rcd. 1664 (Com. Car. Bur. 1993); FONOROLA Corp. and EMI Corp., 7 FCC Rcd. 7312 (1992), on recon., 9 FCC Rcd. 4066 (1994) (authorizing British and Canadian resellers to provide international service upon finding that the foreign country on the other end of the circuit provides equivalent opportunities to U.S. carriers to resell interconnected private lines).

21. See Regulatory Policies Concerning Shared Use and Resale of Common Carrier International Communications Services, 77 FCC 2d 831 (1980). Domestic resale and shared use has promoted competition, lowered rates, and reduced the differences in per-unit charges for large- versus small-volume users.

The FCC renewed its interest in international private line resale by considering it the preferred vehicle for securing lower accounting rates. The FCC ordered USISCs to provide access to foreign resellers of international private lines only if U.S. carriers, particularly non-facilities-based resellers, could secure equivalent opportunities for inbound access. In this way the FCC could leverage inbound private line access to the lucrative U.S. market in exchange for equivalent foreign market access. Such leverage avoids heavy-handed threats to prescribe accounting rates and builds in a requirement for reciprocity of a legitimate form of accounting rate avoidance.

Reliance on proliferating private line resale redirected the FCC from direct confrontation with foreign PTTs over their sovereign right to negotiate accounting rates to "procedural reforms that remove any U.S. regulatory impediments to lower, more economically efficient, cost-based accounting rates" [11]. The FCC assumed that if resale were available on an equivalent basis, inbound and outbound, then the incumbent facilities-based carriers would perceive new incentives to negotiate lower accounting rates to dissuade customers from migrating to private line and resale options.

The FCC also expressed "concern that the present level of certain intraregional accounting rates or other country-to-country arrangements suggest that U.S. carriers may not only be required to pay above-cost accounting rates, but that U.S. carriers are subject to discriminatory treatment in this respect" [12].

Notwithstanding some progress in carrier negotiations and at the ITU to reduce accounting rates, the FCC noted several factors that should have generated even greater accounting rate reductions:

- Rates generally remain excessively high relative to more comprehensive and reliable cost data, including a Common Carrier Bureau study revealing that collection rate charges on some routes to the United States exceed the outbound rate by 80%.
- Intra-European accounting rates are substantially below U.S.-European rates and are subject to heightened downward pressure from user groups.
- Modern fiber-optic cable and satellite transmissions costs are less distance sensitive.
- Rates deviate significantly for routes within the same region.

Rather than prescribe accounting rates, the FCC established benchmark, presumably cost-based, rates as targets achievable on a timely basis through the market pressure from two-way private line resale and heightened vigor by USISCs to negotiate lower accounting rates.[22]

---

22. See Regulation of International Accounting Rates, CC Docket No. 90-337, Phase II, First Report and Order, 7 FCC Rcd. 559 (1992) (hereafter cited as Accounting Rate Phase II First Report and Order). The FCC established "conservative" settlement rate benchmark ranges of 23–39 cents

Facilities-based USISCs, facing competition from resellers,[23] unencumbered by accounting rate liability, may view high accounting rates as imposing a floor on how low they can price end-user rates "to prevent diversion of...customers to a reseller" [13].[24] Presumably, resellers providing outbound services from the United States will acquire market share, thereby reducing the number of IMTS outbound minutes subject to accounting rate settlements. A facilities-based carrier, refusing to negotiate accounting rates closer to cost, would "receive fewer revenues from its IMTS customers and would thus wind up with fewer revenues overall" [14].

9.2.5   Problems in Evaluating Equivalency of Market Access

The FCC decided to link resale opportunities into the United States with an evaluation of whether an equivalent outbound opportunity exists. If the FCC had failed to consider "two-way" resale, it would have given "overseas administrations the incentive and the power to use such [inbound U.S.] resale to evade the international settlements process or otherwise discriminate against competing U.S. carriers" [14]. But the insistence on parity of resale opportunities may require the FCC to examine what constitutes an equivalent opportunity for resale into foreign countries, particularly in terms of the type and quality of access to the foreign country's public switched networks. The FCC may also even face the task of ordering the disconnection of inbound U.S. leased lines where the foreign nation prohibits or does not permit equivalent reseller inbound access.

---

per minute (0.16–0.27 Special Drawing Rights) for European routes and 39-60 cents per minute for Asian routes, and expanded the scope of reports it will require from the five facilities-based USISCs that generated over $25 million in international message telephone service revenues in 1991. (International Accounting Rate Phase II Second Report and Proposed Rulemaking, paras. 8, 16, 25.) The detailed information, which must be filed on 1 January 1993 and 1 year later for all European and Asian routes, includes the accounting rate, status of ongoing negotiations, the number of years since the last accounting rate change, statistics on outpayment, traffic and collection rates, whether the USISC initiated any rate reduction negotiations, a list of problem countries, and the impact of special services on the U.S. outpayment deficit.

23. "Resale [of leased private lines] would bypass the accounting rate mechanism—a major cost to the traditional carrier mode of operation—and increase the feasibility of creating unidirectional traffic channels." (International Telephone Service Imbalances, pp. 107, 116.) If resale remains unidirectional, U.S. facilities-based carriers and consumers will not benefit: resale occurring only in the inbound U.S. direction would increase the U.S. accounting rate deficit. Resale must be bidirectional to have the effect of "expos[ing] the differential between tariffs and accounting rates and ultimately force traditional carriers to renegotiate accounting rates closer to service costs." (Ibid., pp. 116–117.)

24. "To the extent that the accounting rate is above cost, the underlying carrier will face a constraint on how much of a reduction in its revenues it can tolerate." (Accounting Rate Phase II First Report and Order, 7 FCC Rcd. 561.)

Downward pressure on accounting rates from two-way resale currently works only for the relatively few progressive nations such as the United Kingdom, Australia, Sweden, and Canada that permit international resale. It remains to be seen whether additional countries will permit equivalent resale opportunities in view of the impact such access would have on their accounting rate surpluses.

The FCC's two-way resale requirement means that foreign regulatory authorities will have to authorize expanded or first-time inbound resale opportunities to ensure continued outbound access for their resellers, including the incumbent facilities-based carrier interested in routing traffic exempt from accounting rate settlements. For resale to become bidirectional, many nations will have to change prevailing regulatory and licensing policies to permit market entry by both facilities-based carriers and "pure" resellers who simply provide IMTS-type service via leased lines. Heretofore, most nations have limited resale opportunities to IVANs that provide enhanced services over leased lines.

The FCC expects quick results: a 50% reduction in accounting rates in one to five years. It expects most nations to reach this goal in two years, even though after several years of increasingly close scrutiny, only four nations in Asia and Europe as of 1993 had rates within the benchmarks: the United Kingdom, Sweden, Australia, and Singapore [15]. In each of these cases, the existence of facilities-based competition or industrial policy probably explains the low accounting rates.

The FCC continues to support bilateral negotiations among carriers rather than regulatory intervention, despite its assertion of the legal right to prescribe accounting rates for U.S. carriers and to order rate reductions. In support of international comity, the FCC refused to condition authorizations of USISCs to provide service, filed pursuant to Section 214 of the Communications Act, as another vehicle for leverage [16]. However it did express the willingness to consider using the application review process to stimulate facilities-based competition in foreign nations.[25] In 1995, the FCC formulated a proposal to use the Section 214 application review process as a gauge for determining equivalency of investment opportunities by U.S. enterprises in foreign markets.[26]

---

25. See Cable & Wireless Communications, Inc., 7 FCC Rcd. 6855 (Int'l Fac. Div. 1992); BT North America, Inc., DA 94-1257, Int'l Bur. (rel. 14 November 1994) (Sec. 214 review of applications filed by foreign-owned carriers operating in the U.S. for authority to resell international switched services).

26. See Market Entry and Regulation of Foreign-Affiliated Entities, IB Docket No. 95-22, Notice of Proposed Rulemaking, 10 FCC Rcd. 4844 (1995) (proposing to use the Section 214 application review process and the degree of foreign ownership permissible under Section 310 of the Communications Act to promote effective competition, prevent anticompetitive conduct, and to encourage foreign governments to open their communications markets).

9.3 CONCLUSION

In the final analysis "the real problem is not the accounting rate system per se,... [but] regulatory and licensing policies...[that] have inhibited competition... [thereby] buttress[ing] high and static accounting rates" [17]. The incentive to retain high accounting rates can last only so long as the PTT faces no competition at home. Should a foreign government authorize competition for conventional telephone services, the second carrier might project revenue gains by reducing end-user charges and accounting rates to stimulate demand, and possibly to encourage USISCs to route more inbound traffic its way. Despite the keen attention to privatization and liberalization, most countries still reserve a monopoly for switched international telephone services. The PTT may have become privatized and more flexible, but it likely retains a voice services monopoly.

The U.S. government has only begun to find ways to identify and advocate incentives to reduce accounting rates mutually shared by both U.S. and foreign carriers. It has realized the need to demonstrate how facilities-based competition and lower accounting rates will enhance consumer welfare without serious financial harm to incumbent carriers. The FCC has agreed to permit some types of ad hoc carrier initiatives that might deviate from the ISP's core principles of identical accounting rates, proportionate traffic routing, and equal division of the accounting rate.

Accounting rates have remained artificially high, because U.S. and foreign carriers have not identified mutually advantageous financial opportunities for changing the status quo. Unless traffic volumes move toward parity between outbound and inbound directions, one carrier will have every incentive to maintain high accounting rates as the factor used to compensate it for terminating a higher volume of traffic. Heretofore, the United States has disproportionately borne the burden of high accounting rates. But even in the absence of procompetitive initiatives, foreign countries may soon share the financial penalties. Creative routing techniques and facilities-based carrier competition have already begun to remedy the refusal of certain carriers to begin reducing accounting rates.

International carriers have quicker and less messy options available, such as committing to a multiyear transition to cost-based accounting rates. It remains to be seen whether they will pursue them without regulatory agency intervention and self-help measures by large-volume end-users. Likewise, it remains unclear what regulatory incentives can stimulate downward pressure on accounting rates without stifling flexibility and innovation in carrier-to-carrier traffic routing and revenue-sharing negotiations.

References

[1]   FCC, International, Common Carrier Bureau, *Accounting Rates and the Balance of Payments Deficit in Telecommunications Services*, 12 December 1988, p. 29 (hereafter cited as Common Carrier Bureau Report).

[2]   Regulation of International Accounting Rates, CC Docket No. 90-337, Notice of Proposed Rulemaking, 5 FCC Rcd. 4948 (1990).

[3]   Regulatory Policies and International Telecommunications, CC Docket No. 86-494, Notice of Inquiry and Proposed Rulemaking, 2 FCC Rcd. 1022, para. 52 (1986).

[4]   Stanley, K., FCC, Common Carrier Bureau, Industry Analysis Div., *The Balance of Payments Deficit in International Telecommunications Services*, 12 December 1988, p. 19.

[5]   Common Carrier Bureau Report, p. 9.

[6]   Implementation and Scope of the International Settlements Policy for Parallel International Communications Routes, CC Docket No. 85-204, Notice of Proposed Rulemaking, 50 Fed. Reg. 28,418 (1985) (hereafter cited as International Settlements Policy Rulemaking), Report and Order, 59 Rad. Reg. 2d 982 (1986), on recon., 2 FCC Rcd. 1118, on further recon., 3 FCC Rcd. 1614 (1988).

[7]   Implementation and Scope of Uniform Settlements Policy for Parallel International Communications Routes, Notice of Proposed Rulemaking, Docket No. 85–204, 50 Federal Register 28418 (July 12, 1985), Report and Order, 51 Federal Register 4736 (Feb. 7, 1986).

[8]   Regulation of International Accounting Rates, CC Docket No. 90-337, Phase I, Order on Reconsideration, 7 FCC Rcd. 8049, para. 11 (rel. 27 November 1992). See FCC, Common Carrier Bureau, Industry Analysis Division, *Preliminary 1992 Section 43.61 International Telecommunications Data*, Washington, D.C.: FCC, September 1993.

[9]   Regulation of International Accounting Rates, Notice of Proposed Rulemaking, 5 FCC Rcd. at 4951.

[10] Ibid., at 4949.

[11] Regulation of International Accounting Rates, CC Docket No. 90-337, Phase I, 6 FCC Rcd. 3552 (1991).

[12] Regulation of International Accounting Rates, CC Docket No. 90-337, Phase I, 6 FCC Rcd. 3555 (1991).

[13] Regulation of International Accounting Rates, CC Docket No. 90-337, Phase II, First Report and Order, 7 FCC Rcd. at 560 (1992).

[14] Ibid., para. 561.

[15] Ibid., para. 23, note 50.

[16] Ibid., para. 32.

[17] Cheong, K., and M. Mullins, "International Telephone Service Imbalances Accounting Rates and Regulatory Policy," *Telecommunications Policy*, Vol. 15, No. 3, April 1991, p. 117.

International Trade in Telecommunications Equipment and Services

10

The international telecommunications marketplace represents an amalgam of four basic models: government monopoly, private monopoly, private (domestic or multinational) enterprise, and international strategic alliances. The first two models flourish with restricted market access and various barriers to trade, while the latter two thrive with the abandonment of restrictions and expanded opportunities to serve external markets.

Historically, the telecommunications marketplace has emphasized the first model with "national monopolies tied together by an international cartel that legally sanctioned administered prices, equal splits of international revenues, and rules that forbid competition for international traffic" [1]. Even some basic principles of free trade do not apply to international telecommunications because of:

- National security concerns about relying on foreign suppliers;
- Industrial policy grounds where nations strive to insulate indigenous manufacturers and service providers from competition while exploiting market access opportunities in other nations;
- The view that international telecommunications, like aviation, requires diplomatic agreements (e.g., the conferral of landing rights for aircraft and satellite signals), and accordingly cannot rely on marketplace forces.

In the services arena, other factors work against open markets. The cost of providing telecommunication service favored collaborative ventures to achieve economies of scale. While such joint procurements in transmission facilities mean that carriers will have roughly the same per-unit costs, it does not automatically lead to a noncompetitive market. But in the case of international telecommunications, association in satellite cooperatives and submarine cable consortia has encouraged carriers not to compete on price. An industrial structure dominated by government PTTs managed the rollout of technological innovation, controlled access to markets and facilities, established service definitions, institutionalized cross-subsidies, and largely captured or rendered ineffectual the regulatory process.

Recently accelerating privatization initiatives do not by themselves change the trade climate and prospects for market access. In many nations a private monopoly substitutes for one operated by the government. However, change has occurred at an accelerating pace, and some carriers have begun to compete on the basis of price as well as service and other nonmonetary features. The driving force for change lies with:

- Consumers less willing to accept excuses for poor, expensive, or unavailable service and more able to vote with their feet and currency to pursue new options;
- Technological innovations that make it easier for incumbents, newcomers, and users to customize network solutions and provide alternatives to inflexible, expensive, or incompetent service;
- Globalization initiatives of governments that encourage the incumbent to find new markets to compensate for lost domestic market share;
- Increasingly diverse and global service requirements of multinational users and market entry by systems integrators aiming to provide one-stop-shopping and turnkey services.

These forces cannot trigger government to abandon regulatory oversight or to convert telecommunications into a fully competitive marketplace. But they do create the basis for market entry and expanded options.

Improved trading conditions evolve not because governments have opted for a smaller role, but because governments, carriers, and manufacturers have created new market opportunities and have responded to consumer requirements. Government representatives may be instrumental in securing better and more transparent market access opportunities, but the drive for such advocacy primarily comes from users and entrepreneurs. Users are no longer resigned to tolerate service that is expensive, unreliable, and lackluster. Likewise, the telecommunication companies, which have achieved limited market niche opportunities, will lobby for further liberalization. They argue that all markets can support competition and that trade, economic, and regulatory policies designed to insulate incumbents from competition actually serve as a crutch that renders incumbents oblivious to the need for streamlining and efficiency.

Multinational enterprises like banks, airlines, oil exploration firms, and financial service companies need global, integrated, and reliable telecommunications networks. Incumbent carriers, which have lost domestic market share, and new, nimble service providers strive to become full-service operators providing one-stop-shopping convenience to users both domestically and on a regional or global basis. Many companies with global requirements recognize that logistical and cultural challenges support reliance on experts, particularly alliances of several carriers who can provide worldwide service.

Nations that permit corporate alliances and foreign investment in telecommunications recognize the mutual benefit in economic interdependence. By embracing the middle ground between free trade/competition on one hand and monopoly/cartel on the other hand, nations recognize that better service by multinational enterprises and the proliferation of international strategic alliances result, because "it is no longer possible for any country to unilaterally run its domestic market" [2] or to keep it impervious to technological and financial penetration from outsiders.

10.1 TRADE IN TELECOMMUNICATIONS NETWORK SERVICES[1]

Until the middle 1980s, nations did not use trade negotiation machinery to address services like telecommunications. In 1986, the signatory nations to the GATT[2] agreed to consider trade in services in the context of the then upcoming Uruguay Round of multilateral trade negotiations.[3] At the end of 1993, the

---

1. For more comprehensive examination of the GATT and how it began to address trade in services see T. H. E. Stahl, "Liberalizing International Trade in Services: The Case for Sidestepping the GATT," *Yale J. International Law*, Vol. 19, Summer 1994, p. 405; R. A. Brand, "GATT and the Evolution of United States Trade Law," *Brooklyn J. International Law*, Vol. 18, 1992, pp. 101, 117–20; R. N. H. Christmas, "The GATT and Services: Quill and Ink in an Age of Word Processors?" *Fordham International Law J.*, Vol. 10, 1986, pp. 288, 288–89; W. J. Davey, "Dispute Settlement in GATT," *Fordham International Law J.* Vol. 11, 1987, pp. 51, 52 note 1; J. H. Jackson, "GATT and Recent International Trade Problems," *Maryland J. International Law and Trade*, Vol. 11, 1987, pp. 1, 7–9; J. H. Jackson, "GATT and the Future of International Trade Institutions," *Brooklyn J. International Law*, Vol. 18, 1992, pp. 11, 15–17; L. S. Klaiman, "Applying GATT Dispute Settlement Procedures to a Trade in Services Agreement: Proceed With Caution," *University of Pennsylvania J. Business Law*, Vol. 11, 1990, pp. 657, 657 note 2; T. G. Berg, "Trade in Services: Toward a 'Development Round' of GATT Negotiations Benefitting Both Developing and Industrialized Nations," *Harvard International Law J.*, Vol. 28, 1987, p. 2 n. 8; R. A. Cass and E. M. Noam, "The Economics and Politics of Trade in Services," in *Rules for Free International Trade in Services*, D. Friedmann and E. Mestmacker, eds., 1990, pp. 43, 45; S. F. Benz, "Trade Liberalization and the Global Service Economy," *J. World Trade Law*, Vol. 19, 1985, pp. 95, 97.

2. General Agreement on Tariffs and Trade, opened for signature 30 October 1947, 61 Stat. A3, 55 U.N.T.S. 187.

3. The GATT established a Group of Negotiations on Services (GNS) "to establish a multilateral framework of principles and rules for trade in services, including elaboration of possible disciplines for individual sectors [like telecommunications], with a view to expansion of such trade under conditions of transparency and progressive liberalization and as a means of promoting economic growth of all trading partners and the development of developing countries." (Ministerial Declaration on the Uruguay Round, Punta del Este, Uruguay, September 1986.)

GATT formalized a General Agreement on Trade in Services (GATS),[4] with particular attention to a number of sectors including telecommunications, construction-engineering, financial services, transportation, tourism, and professional services, including lawyering.[5]

Increasing interest in applying trade negotiation machinery to services, and telecommunications in particular, results from the growing appreciation that national economies have become integrated and that the ease with which nations can transact commerce in services can significantly affect national welfare. According to Marcellus Snow [3]:

> Today's consensus points to the necessity of information and telecommunications in the more complete establishment of a market economy to replace barter and subsistence agriculture in rural areas of the developing world. There is increasing recognition that information infrastructure is a vital prerequisite for economic development rather than merely a desirable side effect.

The GATT was negotiated in 1947 with an eye toward creating four permanent organizations: (1) the United Nations to address diplomatic and political issues and to resolve conflicts before they result in wars, (2) an international trade organization to establish policies designed to promote commerce between nations and to resolve trade disputes, (3) a vehicle to support trade and development, particularly in developing nations, now generally referred to as the World Bank, and (4) a basis for stabilizing currencies through the International Monetary Fund.

The World Trade Organization (WTO) came into existence in 1994, preceded by years of interim arrangements initially stylized as a Protocol of Provisional Application (PPA).[6] The PPA became the de facto permanent basis for the GATT, and over time the GATT became both an agreement between nations and an organization with a permanent staff and the legitimacy to schedule periodic rounds of tariff reduction negotiations.

The WTO was established to implement the provisions of the GATT, including those in the area of intellectual property protection.[7] The WTO is avail-

---

4. General Agreement on Trade in Services, 15 December 1993, 33 I.L.M. 44 (GATS). GNS, conducted in parallel with the Uruguay Round of GATT negotiations, addressed issues on how to extend the GATT approach to services sectors.

5. See Trade Negotiations Committee Meeting at Ministerial Level, Montreal, December 1988, MTN.TNC/7 (MIN); and Organization for Economic Co-Operation and Development, *Elements of a Conceptual Framework for Trade in Services*, Paris, 1987.

6. Protocol of Provisional Application of the General Agreement on Tariffs and Trade, 30 October 1947, 55 U.N.T.S. 308.

7. For an introduction to the WTO and its functions, see President Clinton's Submission to Congress of Documents Concerning Uruguay Round Agreement, 15 December 1993, Chap. 17, "World Trade Organization."

able only to those countries that: (1) are signatories of the GATT, (2) agree to adhere to all of the provisions of the Uruguay Round of negotiations, and (3) submit schedules of market access commitments for industrial goods, agricultural goods, and services. The WTO agreement sets forth rules governing all of the Uruguay Round accords, including a formal basis for settling disputes, including the possibility of imposing trade sanctions if GATT members violate their obligations.[8]

The trade principles articulated in the GATT, whether applied to goods or services, fall into two major categories:

1. Market access;
2. The question of equivalency and reciprocity of such access and the classifications used by nations to identify the degree to which they will support open markets.

The former typically involves bilateral negotiations among nations with an eye toward reducing market access barriers for particular products or services exchanged between two nations. The latter involves a multilateral concepts like most favored nation (MFN) treatment, the commitment to apply the most open and liberal trade policies to GATT/WTO signatories.

10.2 BASIC TRADE CONCEPTS

10.2.1 Comparative Advantage

The GATT supports the fundamental view that nations should engage in open and free trade to exploit their comparative advantages (i.e., a country's strengths relative to that of other countries). With an unfettered ability to trade goods and services, nations theoretically can enhance overall welfare and wealth by trading products and services in which a nation has a comparative advantage for other goods and services in which a nation has no comparative advantage.[9] For example, the United States can trade aircraft, satellites, and services in the con-

---

8. See Memorandum of December 15, 1993 for the United States Trade Representative: Trade Agreements Resulting from the Uruguay Round of Multilateral Trade Negotiations, 58 Fed. Reg. 67,263, 67,289 (1993); and Understanding on Rules and Procedures Governing the Settlement of Disputes, GATT Doc. MTN/FA II-A2, 15 December 1993.

9. The GATT preamble provides the following statement: "Recognizing that their relations in the field of trade and economic endeavor should be conducted with a view to raising standards of living, ensuring full employment and a large and steadily growing volume of real income and effective demand, developing the full use of the resources of the world and expanding the production and exchange of goods."

sulting, engineering, and financial arenas, where it has a comparative advantage, in exchange for labor-intensive products, like clothing, from nations with cheaper labor costs.

The comparative advantage view considers the exchange mutually advantageous rather than unfair exploitation of cheaper labor and lower standards of living. On the contrary, a nation that does not trade for items and services in which it lacks a comparative advantage risks collectively lowering its national wealth and welfare. For example, in the 1970s Brazil severely restricted information-processing technology, with an eye toward jump starting an indigenous "informatics" industry. Its refusal to trade coffee for software may have retarded growth in its gross national product. Likewise, Japan's insistence on maintaining a domestic rice production capability on land, perhaps better suited for reducing a national housing shortage, arguably elevated concerns for national security and market insulation over the welfare of its citizens.

In economic terms, the concept of comparative advantage stimulates economies of scale and efficiency. Trade can stimulate growth in the demand for products and services, leading to their generation at the lowest per-unit costs. Likewise, a nation might depart from pursuing markets in which it lacks a comparative advantage and with which it could never reach an efficient level of output. Presumably the workers and inputs committed to inefficient production could be redeployed to more efficient outputs, but the transaction costs, dislocation, and human suffering in such transitions cannot be discounted, as shown by the sometimes painful transition in the United States from an economy dominated by agriculture and heavy industrial production (e.g., steel manufacturing) to one dominated by information services and light industry (e.g., semiconductors).

10.2.2 Most Favored Nation Treatment

The MFN clause of the GATT provides:

> Any advantage, favour, privilege or immunity granted by any contracting party to any product originating in or destined for any other country shall be accorded immediately and unconditionally to the like product originating in or destined for the territories of all other contracting parties.

In application this means that once a nation decides to confer a trade benefit, privilege, or concession to one nation, it must do so unconditionally to all signatories to the GATT and also by and for signatories to the GATS [4].

MFN treatment provides a strong incentive for nations to become signatories to the GATT because they can accrue market access benefits secured by another nation. This "free rider" opportunity can level the trade negotiation playing field in the sense that regardless of negotiating strength, a nation can acquire new market access opportunities. On the other hand, the very fact that a market access initiative will accrue to every nation tends to reduce incentives for proposing new access opportunities. With MFN treatment, it becomes difficult for a nation to use a market access opportunity as leverage for negotiating a reciprocal opportunity.

For example, if the United States agrees to increase the percentage investment of British individuals or companies in U.S. broadcasting, MFN treatment theoretically would require the United States to confer this expanded ownership opportunity to all GATT signatory nations. U.S. trade negotiators lose a major bargaining chip if they cannot offer a market access opportunity on a nation-specific basis. Additionally, they risk causing U.S. firms to lose revenues and investment opportunities without offering a reciprocal opportunity for investment abroad if other nations do not similarly expand broadcast investment opportunities for U.S. citizens. The MFN concept may create incentives for trade negotiators to settle on a least common denominator composed of ambiguous language with specific exceptions for particular industries and market segments (e.g., procurement activities of government agencies, including PTTs).

The GATT also institutionalizes the national treatment obligation requiring that foreign suppliers of goods and services be treated no less favorably than what national laws, regulations, and administrative practices provide domestic enterprises. The national treatment concept attempts to prevent nations from nullifying MFS status through the use of discriminatory internal measures, such as local content legislation and other non-tariff barriers (NTBs), as well as discriminatory taxes. However, national treatment as applied by the GATT and GATS recognizes the need to maintain nonparity, as might exist in national regulations (e.g., a cap on foreign investment in domestic broadcast facilities). Whether for legitimate concerns or simply to maintain trade barriers, regulatory exemptions to national treatment mitigates the effectiveness of the national treatment concept.

A nation that conscientiously embraces the concept risks creating unilateral market access opportunities that go unreciprocated, because nations can invoke national security or other regulatory requirements as sufficient reasons for not matching the initiative. For example, the United States might perceive its national treatment obligation to include elimination of barriers to foreign investment in telecommunications, including opportunities to market service and accept overseas traffic from locations within the United States. Other nations would not have to provide a reciprocal opportunity for U.S. businesses to operate incountry, because regulations prohibit such a presence and the GATT/GATS would not require repeal of such regulations. Arguably, a nation that confers a telecommunication equipment or services monopoly does not violate the national treatment concept, because it discriminates equally between domestic and foreign businesses by denying a market access opportunity to both types. Similarly, a country may appear to support national treatment, but the empirical evidence shows a buy domestic cultural heritage. Likewise, technical and regulatory requirements may favor the domestic incumbent or prevent foreign enterprises from meeting all certification requirements needed prior to market entry.

10.2.3   Other Trade Concepts and Requirements in the GATT and GATS

Other fundamental concepts in the GATT and GATS include unrestricted transit of goods and services whereby nations agree not to impede trade involving third parties that merely crosses through national territory. The parity of legal and regulatory treatment between domestic and foreign entities also includes the concept of transparency: establishing trade regulations, practices, procedures, and restrictions in the public eye, because increased visibility and accessibility can simplify commercial transactions and discourage trade restrictions.

When trade disputes arise, the GATT and GATS provide for institutionalized dispute resolution. The efficacy of such a process depends on the willingness of signatory nations to relinquish a degree of sovereignty in recognition that the WTO forum will lend its "good offices" in a fair manner aimed at enhancing trading opportunities. A member state triggers the dispute resolution mechanism by identifying how benefits due it under the GATT are being "nullified or impaired" [5]. The mechanism involves a number of procedures designed to secure a remedy through consultation and coordination as opposed to retaliation and trade sanctions. The complaining nation first attempts to secure informal support for a proposal designed to remedy the problem. If this route proves unsuccessful, the nation may pursue a more formal liaison with other members of the GATT, its director-general, council, or any appropriate intergovernmental organization [6]. The nation may also petition the GATT Council to appoint a panel consisting of individuals drawn from member states to investigate the issue and recommend a solution that might involve a directive to the offending nation to change policies and perhaps to compensate the offended nation. The GATT Council must ratify the report and may amend it or require further negotiation and analysis. Since the GATT has no enforcement mechanism, the panel report can only generate nonbinding recommendations that the offending nation can implement, ignore, or use as the basis for seeking an exemption from an otherwise applicable provision of the GATT [7].

10.3 CHANGE IN THE U.S. BALANCE OF TELECOMMUNICATIONS TRADE

Until 1984, the United States enjoyed a sizable trade surplus in telecommunication equipment and services. The surplus quickly evaporated for a number of reasons:

- Technological innovations promoted heightened competitiveness worldwide and the recognition of the need to globalize (i.e., to serve foreign markets).[10]
- Domestic telecommunications policy initiatives, ostensibly to promote efficiency and competitiveness and enhance consumer welfare also made the U.S. market "vulnerable to international competitors without guaranteeing reciprocity" [8].
- Many foreign nations targeted telecommunications for special concentration of industrial policies, such as tax incentives to promote investments or exports, direct or indirect subsidies, special financing arrangements, protection against foreign competition, worker training programs, regional development programs, assistance for research and development, and measures to help small-business firms.[11]

Additionally, high valuations of the U.S. dollar favored imports, toll revenue settlements resulted in a multibillion dollar outpayment from U.S. to foreign carriers, and the AT&T divestiture freed the BOCs from near-mandatory purchases of Western Electric equipment.

The magnitude of the shift is reflected by the statistic that in 1982 the United States had an $81.7 million trade surplus in telecommunications equipment sales, while in 1988 it had a trade deficit of $2.7 billion [9]. This abrupt change provides evidence of a relatively open U.S. marketplace and a trade policy supporting national treatment, the commitment to apply the same rules, regulations, and classifications to foreign enterprises as applied to domestic

---

10. Aronson and Cowhey have identified several technological and marketplace factors that make trade in telecommunications more essential. These factors include demand by multinational users for functionality anywhere in the world like that available in the United States, the need by PTTs and incumbents to co-opt and accommodate large users, the merger of telecommunications and data processing, and the resulting intertwining of equipment and service markets as well as national and international networks. (J. Aronson and P. Cowhey, *When Countries Talk-International Trade in Telecommunications Services*, Washington, D.C.: American Enterprise Inst./Ballinger Publication, 1988, p. 34.)

11. Historically, the term *industrial policy* has been associated with at least some degree of centralized economic planning or indicative planning, but this connotation is not always intended by its contemporary advocates. ("A Glossary of International Trade Terms," *Business America*, Vol. 9, No. 20, 29 September 1986, pp. 10, 14.)

firms. Perhaps as well the trade deficit demonstrates the lack of policies and procedures designed to thwart market access by foreign manufacturers, unless they seek market share by deliberately underpricing goods and services or pursue other predatory practices.

The ease of access to U.S. consumers in many cases contrasts with the difficulty in achieving a level competitive playing field for U.S. telecommunication equipment manufacturers and service providers in foreign markets. Financial, regulatory, and licensing policies can frustrate market access even where a comparative advantage would favor imports over domestically produced goods. Domestic producers may benefit from closed procurements, close affiliation with or ownership by the government, antitrust law exemptions, government grants for research and development, below-market-rate loans, and equipment testing and procurement that favor domestic producers. A national commitment to making telecommunications a flagship industry and bellwether for a nation's competitive posture can result in government subsidization to achieve shares in the international marketplace. Consuming domestic products becomes a patriotic gesture in some nations.

The market intervention of some nations juxtaposes with a conditional laissez faire philosophy in other nations where national treatment applies unless evidence shows deliberate underpricing of goods (i.e., dumping and substantial asymmetry in market access). U.S. economic and political philosophy militates against the coordination and collaboration that occurs in many foreign nations among government, industry, the financial community, and academia. Foreign efforts to close access to domestic markets in conjunction with affirmative steps to coordinate an industrial policy seeking market share elsewhere have succeeded, despite the risk for retaliation that could trigger a mutually painful trade war.

Set out below is a list of tactics and policies that nations can pursue to tilt the competitive playing field in favor of trade surpluses and increased market share opportunities for domestic manufacturers and service providers.

10.4 ANTICOMPETITIVE TACTICS IN TELECOMMUNICATIONS

10.4.1 Equipment Markets

Closed Procurements

Nations with a single PTT may rely heavily on one domestic equipment manufacturer often owned by or affiliated with the PTT. The national hero may constitute the only enterprise capable of delivering equipment that meets technical specifications and other requirements it may have helped prepare.

Discrimination in Equipment Certification

Foreign manufacturers may have to secure costly and time-consuming equipment testing and certification to qualify for market access. The foreign nation may reject test results from foreign labs, in effect facilitating exclusive market access for domestic enterprises.

National or Regional Standards

Rather than support universal standards as promulgated by the ITU, nations or regions may establish their own, perhaps with an eye toward relying on domestic manufacturer-supplied technical standards and interfaces.

Import Delays and Difficulties

Few nations fully implement the concept of national treatment and may resort to customs, duties, domestic content requirements, and informatics policies (i.e., a requirement that software and other types of intellectual property be created incountry) to create an absolute barrier to entry or price disadvantage for imports.

Barriers to Foreign Investment

Nations may thwart efforts by foreign nations to establish a manufacturing or marketing presence in the nation, or may refuse to consider such installations as domestic when evaluating domestic content requirements, particularly if assembly rather than manufacturing takes place.

Absence of Forum for Dispute Resolution

Foreign manufacturers and distributors may lack opportunities to dispute discriminatory practices in domestic regulatory agencies and courts.

10.4.2 Telecommunication Service Markets

Refusal to Grant an Operating Agreement

Foreign governments and PTTs are under no legal or treaty obligation to confer an operating agreement to market entrants. Without such an agreement, the newcomer cannot access the nation's domestic telephone network to complete calls and in effect is deprived of access. Reciprocity of access is nonexistent in most nations, and even where nations will accept inbound traffic, few will guarantee a commensurate return flow. Fewer still will allow foreign carriers to establish a marketing presence within the nation.

Discriminatory Network Access

Access terms and conditions favor PTT-delivered services, particularly those on a metered, usage-sensitive basis. Leased private lines and other options, including independent transmission facilities, are discouraged, particularly those options that bypass facilities of the PTT or generate less revenues for it.

Impediments to Network Usage

Governments may prohibit shared use and resale of leased lines, impose restrictions on data transmission speed, limit access to leased private lines forcing migration to usage-sensitive public switched service, restrict interconnection to the PSTN, and/or determine who qualifies to construct and operate satellite Earth stations.

Equipment Interconnection Limitations

Governments may establish policies and equipment testing and certification restrictions that limit the ability of carriers and customers to attach equipment providing tailor-made functions unavailable from the PTT or generated by the PTT central office switches at far higher prices.

Lack of a Commitment to Cost-Based and Nondiscriminatory Pricing

The PTT may deliberately overprice some services to subsidize postal and other favored services; pricing strategies may migrate users of private lines and less profitable services to more profitable switched offerings.

Uncertain Regulatory Relief

PTT operations may not be separate from the regulatory/policy-making process; some governments provide no procedures for fairly adjudicating user or PTT competitor complaints.

Prohibitions or Limitations on Transborder Data Flow

Some governments impose restrictions on where information can be processed and where databases can be located.

10.5 LEVERAGE IN TRADE NEGOTIATIONS

While possibly reluctant to resort to retaliatory action, nations do have significant domestically authorized and internationally accepted weapons for remedying trade disputes. The reluctance to use tactics involving the application of financial penalties, like countervailing duties that impose additional, ostensibly parity-fostering costs on imports, stems from the prudent concern that a spiral of sanctions and countersanctions would render all parties worse off from the confrontation.

Accordingly, the most frequently relied-upon weapon is negotiation and "jawboning." This involves forceful advocacy requiring the parties to reach a settlement lest harsher unilateral actions be legislated or promulgated in a far less sympathetic domestic forum like the legislature,[12] than through multinational dispute resolution, like that established under the GATT. In addition to dispute resolution under the aegis of the GATT, nations engage in fact finding and negotiations aimed at addressing the foreign country's industrial structure (e.g., whether and how the PTT might be subject to competition or at least procure equipment on an open and nondiscriminatory basis).

10.6 SECURING MARKET ACCESS PARITY IN TELECOMMUNICATION SERVICE

The United States and other nations supporting free and open trade have attempted to achieve closer parity in market access under the auspices of the existing trade and telecommunications policy forums like the GATT and the ITU. However, the GATT has only begun to incorporate its negotiation and trade machinery for services, in addition to goods,[13] and the ITU has traditionally ad-

---

12. See The Omnibus Trade and Competitiveness Act of 1988, Pub. L. No. 100-418, 1988 U.S. Code Cong. & Adm. News (102 Stat.) 1107 (1988), codified at 19 U.S.C. Secs. 3101–3111 (1990); the Trade Reform Act of 1974, codified at 19 U.S.C. Sec. 2411 et seq., Sec. 301 of the Trade Act of 1974, 19 U.S.C. Sec. 2411 (1990) requires the U.S. Trade Representative (USTR) to take mandatory remedial action if it is determined that U.S. rights under a trade agreement are being denied. Further amendments to Sec. 301, 19 U.S.C. Sec. 2420 (1990), commonly known as "Super 301" required the USTR in 1989 and 1990 to identify U.S. trade liberalization priorities and practices, which if eliminated would significantly increase exports, and to identify priority countries to engage in dialogue to reduce trade barriers. The Office of the U.S. Trade Representative was established as an agency in the Executive Office of the President in 1963. Exec. Ord. 11075, 15 January 1963, 28 Fed. Reg. 473, as amended by Exec. Ord. 11106, 18 April 1963, 28 Fed. Reg. 3911, and Exec. Ord. 11113, 13 June 1963, 28 Fed. Reg. 6183. Pursuant to 19 U.S.C. 2171(c)(1) of the Trade Act of 1988, the USTR, inter alia, has primary responsibility to develop and coordinate the implementation of U.S. trade policy.

13. For an analysis of international trade in services and an assessment of whether the existing trade negotiation apparatus can be adjusted to incorporate services, see G. Feketekuty, *Interna-*

dressed standards and operational rules of the road, not trade policy per se.[14] Neither organization has an automatic right to serve as the primary forum, and accordingly they may "find themselves in the force field of a jurisdictional battle that may not be resolved in the near future" [10].

A key impediment to achieving reduced barriers to market entry in telecommunications lies in the almost metaphysical characteristics of the process (i.e., the difficulty in defining the transaction or set of transactions in a trade context). Accordingly, both the WTO and the ITU need to make significant adjustments in the way they address telecommunications trade issues. Like broadcasting, which transmits programming into the "ether," telecommunication services can involve a number of transborder transmissions with a variety of intermediate processing.[15] Nations created the GATT initially to address concrete, tangible goods, while the ITU has considered telecommunications from a policy or technological perspective aiming to optimize efficiency, innovation, and social objectives, factors somewhat ancillary in a conventional trade policy assessment.

"The fundamental difficulty with telecommunication and data services as a trade-in-services issue is that it is both a telecommunication-policy issue *and* a trade-policy issue *simultaneously and interactively*" [11]. Trade policy officials serve their nation's commercial interests while pressing for a more level competitive playing field elsewhere, including the reduction of barriers or impediments to market access. Telecommunications policy officials ostensibly

---

tional Trade in Services—An Overview and Blueprint for Negotiations, Cambridge, MA: American Enterprise Inst./Ballinger Publishing Co., 1988; and P. Robinson, K. Sauvant, and V. Govitrikar, eds., *Electronic Highways for World Trade Issues in Telecommunication and Data Services*, Boulder, Co: Westview Press, 1989.

14. However, the 1988 WATTC convened in Melbourne, Australia, addressed a number of issues with a trade component, such as establishing a mechanism for "special arrangements" whereby nations can agree on a bilateral basis to permit the provision of new services like value-added networks, even though such services might conflict with conventional rules of the road. WATTC definitely placed the ITU in the mix of legitimate policy-making forums along with the GATT, United Nations, and regional organizations like the European Economic Community and the Organization for Economic Co-operation and Development.

15. William Drake and Kalyso Nicolaidis present six problem areas: (1) Not all service transactions fit under the traditional definition of trade as products produced entirely in one country and purchased in another; (2) the boundary between trade and foreign direct investment is fuzzy and if efforts were undertaken to liberalize investment flows, then developing nations would seek liberalization of labor flows; (3) the boundary line between illegitimate nontariff barriers and legitimate regulations is fuzzy as well; (4) comparative advantage in services sometimes is not based on the same natural factors and endowments as would be the case for goods; (5) applying most-favored nation treatment in telecommunications could erode network integrity, rob incumbents of scale economies, and result in duplication of investment; and (6) most nations would strongly object to simply adding services to the GATT treaty. (W. Drake and K. Nicolaidis, "Ideas, Interests and Institutionalization: 'Trade in Services' and the Uruguay Round," *International Organization*, Vol. 46, No. 1, Winter 1992, pp. 37, 62–63.)

seek to optimize the rules of the road and standards to promote network connectivity, although industrial policy may color their judgment and make them advocates for national heroes and parochial policies.

A number of trade concepts have a direct translation into concrete telecommunications policy proposals. The multilateral rounds of negotiations in the GATT have developed support for such basic principles as:

- Supporting commerce between nations through freely working markets unfettered by discriminatory and nontransparent rules and regulations;
- Reducing barriers to trade on both a bilateral and multilateral basis with due regard to the need to respect national sovereignty, the legitimacy of domestic regulatory goals, and the need to move incrementally to avoid social, political, or economic dislocation;
- Conferring the same market access opportunities on foreign enterprises as available to domestic ones;
- Accepting the GATT mechanism for settling disagreements and for consulting other nations with an eye toward reducing trade barriers.

International telecommunications policy forums may emphasize sovereignty over such trade concepts. Matching half circuits requires close cooperation, and nations may have greater concerns about the transborder inflow of information and entertainment than for the importation of tangible products. Nations relinquish a degree of their sovereignty to accept telecommunication rules with an eye toward promoting beneficial enhancements in efficiency and reduction of transaction costs. While trade policies attempt to achieve similar goals (e.g., to enable nations to specialize based on comparative advantage in the production of goods and services), the debate has little to do with the relinquishment of sovereignty. The sovereignty of a nation is considered a given: the debate focuses on the willingness of nations to allow enterprises of other nations to enjoy MFN treatment, that is, nondiscriminatory market access, including perhaps the right to establish a commercial presence in the foreign country and to benefit from transparent, nondiscriminatory application of regulatory rights and responsibilities. Nations typically leverage threatened reduction of market access opportunities or increased cost of imported goods to achieve closer parity of access. Such tactics include selective reregulation, delays in further deregulatory initiatives like expanded foreign investment opportunities, imposing expectations for market access reciprocity, and limited deviation from national treatment. The prospective loss of market share creates incentives for foreign nations to agree to language that appears to move them toward easing market access restrictions at home, although the timetable and procedures may be vague.

On the other hand, nations aggressively pursuing market access initiatives have faced backlash and retaliation when insisting on too strict a timetable or unilaterally announcing what must transpire. Many nations favor the multilateral GATT system instead of bilateral negotiations for resolving disputes involving telecommunication equipment procurement, network access, interconnection, service pricing, treatment of foreign investment, and regulation of services that add value and enhance basic leased private lines.

10.7 ITU EFFORTS TO ADDRESS TELECOMMUNICATIONS TRADE-IN-SERVICE ISSUES

In 1988, the ITU convened a World Administrative Telephone and Telegraph Conference (WATTC-88) with an eye toward developing new regulations that could address technological innovations and growing conflict on how best to respond to trade disputes. Some nations viewed WATTC as a last chance for the ITU to carve out a legitimate trade portfolio and to accommodate increasing numbers of commercial enterprises, without traditional common carrier certification or service requirements, seeking access to the facilities of incumbent carriers to provide new enhanced services.[16]

The debate at WATTC-88 centered on the scope of international telecommunications and information-processing services that should be subject to fundamental rules of the road set out in the International Telecommunication Regulations, and the instances where incumbent carriers and new enterprises could elect to establish their own special arrangements. While the ITU Convention already contained a section affording administrations this option,[17] the United States and the United Kingdom favored expanding its scope and the frequency of its use, while most nations expressed concern about the potential for special arrangements to become a large loophole for avoiding necessary limitations and for deviating from the status quo.

"Many of the regulatory issues raised by new interconnection relationships or by [new] hybrid offerings do not fit neatly within the service categories of telephony and telex, which have been the traditional subjects of ITU overseeing and regulations" [12]. Similarly, the IVANs and other enterprises providing new services were often market entrants authorized to provide service through the resale of lines leased from incumbent carriers. While the ITU had a category for identifying these types of enterprises, RPOAs,[18] it did not yet have an ac-

---

16. For an extensive assessment of WATTC-88, see W. Drake, "WATTC-88: Restructuring the International Telecommunication Regulations," *Telecommunications Policy*, Vol. 12, September 1988, pp. 217–234 (hereafter cited as WATTC Review).

17. See ITU, *International Telecommunications Convention*, Article 31, "Special Arrangements," Nairobi, 1982.

18. See Constitution of the ITU, Annex, Definition of Terms Used in This Constitution, the Convention and the Administrative Regulations of the International Telecommunications Union, CS/An. 1008, in ITU, *Final Acts of the Plenipotentiary Conference*, Nice, 1989, Geneva: ITU, 1990, p. 65.

ceptable mechanism for allowing them to operate substantially free of the regulations applicable to facilities-based carriers providing the underlying transport functions.

Many nations were content for WATTC-88 to incorporate new services into existing definitions, a decision that would bolster campaigns by incumbent carriers to expand their operations to include such enhanced services. William Drake reports [13]:

> For the USA, the UK and their corporate supporters, a detailed list of services to be covered by the ITU Regulations would be counterproductive. They argued that any list that might be drawn up would quickly become obsolescent, and could raise regulatory questions about the status of future offerings which would not be included. Moreover, given their sometimes less than muted distrust of PTT ambitions, they feared that the exercise could lead to a list incorporating such items as intra-corporate services based on [the use of] leased lines.

WATTC-88 finally adopted a compromise that would apply the regulations to services generally available to the public while exempting specialized services. The ability to reach a compromise reflects enlightened national self-interest in allowing telecommunications to evolve and the ITU to remain effective. But the compromise also reflects how the ITU had to devise solutions to telecommunications trade matters in addressing whether and how nations should accept traffic from new enhanced-service providers or allow such enterprises to operate domestically.

10.8 FRUSTRATION WITH MULTILATERAL DISPUTE RESOLUTION

Many commentators believe that international forums like the WTO and the ITU have declining prospects for resolving disputes and forestalling bilateral negotiations.[19]

---

19. "The consultative phase could be unproductively long; the party whose measures were challenged could simply drag out the bilateral discussions. Even after parties agreed to the establishment of a panel...the GATT had no mechanism for appealing it. Even if a panel finally issued a report, a single contracting party—including the disputant to whom it was adverse—could block its adoption by the GATT Council...Finally, no procedure existed to ensure that the GATT Council would monitor the action or inaction of a party whose measures were found to be GATT-inconsistent, unless prompted by the initiative of the party that successfully challenged those measures." (J. H. Bello and A. F. Holmer, "U.S. Trade Law and Policy Series No. 24: Dispute Resolution in the New World Trade Organization: Concerns and Net Benefits," *The International Lawyer*, Vol. 28, No. 4, Winter 1994 (hereafter cited as Bello & Holmer).)

Ineffectual dispute resolution in the GATT has motivated the U.S. Congress to legislate self-help procedures implemented on a bilateral basis. For example, Section 301 of the Omnibus Trade and Competitive Act of 1988[20] provides for expedited investigation and retaliation involving the imposition or increase of tariffs or quotas in response to unfair and injurious trade practices of a foreign government. Congress authorized the USTR to fashion remedies when a foreign government breaches, nullifies, or impairs benefits to the United States from a trade agreement. Domestic legislation creates a unilateral basis for enforcing a trade agreement by authorizing the USTR to take a more aggressive posture in trade negotiations in which the United States perceives an unjustifiable, unreasonable, or discriminatory trade practice that restricts U.S. commerce.

The aggressive use of Section 301 by the United States had the ironic outcome of generating increased support for previous U.S. proposals to make dispute settlement within the GATT more effective and timely. The new rules established at the Uruguay Round of Trade Negotiations[21] make fundamental changes to the GATT dispute settlement process by providing for:

- Automatic establishment of a panel and automatic adoption of a panel report (unless the council, by consensus, decides to the contrary);
- An exceptional opportunity for appellate review of panel reports;
- Rigorous surveillance of the implementation of adopted panel reports;
- Compensation for, or WTO authorization of, the suspension of concessions if a report is not implemented in a reasonable period of time;
- Expeditious arbitration in the event of disputes about a reasonable period of time for implementation or the appropriate level of compensation or suspension;
- Recourse to these procedures for practices considered as violating the WTO or nullifying or impairing WTO benefits [14].

10.9 REGIONAL TRADE PACTS

The multinational nature of the GATT can result in delayed decision making and a least common denominator ill-fitted for particular nations or regions.[22]

---

20. Pub. L. No. 100-418,. § 1301, 102 Stat. 1107, 1164; see J. H. Bello and A. F. Holmer, "The Heart of the 1988 Trade Act: A Legislative History of the Amendments to Section 301," *Stanford J. International Law*, Vol. 25, 1988, p. 1.

21. See General Agreement on Tariffs and Trade: Final Act Embodying the Results of the Uruguay Round of Trade Negotiations, 15 December 1993, 33 I.L.M. 1 (1994).

22. Bello & Holmer do not exclude the United States from the list of nations that contribute to the GATT's ineffectuality: "Like most nations, the United States tends to be schizophrenic; it wants the benefits of free trade, but also the freedom to act on its own without regard to international

Increasingly, nations have resorted to bilateral or regional trade negotiations. The North American Free Trade Agreement (NAFTA) provides a 2,000-page case study of what three nations, the U.S., Canada, and Mexico, can produce,[23] if they fear the consequences of not responding to free trade zones in other areas of the world. NAFTA affects $6 trillion of trade among nations totaling 360 million people, making North America the largest free trade area in the world [15].

NAFTA provides concrete rules for achieving open and free trade between nations, substantially free of distortions and inequality. As stated in the NAFTA preamble, the goal is to "strengthen the special bonds of friendship and cooperations...contribute to the harmonious development and expansion of world trade...reduce distortions to trade...foster creativity and innovation and promote trade in goods and services that are the subject of intellectual property rights...and preserve their flexibility to safeguard the public welfare" [16].

NAFTA establishes a free trade area, with an objective of eliminating trade barriers, promoting fair competition, increasing investment opportunities, protecting and enforcing intellectual property rights, creating effective procedures to implement and administer the agreement, and establishing a framework to expand and enhance the benefits of the agreement. NAFTA affirms each party's rights under previous agreements to which they are parties such as the GATT. However, to the extent there exist any inconsistencies between NAFTA and other agreements, NAFTA will prevail unless it indicates otherwise. However, preexisting environmental and conservation agreements and annexes to NAFTA take priority over the basic document to the extent there are inconsistencies.

NAFTA's primary objective lies in the elimination of trade barriers such as import duties on goods that originate within Canada, the United States, and Mexico. The agreement establishes a timetable for the reduction of existing duties and prohibits any party from increasing or adopting new customs or duties. NAFTA also addresses nontariff restrictions such as import licenses and quotas. It specifically requires parties to accord national treatment, in accordance with GATT, to the other's goods, but the agreement does allow standards designed to safeguard human, animal, or plant health and the environment.

Chapter 13 of NAFTA contains specific provisions addressing telecommunications, ranging from basic public telecommunications transport networks or services to enhanced or value-added services. The agreement includes a requirement that the parties work on establishing open and nondiscriminatory access to such networks by carriers, service providers and equipment located on

---

restraints. It remains the chief champion and cheerleader for the international rule of law, yet it prizes the sovereign right to act in disregard of such law in exceptional circumstances." (Bello & Holmer, pp. 1103–1104.)

23. The U.S. Congress ratified the NAFTA accord on 20 November 1993. (See North American Free Trade Agreement, December 17, 1992, U.S.-Can.-Mex., Pub. L. No. 103-182, 107 Stat. 2057, reprinted in 32 I.L.M. 605 (1993).)

the premises of end users. The chapter also addresses issues of pricing, standards-related measures, monopolies, transparency, relationships with international organizations and agreements, and technical cooperation for increased compatibility.

Despite progress in promoting free trade, however, NAFTA does include a Cultural Industry Exemption Clause that exempts certain industries from applying the agreement.

NAFTA does not apply where a nation views its cultural identity and sovereignty at risk as a result of market access. For example, concerns about cultural imperialism has led Canada and the EU[24] to restrict both the extent of foreign investment in broadcast enterprises and the scope of foreign programming that can be broadcast.

The telecommunications provisions of NAFTA exempt basic voice telephony services from coverage. This means that NAFTA is limited to value-added or enhanced services, while basic services must be addressed by a separate bilateral agreement. As to enhanced services, NAFTA requires nondiscriminatory licensing, meaning that foreign signatories to the agreement should have the opportunity to partial or complete ownership of such an enterprise in another NAFTA nation.

References

[1]   Aronson, J., and P. Cowhey, *When Countries Talk: International Trade in Telecommunications Services*, Washington, D.C.: American Enterprise Inst./Ballinger Publication, 1988, p. 218 (hereafter cited as Aronson and Cowhey).

[2]   Ibid., p. 229.

[3]   Snow, M. S., "Trade in Information Services in Asia, Asean, and the Pacific: Conceptual Issues and Policy Examples," *California Western Law Review*, Vol. 28, 1991–1992, p. 329 (hereafter cited as Snow).

[4]   GATS, art. II.

[5]   GATT, art. XXIII(1).

[6]   GATT, art. XXIII(2).

[7]   GATT, art. XXIII(2).

[8]   Aronson and Cowhey, p. 32.

[9]   Department of Commerce, National Telecommunications and Information Administration, *NTIA Telecom 2000: Charting the Course for a New Century*, 5 NTIA Spec. Pub. 88-21, Octo-

---

24. For a discussion of Europe's "Television Without Frontiers" policies and how they nevertheless still restrict market access, see L. G. C. Kaplan, "The European Community's 'Television Without Frontiers' Directive: Stimulating Europe to Regulate Culture," *Emory International Law Review*, Vol. 8, Spring 1994, p. 255. See also Council Directive of 3 October 1989 on the Coordination of Certain Provisions Laid Down By Law, Regulation, or Administrative Action in Member States Concerning the Pursuit of Television Broadcasting Activities, 1989 O.J. (L 298) 23, reprinted in 28 I.L.M. 1492 (1989); and Television Without Frontiers: Green Paper on the establishment of the Common Market for Broadcasting, especially Satellite and Cable, COM (84) 300 Final (1984).

ber 1988. See also Duane, J., and W. C. Edgar, "Sectoral Reciprocity in Telecommunications: The Telecommunications Trade Act of 1988," *George Washington J. International Law and Economics*, Vol. 22, 1988, p. 195.

[10] Snow, reprinted by West Law at p. 6.

[11] Woodrow, R. B., "Trade in Telecommunication and Data Services," in *Electronic Highways for World Trade*, p. 23 (emphasis in original).

[12] Renaud, J., "The Role of the International Telecommunications Union: Conflict, Resolution and the Industrialized Countries," in *The Political Economy of Communications—International and European Dimensions*, K. Dyson and P. Humphreys, eds., New York: Routledge, 1990, pp. 33, 43–44.

[13] Drake, W., "WATTC-88: Restructuring the International Telecommunication Regulations," *Telecommunications Policy*, Vol. 12, September 1988, p. 219.

[14] Bello, J. H., and A. F. Holmer, "U.S. Trade Law and Policy Series No. 24: Dispute Resolution in the New World Trade Organization: Concerns and Net Benefits," *The International Lawyer*, Vol. 28, No. 4, Winter 1994, p. 1099.

[15] Koniigsberg, S. R., "Think Globally, Act Locally: North American Free Trade, Canadian Cultural Industry Exemption, and the Liberalization of the Broadcast Ownership," *Cardoza Arts & Entertainment Law J.* Vol. 12, 1994, p. 281.

[16] North American Free Trade Agreement, December 17, 1992, U.S.-Can.-Mex., Pub. L. No. 103-182, 107 Stat. 2057, reprinted in 32 I.L.M. 605 (1993), preamble.

Telecommunications in International Development

11

Rapid advances in telecommunications have not accrued the same financial and social dividends throughout the world. While the world economy has grown increasingly integrated and interconnected, developing nations continue to lag behind both in terms of statistical indexes of technology diffusion, such as number of telephone lines per 100 inhabitants,[1] and the economic stimulation that an efficient and modern telecommunication infrastructure generates.[2]

The telecommunications infrastructure in many developing countries suffers from a combination of underdevelopment, underfinancing, and overuse where available. Accordingly, many of the nations most apt to benefit from telecommunications development have inferior opportunities to improve national cohesion, achieve equitable access to information and social discourse, reduce disparity between urban and rural telecommunications options, and enhance the quality of life. Ironically, the nations least able to finance telecommunications development are the ones most in need of it. These nations may not easily afford such development, but they can ill afford not to commit scarce hard currency to make them more attractive targets for foreign investment.

---

1. The number of telephone lines per 100 inhabitants constitutes a frequently used statistical index of market penetration in a nation. In 1993, Sweden had 68.3 telephone lines per 100 inhabitants, Canada 59.7, the United States 57.8, Australia 47.8, the United Kingdom 46.1, Thailand 3.7, China 1.5, Indonesia 1.0, and India 0.9. (G. Staple, ed., *TeleGeography 1994, National Telecommunications Indicators*, Washington, D.C.: International Institute of Communications, 1992, pp. 164-167). Statistics on teledensity and growth rates are provided at the end of this chapter. More extensive statistics are available in ITU, *Yearbook of Common Carrier Telecommunication Statistics*, Geneva: ITU, 1994.

2. See, for example, R. Saunders, J. Warford, and B. Wellenius, *Telecommunications and Economic Development*, Baltimore: Johns Hopkins University Press, 1983 (hereafter cited as Saunders, Warford, and Wellenius); and B. Wellenius, P. Stern, T. Nulty, eds., *Restructuring and Managing the Telecommunications Sector*, Washington, D.C.: World Bank, 1989 (hereafter cited as *Restructuring and Managing the Telecommunications Sector*).

Ithiel de Sola Pool considered development in three different contexts:

1. Economic measurements such as gross national product (GNP), productivity, and living standard indexes;
2. Multiplication of centers of initiative (i.e., diversification and increased complexity in organization and division of labor in terms of where and how initiatives may occur);
3. Modernization through the transfer of technology (i.e., the acquisition of technical skills to adapt and use innovations, thereby enhancing social welfare).[3]

Professor de Sola Pool identified a number of characteristics that a telecommunication system should have for effective diffusion. It should be [1]:

[A]n adjunct to expression by those persons who have credibility in the culture. It should use the language and the symbols of the culture. Its contents should be capable of local adaptation. It must be cheap. It should require as little foreign exchange as possible. It must also be reliable and relatively rugged, and not require highly sophisticated maintenance and operating personnel. It must operate even in the absence of an elaborate infrastructure...[and be capable of] link[ing] the underdeveloped region.

The lag in telecommunications development persists notwithstanding the commonly held view that an expanded infrastructure and more accessible telecommunications networks will stimulate demand, increase revenues, and help jump-start other sectors of a nation's economy. On the other hand, one should recognize that telecommunications must compete for scarce financial resources, including foreign currencies, with life-maintaining infrastructures like sanitation, health care, water, housing, transportation, and electricity. Even with heightened attention to the benefits of a viable telecommunications infrastructure, only about 3% of the World Bank's loan portfolio finances telecommunication projects [2].

While intuitively pleasing, the favorable contribution made to a nation's gross domestic product from telecommunication investment is not easily measured.[4] Studies have shown a statistically significant positive correlation between GNP and the level of telephone service penetration; that is, nations with

---

3. See I. de Sola Pool and E. Noam, eds., "Communications for Less Developed Countries," in *Technologies Without Boundaries—On Telecommunications in a Global Age*, Cambridge, MA: Harvard University Press, 1990, pp. 167–204.

4. See, for example, S. Pitroda, "Development, Democracy and the Village Telephone," *Harvard Business Review*, November-December 1993, pp. 66–79.

higher GNPs also have more telephone lines per 100 inhabitants.[5] However, most studies have not determined definitively the cause and effect resulting in such a correlation. Does improved telecommunications stimulate economic performance, or do improved economic conditions, measured by increases in GNP, lead to the ability of a nation to finance development, thereby making the telecommunications infrastructure more accessible and reliable?

Once governments decide to invest in telecommunications, they must consider how best to allocate funds. This kind of decision making blends economic concerns for maximizing utility and value with equity and political concerns (e.g., achieving a balance between serving the needs of urban and those of rural residents and between the complex requirements of corporate users and the need for POTS for residential users).

Often policy makers must decide whether to allocate funds to maximize the number of lines and population served or to upgrade the array of features available in locales that already have access to POTS. Service to rural and residential users typically costs substantially more on a per-user basis than service on dense routes and in populated areas. Therefore, government must decide the public welfare merits of extending for the first time a basic lifeline service to the hinterland versus upgrading facilities and replacing deteriorating equipment to improve unreliable or overtaxed service to urban users.[6]

Governments must also respond to the requirements of corporate users who typically need more than POTS. These users have the financial wherewithal and possibly the political clout to erect their own internal telecommunication facilities should the incumbent operator fail to respond adequately and quickly. If a large corporate or government user exits public network facilities, the incumbent carrier may have substantially less revenue available to under-

---

5. See, for example, H. Hudson, *When Telephones Reach the Village*, Norwood, N.J.: Ablex, 1984; H. Hudson, *A Bibliography of Telecommunications and Socio-Economic Development*, Norwood, MA: Artech House, 1988; and ITU/Organization for Economic Cooperation and Development, Project "Telecommunications for Development," Geneva, 1983.

6. Telecommunication entrepreneurs eager to tap developing markets like to think that half of a nation is waiting for a phone line, while the other half is waiting for a dial tone. "The shortage of sets and lines creates two situations that encourage over-use [of existing facilities]. First, because individuals and businesses cannot readily add telephones, each installed station serves a much larger number of people than is the case in the industrialized world. Thus, telephones are, to exaggerate a bit, seldom at rest during the business day. Second, over-utilization, in turn, breeds...sequential dialing in an effort to seize the [often busy] local line once it becomes disengaged. These repetitive calling attempts, but not completions, also generate busy signals that frustrate others attempting to reach the original callers. In an attempt to get through to the party called, some subscribers resort to automatic dialing equipment which further exacerbates the number of busy signals in the system." (R. Bruce et al., "Telecommunications Structures in the Developing World: An Essay on Telecommunications and Development," Chap. 6 in *The Telecom Mosaic—Assembling the New International Structure*, Kent, England: Butterworths, 1988, p. 415.)

write improvements, since large-volume users contribute disproportionately to the revenues generated. Once such users opt for a private facility, they may seek to share or resell its excess capacity. While such facilities may provide sophisticated options not available from the incumbent operator, they can also duplicate available options, possibly resulting in underuse of incumbent carrier facilities, also known as *stranded investment*.

11.1 STATING THE CASE FOR TELECOMMUNICATION DEVELOPMENT ASSISTANCE

In 1984, the Independent Commission for Worldwide Telecommunication Development reported to the ITU[7] substantial disparity in access to telecommunications between nations [3]:

> While telecommunications is taken for granted as a key factor in economic, commercial, social and cultural activity in industrialized countries and as an engine of growth, in most developing countries the telecommunication system is not adequate even to sustain essential services.

The commission noted that nine nations accounted for 75% of the 600 million telephones in the world: the United States, Japan, Germany, France, the United Kingdom, the Soviet Union, Italy, Canada, and Spain. It recommended that "by the early part of the next century virtually the whole of mankind should be brought within easy reach of a telephone, and in due course, the other services telecommunications can provide" [4]. The commission estimated that an annual investment of $12 billion was necessary to achieve even incremental progress in reaching that goal [5]. It asserted that profitability alone constituted an "inappropriate criterion for assessing the merits of telecommunication facility loans and investments, because indirect benefits have to be taken into account" [6].

11.1.1 Failed Strategies

Individual nations and international organizations have pursued different strategies to achieve improved telecommunications in lesser developed nations. Immediately following the Second World War, telecommunication infrastruc-

---

7. *The Missing Link*, Report of the Independent Commission for Worldwide Telecommunications Development, Geneva: ITU, 1984. Sir Donald Maitland chaired the commission created by the Administrative Council of the ITU as directed by the 1982 Plenipotentiary Conference. For a review of the commission's findings, see Rowan, F., and B. L. Waite, "International Communications Law, Part I: Maitland Commission, Economic Development and the United States," *International Law*, Vol. 19, 1985, p. 1339.

ture development assistance constituted a small part of the technology transfer rewards for aligning on either side of the East-West political axis. By the late 1960s, "many Third World nations began to see Western dominance of the international economic and communication system as a source of their 'underdevelopment'" [7].

To bring about greater parity to the international environment, Third World leaders have called for a New World Information Order (NWIO),[8] an environment where nations of all types would have relative parity of access to telecommunication and information resources. While the concept has varied in both definition and application among advocates, many strategies for achieving parity of access were perceived by officials in developed nations as threatening commercial transactions involving intellectual property and even freedom of the press in the face of an authoritarian political regime.

Scholars have identified several new approaches that avoid the stridency of calls for absolutely free access in a short period of time. New viewpoints embrace "grassroots or participatory communication [and political] policies for achieving economic growth" [8], "shaped by...the widespread proliferation of new information/communication technologies" [9]. In application, this means that because development funding will remain inadequate, countries in need must achieve greater self-reliance based on available funding and training while striving for more liberal technology transfer, new sources for loans, and a greater decision-making role in lending and development organizations such as the ITU. A World Bank technical paper notes a rise in entrepreneurship, a "trend deserv[ing] all the domestic and international support it can possibly get, because its strengthening promises to greatly contribute toward the 'normalization' of industrial and business conditions...[where] moderate incentives can be very effective" [10].

11.1.2 The Payoff From Telecommunication Investments

Telecommunications joins with energy and transportation as an essential component in social and economic development, according to [11]:

> The needs...[for a] telecommunication infrastructure to guarantee the
> success of any capital investment is very obvious. The relocation of
> developed countries' industrial estates to the developing countries

---

8. See M. Messmate, "The New World Information Order," in *Crisis in International News*, J. Richstad and M. Anderson, eds., 1981 p. 77. For a review of how the United Nations Educational, Scientific and Cultural Organization (UNESCO) addressed the subject, see F. Cate, "The First Amendment and the International 'Free Flow' of Information," *Virginia J. International Law*, Vol. 30, No. 2, 1990, pp. 373–420. For a review of the politics of transborder delivery of information, see M. Jussawalla and C. W. Cheah, *The Calculus of International Communications*, Littleton, Co.: Libraries Unlimited, 1987; and J. Rada and G. R. Pipe, eds., *Communications Regulation and International Business*, New York: North Holland, 1984.

will require a high quality and reliable telecommunication [network] for control and transfer of information between the new industrial location and the head office.

These projects usually turn a profit[9] and enhance social welfare as measured by statistical methods[10] and economic analysis, as noted in [12]:

> In a macroeconomic sense, the large unsatisfied demand for telecommunications services in developing countries and the high returns on new investment is evidence that the perceived communication needs of both producers and consumers are not being met. Not meeting this demand may worsen the extent to which information is distributed unequally between parties to transactions and may diminish the opportunity for information transfer.

Common sense and anecdotal information confirm the view that a poor telecommunications infrastructure can retard national economic development [13]:

> Inadequate telecommunications reduces efficiency throughout the economy, diminishes the effectiveness of investments in priority sectors and development programs, causes a comparative disadvantage in trade and in attracting investment, and lowers the quality of life in terms of personal access to emergency services and communication with kin and friends. In Uganda in 1983, for example, because of inadequate rural and provincial telecommunications and postal services, trucks collecting coffee and cotton from a large union of cooperatives made trips estimated to be 20 percent ineffective; also about 200 workyears were wasted in otherwise unnecessary administrative travel by senior government officials.

---

9. "The reason for inadequate investment in the telecommunications sector in developing countries is also not that telecommunication entities lose money or require government subsidies. In general, reasonably well managed telecommunications entities can generate large financial surpluses in local currency." (Saunders, Warford, and Wellenius, p. 12.)

10. Even anecdotal evidence confirms the view that properly targeted and managed investments in telecommunications will enhance efficiency and other investments throughout a national economy. "The introduction of long distance telephones in the Amazon region of Peru in the late 1970s resulted in substantial cost savings and increased revenue in river transportation; in another region in 1983, the use of telex for reservations increased average hotel occupancy in a tourist town from less than 50 percent to about 70 percent. In Sri Lanka in the 1970s, telephone access to market information allowed farmers to place their product at 80–90 percent of Colombo prices, as compared with 50–60 percent before; in the 1980s, Colombo food merchants relied on the telephone and telex for operations in international markets." (B. Wellenius, "Beginnings of Sector Reform in the Developing World," in Restructuring and Managing the Telecommunication Sector, p. 91.)

In response to allegations that one cannot quantify with any precision the economic stimulation accruing from telecommunications development, Sir Donald Maitland restated the question [14]:

[H]as economic and social progress occurred in any country without accompanying investment in the communications infrastructure? The evidence available to the Commission left us in no doubt that there was indeed a correlation between economic development and investment in telecommunications. Which was the chicken and which was the egg seemed an esoteric question.

Telecommunications development means more than expanding the number of telephone lines per 100 inhabitants. To become attractive to foreign investment and to operate in an integrated, global economy, nations must also deploy enhancements to POTS, such as high-speed data networks, cellular radio, mobile satellite services, Internet access, and facsimile. Developing nations face the same pressure to upgrade and diversify the telecommunications sector as developed nations, but are far less able to do so when even the basic infrastructure is incomplete, obsolete, or in disrepair.

To take advantage of the development and economic stimulation opportunities presented by telecommunications, developing countries must find solutions to:

- Limits on access to borrowed or self-financed capital;
- The goal of providing universal basic service, while at the same time installing state-of-the-art overlay networks, including cellular radio, digital microwave, and VSAT systems;
- Regulatory limitations on the pricing of services that may translate into mandatory underpricing of certain services;
- Compulsory contributions of telecommunication revenues into the general treasury;
- Institutional limitations that generate inefficiency, lower productivity per employee, and higher costs per unit of output;
- Limits on access to equipment, resulting from domestic trade and procurement policy, or "tied-aid," linking loans and other financial assistance with the obligation to buy equipment only from suppliers in the nation providing the aid;
- The frequent requirement to pay for at least a portion of any procurement with hard currency;
- The diversification of constituencies (e.g., residential versus corporate) and interests (e.g., universal service versus business networks) mandating a new regulatory and management structure that promotes fairness.

11.2   A NEW STRATEGY: PRIVATIZATION AND LIBERALIZATION

Some developing nations have viewed public ownership as a constraint on development rather than a guardian for universal service, social equity, reasonable rates, and growth. Privatization and liberalization present an opportunity for the infusion of more capital, improved productivity, and enhanced efficiency. But selling the concept and maneuvering it through the political process has proved daunting. The legislature must confront the consequences of significant reductions in the number of individuals employed in telecommunications, an industry typically unionized and often an integral part of what Eli Noam has termed the "postal-industrial complex," a coalition of the PTT, its domestic equipment supplier, residential and rural customers, the political left, the newspaper industry, which benefits from rate subsidies, and affiliated experts [15].

In revamping the operational and regulatory structure of the telecommunications sector,[11] governments are tempted to view privatization as a one-time infusion of capital into the public treasury. Support for competition in telecommunication services, rather than maintaining a PTT monopoly, requires governments to acknowledge the potential for greater uncertainty and a bigger role for the private sector, quite probably including foreign players. In the absence of effective regulation and contractual commitments, the loss of government ownership and monopoly control may jeopardize lucrative revenue streams needed to finance infrastructure development projects. On the other hand, privatization can accelerate infrastructure development with an infusion of capital typically from foreign enterprises. In granting the telecommunication franchise, many nations have imposed infrastructure improvement benchmarks (e.g., an increase in the number of telephone lines by a future date), which if not reached will translate into financial forfeitures.

Privatization and liberalization arise as policy solutions principally because the state has failed to provide reliable ubiquitous service and it can no longer ignore consumer complaints or the economic consequences of underdevelopment. "The telecommunications departments of many developing countries have become bloated bureaucracies characterized by inefficiency and unresponsiveness" [16]. These departments often cannot support facilities expansion with existing cash flows even by charging IMTS rates well in excess of cost. "[S]urpluses from telecom operations are frequently diverted to fund postal operations and electric utilities, which often fall under the same ministry, or go toward funding general government expenditures" [16].

---

11. Rather than simply defer to the PTT, the government typically creates an expert regulatory agency capable of implementing new legislation, providing for interconnection with facilities of the incumbent carrier by unaffiliated market entrants, managing spectrum, addressing the question of foreign ownership and national security, overseeing the terms and conditions for service, and ensuring that public policy goals like universal service are pursued.

Simply changing the ownership or corporate charter of the PTT does not guarantee efficiency and public dividends. Governments increasingly deem it necessary as well to stimulate competitive behavior by linking privatization with revamped regulation. To moderate the influence of incumbent carriers and elected officials, governments have modified laws to create an independent regulatory agency authorized to monitor the performance of the privatized telecommunications carrier, regulate rates, enforce quality-of-service standards, and ensure that revenues are fairly shared among service providers. Governments establish liberalized policies to reduce or eliminate barriers to market entry by competitors *and* accord the incumbent carrier, no matter who owns it, greater flexibility to operate under changed and more competitive circumstances.

11.2.1 Will Privatization and Other Initiatives Serve the Interests of Developing Countries?

In most developing countries, telecommunication services evolved "under the auspices of government departments, run by civil servants under government rules and procedures" [17]. This industrial structure worked well enough when telecommunications was not viewed as an integral part of economic development. So long as elites and tourists had access to basic services and nonelites were not demanding improvements, a shaky system could get by, despite the burdens of inadequate investment, bureaucratic waste and delays, and the lack of properly trained staff.

Increasingly, policy makers have viewed telecommunications, which includes both basic telephone service and enhanced services like high-speed data, an essential ingredient for progress and enhanced welfare of the citizenry. But can a developing nation achieve progress simply by targeting telecommunications for expanded investment from the public sector, or should it resort to structural changes and pursue external capital markets and investors? The consensus opinion of decision makers supports reforms to a point, despite the likely opposition by the government-owned PTT.

Even the first step toward institutional change, separating telecommunications from postal operations, can generate virulent opposition in many developing countries. Telecommunication officials do not necessarily welcome this initiative, despite the fact that postal operations typically require cross-subsidization from telecommunication service revenues. A conventional PTT's management is often top-heavy with postal officials who oppose separating telecommunication operations.

Reporting on the situation in India, T. H. Chowdary notes the reluctance of telecommunication employees' unions to depart from their postal colleagues, based in part on the concern that fewer numbers mean a loss of clout with the legislature and executive branch [18]. Incumbent PTT employees worry that a corporate telecommunication service provider presents the allure of a taxable entity even though the government may have to offer investment credits and other favorable policies to secure bidders. Reduced tax liability could result at the same time as the company is restructured and it furloughs substantial numbers of employees. Advocates for privatization counter that some severance payments could be converted into shares in the new corporation and that a more efficient carrier would grow in time, thereby creating new employment opportunities. Some opponents to privatization question the potential for better service and expedited development of hinterland service. The counterargument refers to numerous examples in which private enterprises have operated profitably in either small or widespread geographical areas, with expanded line penetration and improved worker productivity as measured by such indexes as number of access lines per worker.

On the matter of access to capital, Mr. Chowdary notes that PTTs must vie with other government agencies for funds while corporations have direct access to capital through equity and debt offerings, the issuance of which provides a stimulus for efficiency. Mr. Chowdary refutes the view that private ventures will choose to serve only profitable markets by noting that privatization does not result in the elimination of regulations requiring universal service and preventing the corporatized telephone company from neglecting social objectives. To prevent the private company from price gouging and failing to improve performance, a nonexclusive license could be issued reserving the possibility for facilities-based competition in the future, a likely prospect for lucrative niche markets like cellular radio and packet-switched data services.

Corporatization and private ownership of shares in the telephone company need not constitute an option only for developed countries with mature networks. There is no ideal or threshold level of line penetration, social cohesiveness, and national wealth necessary to implement privatization. Mr. Chowdary concludes that [19]:

> [S]ervice provision by government department is not leading to the establishment of a mature network fast enough. Reserving this task for the government bureaucracy is impoverishing and frustrating needy customers [like the software industry in India that has a comparative advantage internationally but no access to a high-speed data network to download the product to foreign points], harming the country and widening the gap between the developed and developing nations. Technology's possibilities, individual enterprise, the skills of trained managers, and the savings of the public are all unused, wasted, and frustrated. The people of India and other developing countries can no longer afford to leave telecommunications to a civil service department.

International development organizations like the World Bank regularly consider lending for telecommunication projects that may cost in the range of $1,000 per installed line, but may generate commercially inadequate annual returns on investment, at least in the short term. "This is obviously not the sort of business anyone would want to be in. This however, is just the sort of arena in which…[the World Bank] and other donors must be involved" [20]. Hence, corporate financing of telecommunications can never be expected to fund all necessary and worthwhile projects. Financial aid, tied to equipment purchases and with an eye toward future growth, will certainly stimulate some corporate interest. But the vast majority of telecommunications development in the most needy nations will occur because development organizations exist to consider more than internal rates of return.

These organizations prioritize projects based on their perceived ability to stimulate economic growth and eventually to lead to a self-sustaining telecommunications sector. While respecting the sovereignty of recipient nations, funding organizations have to provide the kind of advice needed by developing nations in deciding where to allocate financial aid and how to make informed policy decisions. This requires ongoing assistance in technical, regulatory, and operational matters.

11.3 THE ROLE OF THE ITU

The 1989 Plenipotentiary Conference of the ITU recognized the need for greater emphasis on development assistance and created a new permanent organ: the Telecommunications Development Bureau (BTD).[12] The BTD operates in an environment where the developing world still lacks access to basic telecommunications capabilities and where the disparity grows between developed and developing nations in terms of overall access to telecommunication equipment and services. On the other hand, as the dichotomy of technology access worsens between developed and developing nations, the ITU's one-nation, one-vote structure has motivated developing nations to seek more attention to the problem and a commitment of more funding by developed nations.

International development issues tend to arise in a crisislike atmosphere, replete with acrimony and recriminations. The 1982 Plenipotentiary Conference became nearly captive to political issues related to development and equitable access to the satellite orbital slots.[13] At the behest of developing nations, the conference amended the ITU Convention, obligating the organization to in-

---

12. See Constitution of the International Telecommunications Union, Article 14, reprinted in *Final Acts of the Plenipotentiary Conference*, Geneva: ITU, 1989, pp. 17–19. For a summary of the proceeding leading to formation of the BTD, see B. Harris, "The New Telecommunications Development Bureau of the International Telecommunications Union," *American University J. In-*

itiate efforts aimed at serving the special needs of developing countries.[14] The 1982 Plenipotentiary became a politicized forum for various articulations of what constitutes special needs and the scope of obligations developed nations should assume to provide favorable treatment to their less developed counterparts.

In a less politicized environment, the BTD serves as a legitimate forum to address development issues, including access to capital, technical and managerial training, education of policy makers in the importance of telecommunications, and private industry's participation in telecommunications financing. The ITU Convention authorizes the BTD to convene world and regional conferences to address such issues.

References

[1]   De Sola Pool, I., and E. Noam, eds., "Communications for Less Developed Countries," in *Technologies Without Boundaries—On Telecommunications in a Global Age*, Cambridge, MA: Harvard University Press, 1990, p. 177.
[2]   Saunders, R., J. Warford, and B. Wellenius, *Telecommunications and Economic Development*, Baltimore: Johns Hopkins University Press, 1983, p. 67.
[3]   Report of the Independent Commission for World Wide Telecommunications Development, *The Missing Link*, Executive Summary, Geneva: ITU, 1984, p. 3.
[4]   Ibid., p. 4.
[5]   Ibid., p. 13.
[6]   Ibid., p. 7.
[7]   Ayish, M., "International Communication in the 1990s: Implications for the Third World," *International Affairs*, Vol. 68, No. 3, 1992, p. 487 (hereafter cited as "International Communications in the 1990s").
[8]   Jussawalla, M., "Is the Communications Link Still Missing?" *Telecommunications Policy*, Vol. 16, August 1992, p. 485.
[9]   "International Communications in the 1990s," p. 488.
[10] Ivanek, F., T. Nulty, and N. Holcer, *Manufacturing Telecommunications Equipment in Newly Industrialized Countries: The Effect of Technological Progress*, World Bank Technical Paper No. 145, Washington, D.C.: World Bank, 1991, p. 31.
[11] Soerjodibroto, T., *ASEAN and World Telecommunication Development*, paper presented at PTC'92, Honolulu, Hawaii, p. 1.
[12] Mendis, V., "Phased Privatization with Proposed Foreign Participation: The Sri Lanka Experience," in *Restructuring and Managing the Telecommunications Sector,* B. Wellenius, P. Stern, T. Nulty, eds., Washington, D.C.: World Bank, 1989, p. 103.

---

13. For a summary of the 1982 Plenipot, see Probst, S. E., "The Plenipotentiary Conference of the International Telecommunications Union, Nairobi, 1982: A Summary of Results," *American Society of International Law and Practice*, Vol. 77, 1985, p. 354.

14. See, for example, *Id.* at Recommendation No. 2, "Favourable Treatment for Developing Countries," at 343–345 (recommending that "developed countries should take into account the requests for favourable treatment made by developing nations in service, commercial or other relations in telecommunications, thus helping to achieve the desired economic equilibrium conducive to a relaxation of present world tensions."

[13] B. Wellenius, "Beginnings of Sector Reform in the Developing World," in *Restructuring and Managing the Telecommunications Sector*, B. Wellenius, P. Stern, T. Nulty, eds., Washington, D.C.: World Bank, 1989, p. 90.

[14] Maitland, D., "'The Missing Link': Still Missing 8 Years Later?" *Proc. Seminar on Telecommunications and Its Role in Socio-Economic Development*, Eigtveds Pakhus, Copenhagen, 2–3 November 1992, p. 5.

[15] Noam, E., "International Telecommunications in Transition," in *Changing the Rules: Technological Change, International Competition and Regulation in Communications*, R. Crandall and K. Flamm, eds., Washington, D.C.: Brookings Institution, 1989, p. 258.

[16] Ambrose, W., P. Hennemeyer, and J. P. Chapon, *Privatizing Telecommunications Systems—Business Opportunities in Developing Countries*, International Finance Corporation Discussion Paper No. 10, Washington, D.C.: World Bank, 1990, p. 5.

[17] Chowdary, T. H., "Telecommunications Restructuring in Developing Countries," *Telecommunications Policy*, Vol. 16, No. 6, September/October 1992, pp. 591–592.

[18] Ibid., p. 593.

[19] Ibid., p. 600.

[20] Lomax, D., "Telecommunication Projects as Viewed by the World Bank," *Proc. Seminar on Telecommunications and Its Role in Socio-Economic Development*, Eigtveds Pakhus, Copenhagen, 2–3 November 1992, Tab 11, p. 3.

The Technology, Law, and Policy of International Satellites

12

The technical characteristics of satellites and their cost support widespread shared use among nations. When operating in an orbital location approximately 22,300 miles above the equator, satellites move at the same rate as the Earth, thereby presenting a stationary target for signals uplinked to them by Earth stations.[1] From this geostationary location, a single satellite can illuminate about one-third of the Earth, although operators often concentrate the signal to cover smaller regions or single nations.

Because the footprint of a satellite typically covers a wide geographical area, carriers can provide service throughout a region. A new point of communication can be added at low cost, making it possible to spread the cost of satellite service over a number of different routes of different distances and traffic densities. This insensitivity to distance and number of receiving points means that, with proper coordination among governments, carriers, and users, satellites have the potential to achieve three important outcomes:

1. Nations without extensive demand can share access to a single satellite constellation, because once a satellite is deployed to meet dense route requirements, it can also provide sparse route service within the geographical area of coverage.
2. Satellites can provide point-to-multipoint service, such as widespread distribution of a video program to a number of broadcast and cable

---

1. A satellite in geostationary orbit "is not exactly stationary. Rather, it moves in a figure-eight pattern...[requiring] [s]tation-keeping maneuvers...to maintain its nominal position. With current technology, a satellite can be maintained within 0.1 degree of its nominal orbital location on the equatorial plane." (M. Smith III, "The Orbit/Spectrum Resource and the Technology of Satellite Telecommunications: An Overview," *Rutgers Computer & Technology Law J.*, Vol. 12, 1987, pp. 285, 287.)

television outlets, at roughly the same cost as a single point-to-point transmission.

3. Lesser developed nations can achieve satellite access to the rest of the world with minor incremental cost, primarily in ground segment facilities like Earth stations, provided an administrative mechanism exists that allows access by carriers with little or no investment in the satellite.

Satellites provide a primary medium for modern, widespread international telecommunications. While submarine cables provide a conduit for communications between points in relatively close proximity to the coastline, satellites provide the basis for communications between and among any point under its footprint. Accordingly, no single location or route of traffic needs to generate the level of volume needed to justify investment in a dedicated link provided by cables. With the deployment of at least three satellites, one in each of the major ocean regions of the world (Atlantic, Pacific, and Indian) virtually any point on Earth can access any other point. With the introduction of satellites, access to the rest of the world has been made possible for users in remote regions far interior from the coast where a submarine cable makes its landfall and far away from urban corridors with more robust traffic volumes.

12.1 TECHNICAL AND LOGISTICAL FACTORS IN SATELLITE USE

Satellites make ubiquitous telecommunications feasible even as various regions of the world present different requirements and operating conditions. That basic concept often gets obscured when international conferences of the ITU approach gridlock due to strongly differing positions on what spectrum allocations to make and which standards to adopt. Reaching consensus on satellite policies requires special effort because of differing impact on nations, service providers, and end users from technological innovations, climate, geography, financial resources, and market demand.

12.1.1 Key Factors Shaping Regional Differences in Satellite Use

Regional differences in the type of satellite used and the services provided involve much more than whether users must avoid using higher frequencies due to the potential for heavy rainfall to weaken signals. A more complete analysis considers such factors as:

- *Geography and climate* (e.g., a nation's location relative to any particular satellite's *boresight*, the location of its strongest signal);

- *Technology* (e.g., *beam types* and *look angles* of accessible satellites: the ease in pointing a satellite dish above the horizon to receive a usable signal);
- *Politics* (e.g., whether a nation permits satellite news gathering, use of transportable dishes, and reception of direct-to-home satellite broadcasting);
- *Spectrum use and management* (e.g., the extent of terrestrial radio use and interference potential);
- *Marketplace factors* (e.g., consumer demand, demographics, and financial resources available for satellite deployment or transponder leasing);
- *Regulation* (e.g., policies regarding the number of carriers, use of portable terminals, tariff restrictions on end-user access to space segment, and extent of carrier flexibility to satisfy consumer requirements).

12.1.2 Geography and Climate

Geography and climate play an important, if perhaps overemphasized, role in satellite planning and development. Rain can weaken satellite signals, and frequently rainy areas are better served from powerful satellites operating in the C-band—6-GHz transmission uplinks to the satellites and 4-GHz transmission downlinks from the satellite—rather than the more recently activated Ku-band (14-GHz uplink and 11-GHz downlink). Geography, or more specifically population distribution and economics, primarily affects satellite carriers' decisions on which frequencies to use and what locations to serve.

Even established regional and global carriers must balance their mission to achieve widespread access with a pragmatic assessment of how much traffic each route will generate. Stronger and more numerous beams are available for transoceanic routes north of the equator to link developed nations (e.g., the United States to Japan and Europe) that already have generated reliably high demand for service. Such a strategically necessary response results in reduced signal strength for satellite beams accessible from developing countries in equatorial and southern hemisphere locations.

Even as equatorial countries develop economically and nations south of the equator begin to receive access to dedicated international satellite capacity, the vestige of earlier conditions and an installed base of equipment operating on lower frequencies have slowed migration to higher frequencies, irrespective of rainfall totals. While a nation's physical location and climate matter, other factors play an equal or more significant role in the equation.

12.1.3 Technology

Satellite technology development tracks the wider evolution of radio as innovators find cost-effective ways to make use of higher frequencies. The onset of solid-state electronics and, more recently, very-large-scale integrated circuit chips has enabled carriers to meet consumer demand by constructing satellites that operate on higher frequencies. Carriers can conserve the scarce orbital arc by positioning two or more satellite, operating on different frequencies, in the same orbital location. Consumer welfare-enhancing innovations include interconnection between frequency bands (cross-strapping), onboard satellite circuit switching, steerable spot beams to concentrate signal strength as demand dictates, higher powered signals, circuit multiplication through compression, and other capacity-conserving technologies and intersatellite links (i.e., interconnection between two or more in-orbit satellites).

Because consumer requirements are not equally dispersed throughout the world, satellite planners must deploy satellites with features when they are needed and when a user community can afford such innovations. The developed world, with a longer history of satellite use and greater aggregate traffic volumes, typically has initial access to such innovations. However, new generations of satellites may have little value for secondary markets, since the signals at higher frequencies lose strength more quickly as distance increases from the boresight. Likewise, developers of such satellites are not likely to concentrate signal strength for service to secondary markets through spot beams. Users in countries that cannot afford the latest technological innovations (e.g., most of Central and South America and Africa) may find that they will have to make do with larger hemispheric and zone beams of older satellites. Cutting-edge technologies will first serve high-volume routes with an established, sophisticated, and financially qualified user base. Satellite news gathering, direct-to-home broadcasting, teleconferencing, telemedicine, computer file transfer, and other evolving services will first acquire market share in nations with the financial wherewithal to sponsor technological innovation.

12.1.4 Politics

National regulatory policies toward satellites can promote or retard use, demand, and availability. Most nations, with the United States a key exception, have attempted to confer a satellite service monopoly and to restrict available options. Reserving a satellite service monopoly constitutes part of the PTT model and the view that a single enterprise can achieve scale economies while providing essential services ubiquitously. While some nations (e.g., the United Kingdom, Canada, Australia, and Japan) have liberalized policies regarding access to satellite capacity, virtually all nations retain a monopoly for basic switched satellite services.

Growing interest in privatization and liberalizing regulatory policies may encourage nations to rethink the merits of a single monopoly. The cooperative model in international satellite telecommunications has achieved nearly ubiquitous access to space segment at globally averaged rates and the opportunity for developing countries to qualify as a participant with less than a $1 million investment. However, the model may allow monopoly attitudes to persist despite changed circumstances that might support some degree of facilities-based competition. Decision makers have to balance the view that the marketplace cannot support multiple systems that would collectively serve all user requirements with concerns about market failure and the inability or unwillingness of commercial enterprises to serve sparse and probably unprofitable routes and users.

Additionally, some nations may have concerns that private market entry will lead to orbital slot congestion, either by in-orbit satellites or by "paper satellites" in the ITU registration process. An open skies policy may expand the scope of opportunities for people to receive news and to communicate with the rest of the world, an outcome some governments might view as stimulating political instability. Stifled market development and retarded demand result from restrictions on the operation of Earth stations, whether to receive television broadcasts or to transmit internationally. Such restrictions occur for a number of reasons running the gamut from aesthetics, with urban zoning restrictions, to concerns for political stability, with attempts to prohibit individual ownership and operation of even receive-only Earth stations.

Satellite-delivered messages do get through and they can have a significant social impact. Relatively easy reception of commercial satellite television broadcasts may have contributed to the disintegration of the Soviet Union and its bloc of client nations in Eastern Europe. Uncensored views of the West, particularly its advertised products and the perception of wealth and opportunity, may have stimulated the desire for access and freedom.

Even now, many nations have restrictive policies. For example, China has periodically imposed a ban on any residential satellite television reception, and Indonesia has limited receivers for direct-to-home satellite broadcasting to C-band reception, presumably to support its Palapa system. Conservative members of the EU (e.g., Italy and France) have vigorously opposed policy initiatives of the European Commission to liberalize satellite policies, including the elimination of restrictions on use of receive-only Earth stations and user choice of satellite carrier.

Pricing policies can also retard demand, even as a larger percentage of the population discovers the versatility and potential for satellite use. The Gulf War in 1991 provided a case study of the potential for mobile satellite communications to provide access from remote locations. Live news feeds from Baghdad informed us of developments on a real-time basis, even as television networks and interexchange carriers paid exorbitant fees to the Saudi Arabian government for authorization to install and operate Earth stations to provide satellite access for television networks and for soldiers to call home.

12.1.5 Spectrum Use and Management

Spectrum planning and the frequency allocation process at the ITU slowly respond to consumer requirements and frustrations. WRCs painstakingly study and consider the need for and consequences of reallocated spectrum for existing and new services. While technological innovations make it feasible to use existing allocations at higher frequencies, such as the Ka-band (30-GHz uplink, 20-GHz downlink), the ITU's allocation of all spectrum means that incumbents must share allocated spectrum with newcomers and run some risk of losing bandwidth to other, more compelling spectrum requirements.

The slow, incremental process of spectrum reallocation means that new decisions typically build from prior actions and preserve the status quo. When allocating more spectrum for satellites, ITU conferences usually address expansion bands adjacent to existing allocations. In application, this means that the number of satellite bands remains relatively small, with the available bandwidth expanding only somewhat over time.

The volume of terrestrial traffic and potential for interference serve as the primary drivers for frequency band migration. When microwave congestion and Earth station installations reach a point in urban locales at which users cannot install additional C-band Earth stations with a certainty of noninterference, the migration to higher Ku-band frequencies will occur. The convenience of smaller and less bulky terminals constitutes a favorable by-product of a necessary and expensive migration to another frequency allocation.

Migration to the Ku-band also occurs when developed nations exhaust the supply of orbital slots available for providing optimal footprint coverage. Ku- and Ka-band satellites collocated with C-band space stations can double and triple the number of in-orbit resources. This provides a solution to complex parking place negotiations between nations already occupying several slots and adjacent countries with later-evolving satellite requirements and a keen national interest in having their own satellites.

Without such a safety valve, along with regulations of some nations requiring closer spacing of satellites, the ITU would face even greater pressure from developing countries to achieve equity in orbital slot usage by reserving space for future use or sale. Except for certain expansion frequencies, the ITU still relies on a first-come, first-reserved, a priori basis for orbital slot usage. The reservation of 31 satellites in 26 orbital slots by the Kingdom of Tonga, an archipelago nation of 100,000 inhabitants in the South Pacific, and the subsequent transfer for compensation of some slots to carriers from other nations, present another challenge to the status quo. Developing nations have begun to realize the value in staking claims to a scarce resource previously reserved only if a near-term requirement can be demonstrated. If such a trend were to develop, proliferating orbital slot registrations would tax the ITU's resources and likely lead to more expense, controversy, and disputes over who has a rightful claim to a shared global resource.

12.1.6 Marketplace Factors

Despite efforts to ignore or blunt marketplace factors, consumer requirements and financial resources continue to shape the industry. Some nations readily support market-oriented allocations of the orbital arc and determination of the number of carriers. For example, the FCC has articulated an "open skies"[2] policy to encourage all technically, legally, and financially qualified applicants to provide satellite services. On the other hand, many nations view the expense, risk, and cross-border capabilities in satellites as supporting consolidation of resources through a domestic monopoly and through global or regional cooperatives owned by a number of carriers. At the individual user level, particularly in developing nations, one may wish for newer and more versatile functionality, but financial exigencies may deem otherwise. The prevalence of C-band networks in certain equatorial nations may have less to do with climate and more with financial or technology transfer limitations that necessitate second-hand equipment purchases, or with extending the usable life of antiquated equipment.

12.2 STRATEGIC PLANNING IN SATELLITE DEVELOPMENT

The commercial communications satellite marketplace has evolved from a comparatively low-capacity, high-cost medium to one that is price-competitive with submarine carriers and just as versatile. We can no longer consider this industry as the one-by-one manufacture of a few dozen designer satellites a year with services provided by a few predominately government-owned carriers participating in global or regional cooperatives. Instead, a variety of developing secondary markets present the prospect for service diversity and perhaps even greater price competition.

The potential exists for off-the-rack satellite manufacturing rather than the one-at-a time method currently used. Likewise, developments in launch technology may make it possible to secure quick response scheduling rather than requiring an operator to join a cue that may exceed two years. In the near term, mobile satellite service ventures anticipate launching dozens of assembly-line-manufactured satellites to form a low-Earth-orbiting (LEO) constellation.

---

2. Domestic Communications Satellite Facilities, 35 FCC 2d 844 (1970).

12.2.1 Lost Market Share or New Opportunities?

Some incumbent satellite manufacturers may regret lost opportunities to sell new, high-capacity space stations. Likewise, incumbent carriers are not likely to welcome even more competition. Manufacturers may long for the time when satellites had single owners who gave no thought to subsequent resale or life-extending maneuvers and who had no option to consider quickly deploying a gap-filler, smaller capacity satellite to satisfy immediate consumer requirements. Satellite operators may wish they did not have to narrow profit margins in the face of the potential for lost market share.

Such nostalgia parallels the dominant thinking in the domestic U.S. telecommunications arena of 20 years ago. At that time, AT&T could prevent the creation of secondary markets for used Western Electric equipment, because the captive Bell Company customers dutifully returned equipment for centralized salvaging. The AT&T-managed Bell System provided end-to-end functionality with limited competition. Such vertical integration never has dominated the satellite marketplace, nor has brand loyalty seemed particularly strong in this industry.

A more productive response, like that displayed by INTELSAT in the face of increasing satellite service competition, is to view changes as healthy and reasonable market evolution. New opportunities and more market segments have arisen, even as more players enter the game. Collectively, a larger set of players promotes a healthier industry, particularly in light of mergers and acquisitions that have concentrated market share.

12.2.2 New Satellite Services and Market Segments

The proliferation of satellite service options may result in fewer new, high-capacity, multipurpose satellites and in more specialized systems. Such new roles include satellites being:

- Relocated from one orbital slot to another and perhaps from one owner and user base to another;
- Retrieved by a space shuttle from a useless orbit due to launch failure;
- Able to provide discounted service to users who can track a moving target with little remaining station-keeping fuel;
- Streamlined in size, weight, and capacity for quick deployment into low Earth or geostationary orbits;
- Renovated and relaunched;
- Privatized and transferred from government to private ownership;
- Launched to provide gap-filling service to meet capacity requirements that cannot be met by smaller satellites with capacity and cost less than conventional 36-transponder satellites.

These new and reconstituted options speed the development of market segments and a more diversified set of services. But almost from the outset, carriers and manufacturers have provided some degree of product differentiation and price options. Risk adverse customers have paid premium rates for noninterruptible service backed up by protection transponders available for restoration in the event of outage. These users can tolerate no service interruption, regardless of whether their transponder or the complete satellite fails. On the other side of the spectrum, price-sensitive users have agreed to take unprotected, preemptible service in exchange for substantial discounts. These users will consider market entrants, even ones lacking their own in-orbit satellites available to restore service in the event of an outage.

The open skies policy in the United States stimulated facilities-based competition when the satellite industry was in its infancy. Later, the United States advocated transborder use of domestic satellites and conditional international competition from private ventures. The new options presented here identify some of the changes in satellite design, functionality, usable life, and ownership that existing and new carriers will consider when serving an increasingly diverse set of users.

12.2.3 Price as a Function of Both Cost and User Risk Profile

Diversification in user populations means that a new or expanding group will demand vastly discounted service, but will accept a higher risk of service outages and even discontinuation of service from a particular satellite. Satellite operators offer such inferior service based on the determination that it will not result in migration by existing users to more expensive, higher quality service. Such service cannibalization will not likely occur when users, like television broadcasters and telephone companies, have absolute and continuous service requirements and service providers impose substantial termination charge liability for premature discontinuation of long-term service arrangements. Additionally, there is a perception by some that secondary market services represent damaged or inferior goods for use only by those customers who would otherwise physically transport databases and videotapes, or who could not afford better and more expensive options.

Second Leases on Life

Traditionally, satellites had the same owner/operator for the life of the system. Heightened responsiveness to consumer requirements may result in the launch of new satellites, even though a prior generation may still have many years of remaining life. While some carriers may not differentiate the aging satellite capacity, others may sell them to nations unable to afford a new satellite, but quite anxious to acquire in-orbit (i.e., existing) capacity. Carriers holding onto aging capacity may reposition the satellites into a different orbital slot to:

- Accommodate demand;
- Free up a preferred orbital slot for a new satellite;
- Adjust for inaccurate projections or satellite/launch failures.

When carriers transfer ownership of satellites, a new secondary market develops for used satellites, and presumably ITU regulations can accommodate the transfer of an orbital slot registration from one nation to another. The satellite may cost no more than the original investment, less depreciation to reflect the remaining usable life. But it might fetch a premium based on a buyer's keen desire for readily available, in-orbit capacity and an aversion to multiyear construction schedules, launch queues, and other risks. The selling of previously owned satellites is evidence of a maturing and diversifying market.

New Options and New Leases on Life

Satellite operators now more actively consider ways of squeezing out longer life and higher revenues from their assets. Such conservation efforts now extend to finding innovative ways of salvaging satellites that previously would have been written off as space junk. For example, the AsiaSat-1 satellite had a previous brief and unsuccessful life in a useless orbit. Space shuttle astronauts retrieved the satellite for reconditioning on Earth. The satellite was subsequently relaunched. An INTELSAT VI satellite got a new lease on life courtesy of the Shuttle Endeavor's heroic crew who retrieved, retrofitted, and reboosted the satellite into proper orbit without first returning the satellite to Earth.

Government Exits a Market, Privatizes It, or Expands Its Market Horizons

A variety of new opportunities for market access result when a government owner/operator decides to exit a market, allow a private venture to resell capacity deemed surplus, or expand its target market in search of hard currency. Columbia Communications got the opportunity to enter the international satellite business as a result of the reluctant decision by NASA that commercial operations constituted an activity outside the scope of its mission, despite having contracted for the construction of 12 C-band transponders on each of two tracking and data relay satellites. PSN (PT Pasifik Satellite Nusantara) of Indonesia presents a variation of the NASA model: the creation of a private venture to market the remaining usable life of an entire satellite (Palapa B-1), previously owned and operated by the government. Several new private ventures have emerged to participate in the construction or marketing of satellites in the Commonwealth of Independent States.

Squeezing Extra Use out of a Satellite

A new deep discount market segment has evolved for satellites in nonstationary orbits. These satellites, sometimes referred to as *wobblesats*, have exceeded their estimated usable life, but can achieve life extension through conservation of remaining station-keeping fuel. The satellite operating in such "inclined orbits" present a moving target and therefore require Earth-station tracking.

Because these satellites have an uncertain remaining life, carriers must provide attractive financial inducements. The satellites typically serve a developing market niche composed of very sophisticated users capable of investing in the management of satellite-tracking Earth stations or users who might not otherwise have the financial wherewithal to use satellites without a deep discount. These satellites also enable major incumbent operators like INTELSAT to maintain an adequate transponder inventory in the face of launch failures or underestimated demand. Likewise, they allow incumbent carriers to maintain the priority right of occupancy at the orbital slot.

Smoothing Out Capacity Lumpiness/Gap Fillers

A new market has evolved that matches a manufacturer with an extra or partially constructed satellite, and an operator with unsatisfied user requirements. Typically, satellites are customized to meet particular geographical and service requirements. However, early on in the construction process, a satellite can be reconfigured to satisfy the different requirements of a new or replacement customer. INTELSAT jumped at the opportunity to satisfy unmet demand in the North Atlantic region for Ku-band satellite capacity when it secured the option to acquire a partially constructed satellite no longer on order.

Another market niche results from the match between an operator experiencing a capacity glut and one with new or additional requirements. Satellite capacity is deployed in large increments and carriers must build into its procurements a number of satellites to account for launch failures that on average occur 20% to 25% of the time. Should an operator overestimate demand, merge with a competitor, or beat the statistical odds on launch failures, it may experience an inventory glut. Just the opposite may occur when an operator underestimates demand, experiences delays in construction and launches, or suffers particularly bad luck in the launch of its satellites. It reasonably follows that a market should evolve to match operators with different levels of unused inventory.

New Product and Service Lines

A maturing satellite industry promotes new options and market niches in satellite manufacturing. Small-satellite manufacturers can provide lesser developed nations with an opportunity to own and operate a lower cost, smaller capacity network. These manufacturers also furnish a smaller capacity alternative to incumbent operators facing an immediate, but low-volume, capacity shortfall that can be solved on a more timely basis by the launch of a small satellite with fewer than 36 transponders.

Mobile satellites, operating in nongeostationary LEO only a few hundred miles above Earth, present the prospect of ubiquitous mobile satellite service via handheld terminals. The proliferation of both LEO and small satellites will require new ITU procedures to coordination usage between types of satellites and between satellite and terrestrial users. However, consumer demand for mobile telecommunications, as shown by the unprecedented ramp-up of cellular radio service, should put pressure on governments to reach consensus. Likewise, the demand for mobile satellite service may stimulate development of new launch options, particularly ones capable of deploying several satellites at a time and able to launch replacement satellites in a matter of days.

12.3 THE LIGHTSAT ALTERNATIVE IN AN INCREASINGLY CONGESTED ORBITAL ARC

Increasingly fractious orbital slot battles create the potential for a geostationary orbiting *lightsat* alternative to conventional satellites. These smaller and lighter satellites operate with fewer transponders, but can be manufactured to operate on frequencies that will not cause interference to a planned or already operating full-capacity satellite.

The orbital slot registration process, managed by the Radio Regulation Board of the ITU has become more difficult and time-consuming. The process was not designed to ration slots or to assign them with marketplace considerations in mind. By agreeing to participate in a single international forum, nations have sought a mechanism for ensuring interference-free satellite operations. But the effectiveness of this process falls when nations or operators cannot begin operations on a reasonable schedule and have to make costly compromises in terms of geographical coverage, signal strength, and operational frequencies. The matter grows worse when some players take matters into their own hands by operating satellites in unregistered slots and boldly refusing to comply with treaties accepted by their governments. Others flout the unwritten rules of the road by converting the process into a speculative market instead of a process for avoiding interference between operators with actual satellites to launch and real markets to serve.

Historically, nations and satellite cooperatives have recognized that the process can work only if satellite operators limit their orbit registrations to a reasonable level based on realistic expectations of need. Recently, however, many incumbent operators have perceived strategic advantages in warehousing orbital slots (i.e., reserving orbital slots well in excess of their reasonably antici-pated future requirements). Even if an expanded satellite constellation makes marketing sense and serves growing market requirements, it places increasing pressure on the registration and interference avoidance process.

Others, primarily prospective satellite operators, consider the orbital slot reg-istrations an opportunity for entrepreneurialism or the assertion of national sover-eignty. The cumulative effect of these new motivations is a rise in the number of Advanced Publications of satellite deployment plans and a commensurate increase in the complexity, acrimony, and time consumption of the process.

The aggregate demand for orbital slot registrations has reached a point where few proposed uses can go through the registration process unencumbered and on a timely basis. If prospective registrants cannot achieve a priority right of access to an orbital slot, their business plans may suffer as ready-for-service dates pass and market access opportunities dwindle. So far, most satellite op-erators have chosen to fight the battle in recognized domestic and international forums (e.g., the FCC and the ITU). But the potential exists for more disputes over orbital slot registrations as has occurred for Russian, Chinese, Indonesian, and INTELSAT satellites. Likewise, the risk increases that carriers will launch unregistered "rogue" satellites that challenge the legitimacy of the overall or-bital slot registration and coordination process. Contested satellite registration plans hold the risk that a venture—and the venture's chief asset—cannot even get off the ground.

12.3.1 A New Look at Lightsats

Geostationary lightsats have failed to capture market share primarily because of the view that they present a less attractive and more costly alternative. This less attractive perception stems in large part from optimistic business plans that pro-ject transponder requirements beyond the 12 or so a typical lightsat can provide. Satellite operators want to deploy the region's "hot bird," with the goal of hav-ing most Earth stations pointing to their satellite, thereby supporting premium lease rates. Hot birds typically sell out of transponders at or before launch, and the lucky operator wishes it had more in-orbit inventory to sell. Under this ro-bust market scenario, lightsats may not provide sufficient transponder capacity.

The counterpoint to this scenario is the more likely instance in which no single satellite or operator dominates the market. This more realistic and di-verse environment requires more than a one-size-fits-all type of satellite. Hot birds tend to exist only in the video programming market segment. Most satel-lite operators have to start with few if any prelaunch transponder lease agree-ments, and even incumbent operators run the risk of market cannibalization rather than new leases when deploying an additional satellite.

Recent INTELSAT satellite deployment plans support the view that the satellite marketplace should be considered an amalgam of several different market segments. INTELSAT initially deployed "plain vanilla" satellites in middle ocean orbits, primarily to serve east-west, transoceanic applications. The cooperative appeared less interested and equipped to serve intraregional and north-south applications until Pan American Satellite (PanAmSat) proved the viability of these markets in South and Central America. INTELSAT has now recognized the need to develop satellites configured for service to a particular region. It has also launched satellites designed to serve users within a particular land mass, as opposed to the traditional transoceanic role that links users in two or more locales separated by oceans.

Lightsats have also failed to secure market share based on the view that fewer transponders will result in a higher cost per transponder, particularly without a shared launch or lower cost launch option. Lightsat manufacturers need to find ways of achieving absolute transponder cost parity with *heavysats*. Failing that, they must secure cheaper launch options and identify transponder lease ramp-up scenarios, where a lower capacity satellite can just serve a customer's requirement, with larger satellites being too big to meet initial demand. Under such circumstances, a lightsat would provide the right inventory and a heavysat deployment would result in unused capacity and higher carrying costs.

12.3.2 The New Comparative Advantage for Lightsats

Not all satellites are the same when it comes to avoiding orbital slot registration delays, costly compromises, and denied or conditional access to the orbital arc. Lightsats have a comparative advantage over heavysats, because they can share an already occupied orbital slot without causing interference by operating on satellite service frequencies not used by the incumbent heavysat. With the proliferation of advanced publications by both incumbents and newcomers, it will become increasingly difficult for a heavysat operator to avoid an orbital registration battle.

The first-time registrant may have the equities in its favor, but the process emphasizes a first-filed, first-registered procedure. Accordingly, newly industrialized countries and those nations with late-developing transponder requirements face an increasingly contentious process. The most frustrated and vocal of developing country registrants may seek major procedural changes in the process, perhaps going so far as to seek a broader mandate for reserved slots, regardless of near-term requirements, to foster a proper balance of access.

Even without a change in the orbital registration process, lightsats will serve an increasingly important function: the expedited deployment of capacity for nations not yet in space and those with additional requirements. At some point, nations facing impediments to satellite access on account of finances or orbital slot registration problems may consider alternatives to the conventional heavysat option. Candidates include nations with unrealized heavysat aspirations, such as South Africa, New Zealand, and Argentina and regional alliances, such as the Regional African Satellite Communications System (Rascom) and the Condor Alliance of Andean Nations in South America, that have had to settle for leased services in lieu of dedicated satellites.

Lightsats may present more immediate, practical, and less costly solutions. They have the potential for avoiding lengthy launch queues. They can serve nations historically considered a thin route destination with the kind of signal strength and technical parameters (e.g., spot beams) typically reserved for high-density routes. As a lower cost option, lightsats present the potential for bringing to fruition ambitious satellite deployment plans that have languished from lack of capital.

12.3.3 More Than a Flag Carrier

Lightsats have financial and service justifications beyond the exercise in sovereignty that supports many unprofitable national airlines. INTELSAT again provides an instructive example. Its decision in 1993 to acquire three 12-transponder Russian Express satellites shows that it has recognized the merit in quickly deployed satellites with a limited purpose and a lower price tag, the very features lightsats have to offer.

At the low end, lightsats present a cheaper and appropriately reduced transponder inventory that can grow incrementally as requirements emerge and as the ability to pay improves. At the high end, lightsats present the first ever opportunity for just-in-time rapid deployment of capacity to meet unanticipated demand and underserved requirements.

12.3.4 Market Trends Favoring Lightsat Deployment

In addition to the increasingly contentious orbital slot sweepstakes, a number of marketplace developments favor a new look at lightsats. These include:

- More extensive deregulation and privatization that improve the prospects for satellites to serve pent-up demand with greater flexibility and fewer financial burdens such as the obligation to subsidize domestic telephone and postal service;
- New and expanded demand for services with broadband requirements, such as HDTV, file transfer, electronic data interchange, telemedicine, and teleconferencing;

- The ability of quickly deployed, limited-purpose lightsats to eliminate satellite capacity shortages for certain types of applications until such time as INTELSAT deploys more heavysat capacity;
- Expanded spectrum allocations for satellite services;
- Economic growth, particularly in Asia-Pacific;
- Liberalization that supports a conditional "television without borders" policy, particularly for transborder reception of proliferating video programming options;
- A "qualified open skies" policy favoring diversified satellite, Earth station, value-added network, and PTT competition.

12.3.5 Market Trends Moderating Lightsat Attractiveness

A number of factors can work against the lightsat alternative, including:

- The inability to achieve cost parity with heavysats on a per-transponder basis;
- The failure to convince operators of the need to consider how to prepare for the consequences of not having a hot bird to market and the commensurately slower loading of transponders with traffic;
- A PTT backlash to competition, leading to reregulation or less liberalization, resulting in fewer market access opportunities;
- Further innovations in digital circuit multiplication and compression, resulting in reduced aggregate demand;
- The availability of submarine cable networks to offer diverse routing even for broadband video applications;
- Further orbital slot warehousing that reduces the number and availability of satellite parking, thereby creating the perceived need to deploy the highest capacity possible per slot;
- A persistent economic downturn and trade protectionism.

Heightened concerns about the time and effort needed to register satellite orbital arc usage may improve the marketplace prospects for lightsats. Few incumbent or prospective satellite operators have considered lower capacity satellites. Perhaps this oversight stems from the lack of cheaper launch options and the optimism articulated by business plan projections necessitating the conventional 24- to 36-transponder inventory in heavysats. The lightsat alternative serves markets with less robust demand. It provides a just-in-time, incremental solution that conserves capital and avoids many contentious orbital registration scenarios.

12.4 FUTURE TRENDS

Telecommunication service providers and regulators increasingly recognize the need for responding to consumer sovereignty, free marketplace ideology, competitiveness, globalization, liberalization, and privatization of government-owned enterprises. They recognize that telecommunications plays a key role in a strategy for economic and social revival.

However, in the satellite arena, a number of factors work against total reliance on marketplace resource allocation as the vehicle to promote innovation, investment, and improved service: the multinational nature of the business, the shared use of a scarce spectrum and orbital arc resources, and the reliance on satellite revenues to subsidize postal and other favored telecommunication services. Consumers will have an increasingly significant impact, particularly in regions served by many carriers and in countries where customers can legally opt to do business with competitors of the incumbent national carrier.

12.4.1 Dismantling Service-Specific Spectrum Allocations

Consumer demands for flexibility in satellite use may create pressure on spectrum block allocations based on narrow service definitions. With the onset of portable, even handheld satellite transceivers, users will assume the right to communicate using the same terminal operating on the same frequency whether on land, aboard a ship, or in the air. However, international and domestic spectrum allocations have created service-specific constituencies that fight mightily to preserve existing allocations. While spectrum management should emphasize efficiency, it cannot avoid politics and the ability of special interest groups to protect existing allocations.

12.4.2 Responding to the Demand for Mobile Communications

Consumers have recognized that "tetherless" communications enhances efficiency and productivity. Wireless communications from terrestrial networks will include a satellite component, primarily from LEO satellites and more cheaply launched lightsats with lower capacity but also faster timetables for construction and launch. The scope of such reliance on satellites depends on the extent to which national carriers will permit usage and agree to provide interconnection with the PSTN. Whether satellites will play a major role in satisfying consumer demand for ubiquitous mobile communications depends more on regulatory and industrial policy and less on technological capabilities or the ability of entrepreneurs to raise funds for such ventures. Table 12.1 outlines the major global mobile satellite ventures.

Table 12.1
Comparison of Major Mobile Satellite Systems

Organization	Investors	Cost	No. of Satellites and Orbit	Description
Iridium, Inc.	Motorola; China Great Wall Industry; Krunichev State Research; Lockheed, Raytheon, STET, Sprint, Thai Satellite Co., Veba, and consortia in Japan, Africa, South America, Canada, India, Asia/Pacific, and the Middle East	$3.4 billion	66 in LEO	Satellites directly access each other via intersatellite links; system architecture involves time-division multiple access (TDMA) operating bidirectionally; expects to 2–3 million subscribers with global coverage by 1998
Inmarsat-P	Inmarsat, a cooperative of signatories from 77 nations; many of the signatories individually	Approximately $2 billion	Design not finalized, but likely to be 10–20 in an intermediate circular orbit	Inmarsat created a separate corporate entity to pursue land mobile service markets; while Inmarsat and many of its signatories have invested
American Mobile Satellite Corp. (AMSC)	Hughes, AT&T (McCaw Cellular), Mtel, Singapore Telecom, and publicly traded	$550 million	2 geostationary orbit satellites serving North and Central America	The FCC forced competing applicants to form a single consortium; the FCC granted AMSC a geostationary orbit satellite service monopoly

Organization	Investors	Cost	No. of Satellites and Orbit	Description
Globalstar	Space Systems/Loral, Qualcom, Alcatel, Deutsche Aerospace, Air Touch, Dacom, Hundai, Vodaphone, and publicly traded	$1.6 billion	48 in LEO	Globalstar emphasizes a "carriers' carrier" role with numerous gateway facilities to terminate traffic
Odyssey	TRW, Teleglobe	$1.3 billion	12 in middle Earth orbit (MEO)	Odyssey reduces the number of satellites required by raising the orbital location with an eye toward offering the least-cost service
Teledesic	Craig McCaw, William Gates, Edward F. Tuck	$6.3 billion	840 in LEO	Teledesic plans on providing ubiquitous broadband access primarily from fixed locations; the network design borrows from technology developed for the Strategic Defense Initiative

12.4.3 Wise Deployment of New Technology

Satellite carriers do not willingly add risk to an industry that faces significant risk with every launch. Carriers and users will tap new technologies if they provide efficiency, customized services, and lower costs. Carriers will apply maneuvers to prolong the life of usable satellites, and consumers will add circuit multiplication equipment to derive more throughput. Deployment of expensive and complex Ka-band systems and other new technologies will occur when they prove to be qualitatively better and cost-effective or when bandwidth requirements and congestion in the existing bands necessitate the inevitable migration up in frequency.

12.4.4   Uniformity and Global Coordination Remain Elusive Goals

ITU conferences will continue to have plenty of issues to resolve because nations continue to have different agendas, industrial policies, and spectrum allocation plans. Perhaps in recognition of international comity, advocates for new services must identify global service and financial benefit in order to support requirements that incumbent users share spectrum or perhaps even move to another frequency band. Developed nations will find it increasingly necessary to prove to the numerous developing and nonaligned nations, who command a majority of votes at the ITU, that innovations will accrue more than profits for manufacturers and carriers. Industrial policy and the consequences of previous domestic spectrum allocations inconsistent with an ITU vote mean that consensus decisions will remain elusive.

12.5   CONCLUSION

Satellite communication remains an essential vehicle for promoting commerce and world peace. The strategic planning process requires consideration of a diverse array of factors, many of which working at cross-purposes. The satellite service provider, finally convinced of the need to satisfy user requirements, may find that the ITU spectrum planning process frustrates such service. But on the other hand, the service provider content to rely on the ITU consensus building process and tolerant of delay, may find consumer requirements satisfied by innovators with terrestrial solutions or may find creative ways to maneuver around regulatory solutions.

   With digital audio, HDTV, and diverse mobile services clearly on the horizon, advocates may find a global consensus on spectrum allocations and operational rules elusive.

References

[1]   P.L. 87-624, 76 Stat. 419, 31 August 1962, codified at 47 U.S.C. Sec. 702(A).

[2]   *Agreement Relating to the International Telecommunications Satellite Organization "INTELSAT" (INTELSAT Agreement)*, 20 August 1971, entered into force 12 February 1973, 23 U.S.T. 3813, T.I.A.S. No. 7532, *Operating Agreement Relating to the International Telecommunications Satellite Organization "INTELSAT" (INTELSAT Operating Agreement)*, 20 August 1971, entered into force 12 February 1973, 23 U.S.T. 3892, T.I.A.S. No. 7582 at Preamble.

Case Studies in Satellite Policy 13

13.1 SPACE COOPERATIVES

Governments first tapped the social, tactical, national security, and strategic benefits accruing from satellites for military and intelligence gathering. A developmental lead enjoyed by the former Soviet Union presented a challenge to the United States that stimulated a race for global leadership in satellite and spacecraft technology. The U.S. taxpayer underwrote billions of dollars in space research and experimentation, with an emphasis on landing astronauts on the Moon.

Even as nations pursued space on national pride and security grounds, some officials recognized the benefits of commercializing space or at least of providing the basis for making space technology available for civilian applications. In 1962, the U.S. Congress passed the Communications Satellite Act[1] with an eye toward establishing a private enterprise, the Communications Satellite Corporation, that would "establish, in conjunction and in cooperation with other countries...a commercial communications satellite system" [1].

Comsat was instrumental in creating INTELSAT, the primary provider of satellite capacity throughout the world. In 1964, the United States and 10 other nations set the foundation for creating INTELSAT.[2] Twenty years later, the U.S. government officially endorsed conditional competition with INTELSAT and

---

1. P.L. 87-624, 76 Stat. 419, 31 August 1962, codified at 47 U.S.C. Sec. 701 et seq.

2. The INTELSAT Interim Agreement was executed by Australia, Canada, Denmark, France, Italy, Japan, the Netherlands, Spain, the United Kingdom, the United States, and the Vatican City. See *Agreement Establishing Interim Arrangements for a Global Commercial Satellite System (Interim Agreement)*, signed 20 August 1964, 15 U.S.T. 1705, T.I.A.S. No. 5646; see *Agreement Relating to the International Telecommunications Satellite Organization "INTELSAT" (INTELSAT Agreement)*, 20 August 1971, entered into force 12 February 1973, 23 U.S.T. 3813, T.I.A.S. No. 7532, *Operating Agreement Relating to the International Telecommunications Satellite Organization "INTELSAT" (INTELSAT Operating Agreement)*, 20 August 1971, entered into force 12 February 1973, 23 U.S.T. 3892, T.I.A.S. No. 7582.

currently supports the privatization of satellite cooperatives as a means to support a level competitive playing field.

INTELSAT provides the basis for pooling financial resources and sharing risk. At the introduction of civilian satellite service, a single cooperative was essential for achieving economies of scale and for deploying one or more satellites in each of the world's three ocean regions to make global service possible. Nations achieved favorable network externalities by consolidating traffic demand and financial resources, making it possible for a single global satellite constellation "to provide expanded telecommunications services to all areas of the world…which will contribute to world peace and understanding" [2].

Establishing a global satellite network so early in the product life cycle of commercial satellites required government involvement. A private, commercial venture would not average prices across dense and sparse routes, thereby promoting global access even by nations lacking significant demand and financial resources to make anything more than a token investment in the enterprise. A cooperative structure could aggregate resources and demand to achieve the critical mass sufficient to support satellite construction in the same way that farmers form investment pools to underwrite a shared facility for processing and marketing crops.

INTELSAT has twin and perhaps conflicting missions:

1. Operate efficiently and in a businesslike way, particularly because nations will have to abandon or temper the extent to which they pursue individual satellite options;
2. Serve social goals that promote the widespread, affordable use of satellites, perhaps at the expense of profitability, lower rates on dense routes, and a proliferation of service options for high-volume users.

On one hand, the satellite cooperatives operate like a business with a chief executive officer and a professional staff. Investors, known as *signatories*, provide oversight of the permanent staff and set general policies through a board of governors composed of signatories with the highest single investment stake, along with other signatories representing a group whose collective investment share qualifies quantitatively or on the basis of geographical diversity.[3]

On the other hand, nations execute a treaty-level document to become a party to the INTELSAT Agreement. The governance structure includes the assembly of parties, where governments confer on broad policy matters affecting the cooperative.[4]

---

3. See INTELSAT Agreement, Art. IX., Board of Governors: Composition and Voting.

4. "It shall have the power to give consideration to general policy and long-term objectives of INTELSAT consistent with the principles, purposes and scope of activities of INTELSAT as provided for in this Agreement." (INTELSAT Agreement, Article VII(b).)

13.2 INTERNATIONAL SATELLITE COMPETITION

In 1984, a U.S. "Presidential Determination" deemed international satellite systems separate from the INTELSAT cooperative, to be "required in the national interest" [3]. This one-paragraph document triggered a substantial change in telecommunications policy that required extensive coordination within the U.S. government and a major foreign relations campaign to demonstrate that the policy blended procompetitive initiative with adequate safeguards for INTELSAT. Despite prior efforts to create a single global cooperative largely protected from separate system competition, U.S. policy makers now believe that the international telecommunications marketplace can support limited competition. Officials in many other nations expressed skepticism or resentment over what they considered a heavy-handed attempt by the United States to dismantle the cooperative model for providing the ubiquitous international satellite capacity it was instrumental in creating.

Both sides agreed that INTELSAT, a cooperative that had investors from 135 nations in 1995, demonstrated how global cooperation can achieve ubiquitous satellite service at rates reflecting favorable economies of scale. Furthermore, the United States was instrumental in creating and had executed treaty-level documents requiring it to avoid actions such as approving separate satellite systems that might cause technical or economic harm to the cooperative. Article XIV of the INTELSAT Agreement, requires parties to the agreement (i.e., governments and their signatories who provide international services) to consult with INTELSAT prior to operating or using a separate satellite system to "ensure technical compatibility of such facilities...and to avoid significant economic harm to the global system of INTELSAT" [4].

The United States proposed the consultation requirement as a way to ensure that INTELSAT could operate free of interference and largely insulated from competition. In this way, INTELSAT and its constituency could achieve scale economies, thereby making satellites cost-competitive with established submarine cable systems. Also, perhaps, U.S. satellite manufacturers, launch companies, and taxpayers could enjoy greater certainty of benefiting from the commercialization of space. An international cooperative might have open tenders, as opposed to national systems that might have closed procurements of equipment and services from domestic companies.

Nevertheless, the U.S. government, through the executive branch and the FCC, implemented the Presidential Determination with the view that changed philosophy and circumstances now supported conditional international satellite competition that would not harm INTELSAT. In a joint "White Paper on New International Satellite Systems,"[5] the Departments of Commerce and State

---

5. Senior Interagency Group on International Communication and Information Policy, "A White Paper on New International Satellite Systems," (February 1985).

outlined how the United States proposed to balance the two objectives. The White Paper suggested criteria that the FCC subsequently incorporated as conditions to licensure [5], placing limits on the scope of direct competition with INTELSAT. The White Paper proposed that U.S. separate systems must:

- Not interconnect with the PSTN in either the United States or any foreign locale;
- Establish long-term contracts for the sale or lease of capacity;
- Secure a foreign operating agreement and complete the INTELSAT consultation process before commencing service.

In 1985, the FCC accepted the executive branch's recommendations when it licensed PanAmSat to provide international satellite service [6]. Upon granting PanAmSat's application to construct, launch, and operate a separate system, the FCC and its executive branch counterparts had to undertake the required consultation with INTELSAT to demonstrate the absence of technical and economic harm. This process first generated opposition and delay, but subsequently resulted in a mutually acceptable compromise.

The consultation process provides an interesting insight into the awkwardness and mixed agenda resulting when nations blend private enterprise and quasipublic undertakings. For example, the U.S. government had to ensure that publicly traded Comsat Corporation, the nation's sole signatory to INTELSAT, would fairly perform its task as emissary in the consultation process, even though a favorable finding of no technical or economic harm would set a precedent on the matter of competition that over time could adversely affect Comsat's profitability.

Comsat was established by an Act of Congress [7] to concentrate resources and expertise at the beginning of commercial satellite telecommunication. The Communications Satellite Act of 1962 requires it to serve the public interest as articulated by the U.S. government, even as its corporate status obligates it to pursue the best interests of its shareholders. Comsat's sole signatory status requires it to adhere to U.S. government–issued instructions on how to vote and what to say on matters involving the public, as opposed to Comsat's private interest.

The FCC licenses Comsat as a "carrier's carrier," in effect authorizing it to serve as the middleman between INTELSAT and U.S. international telecommunication service providers [8]. The FCC regulates Comsat as a common carrier, subject to Title II of the Communications Act of 1934.[6] Common carriers cannot

---

6. 47 U.S.C. Sec. 151 et seq. (1986). Beginning in 1982, the FCC authorized domestic satellites to provide transborder services, taking advantage of the fact that their footprints traverse nearby nations. (See Transborder Video Services, 88 FCC 2d 258 (1981); Hughes Communications Galaxy Corp., 6 FCC Rcd. 297 (C.C.B. 1991).) Transborder, international services of domestic U.S. satellites required consultation with INTELSAT and addressed many of the economic harm

discriminate, engage in unreasonable practices, and must file tariffs that specify service terms and conditions. Because Comsat uses spectrum to provide service, the FCC imposes additional requirements relating to Comsat's operation of Earth stations under Title III of the Communications Act.

Prior to the introduction of separate systems, Comsat enjoyed a monopoly on international satellite service to and from the United States [9]. Because Comsat must comply with official instructions of the U.S. government when voting its approximately 25% ownership share in INTELSAT, occasionally it must work against its financial self-interest, as occurred when ordered by the FCC, State Department, and NTIA to push for the favorable consultation of separate systems. Comsat had to lobby its fellow signatories to act in a manner that supported intramodal competition (i.e., two or more satellite facility providers). As their nations' only international carriers, most INTELSAT signatories also invest in submarine cable ventures, thereby supporting intermodal competition between cables and satellites. However, these same enterprises object to intramodal satellite competition because of:

- The more direct potential for adverse financial impact in terms of lost revenues;
- The reduced value of their INTELSAT investment;
- The view that separate systems would creamskim (i.e., serve only the most profitable, dense routes leaving INTELSAT as the only carrier with the mission to serve remote locations at rates designed to average both dense and sparse routes).

In favoring international separate systems, the U.S. government required Comsat to lobby its counterparts on the need for favorable consultations. Previously, INTELSAT had evaluated a number of domestic transborder and regional systems, finding no significant technical or financial harm. These include dozens of U.S. domestic satellites providing some services to Canada, the Caribbean, Central America, and parts of South America. In addition, INTELSAT had favorably consulted major regional systems operating in Europe, Southeast Asia, India, Brazil, the Middle East, and elsewhere.

13.2.1 The INTELSAT Consultation Process

In addition to its management and operational duties, the INTELSAT executive organ provides the cooperative with expert analysis of the extent of separate

---

questions raised by separate systems. In 1995, the FCC proposed to merge its transborder and separate system policies. See Amendment to the Commission Regulatory Policies Governing Domestic Fixed Satellites and Separate International Satellite Systems, ID Docket No. 95–41, FCC 95–146, 1995, WL 240658 (FCC) (rel. April 25, 1995).

system technical interference and prospective traffic and revenue diversion. The executive organ staff advises the board of governors, which in turn prepares a recommendation to the assembly of parties of whether to consult favorably on any separate system. The assembly of parties then issues an official finding, stylized as a recommendation to respect sovereignty of the consulting party. Nevertheless, the pursuit of separate system plans in the face of a negative recommendation would violate the spirit of the INTELSAT Agreement and would likely result in no other nation conferring an operating agreement.

While INTELSAT initially faced the prospect of more than five large-capacity U.S. separate systems, only three operators, PanAmSat, Columbia Communications, Inc., (using capacity on two NASA tracking and data relay satellites [10]), and Orion Satellite[7] had become operational by 1995. Separate systems may adversely affect the ability of INTELSAT to provide ubiquitous satellite service at globally averaged rates.[8] Global averaging means that rates for high-volume routes such as the United States to the United Kingdom are generally priced the same as service on less dense routes such as Nigeria to Thailand. Proponents of separate systems object to special efforts to safeguard INTELSAT, claiming that they too will serve lesser developed nations with innovative services at rates equal to or lower than what INTELSAT offers.

To safeguard its ability to average rates over a large global network and perhaps because of more parochial concerns about losing its near monopoly, INTELSAT has used a worst-case analysis of technical and economic harm. INTELSAT recently expressed the intention to eliminate the economic consultation requirement and has streamlined the process for separate systems with initial plans to launch satellites with relatively small capacity. Nevertheless, it continues to use a technical harm assessment that evaluates operational and

---

7. On the other hand, the U.S. separate system restrictions limited the potential for INTELSAT to lose large traffic volumes. By initially prohibiting services that access the PSTN, the U.S. insulated over 80% of INTELSAT's revenues. INTELSAT could continue to average costs for global switched services without fear of significant traffic diversion. Separate systems would concentrate on the growing market for customized, nonswitched offerings like video. Accordingly, the few U.S. separate systems that entered the marketplace could be considered niche players in no position to threaten INTELSAT, impair spectrum or orbital slot management, harm developing nations, jeopardize national security concerns, or hinder the U.S. balance of trade. The potential for enhanced consumer welfare, service diversity, some downward rate pressure, and innovation was too good to pass up based merely on an unscientific and biased internal INTELSAT analysis. (See Orion Satellite Corp., 101 FCC 2d 1302 (1985) (conditional authorization), 5 FCC Rcd, 4937 (1990) (granting authority to construct, launch, an operate subject to filing an expanded explanation of its financial plan), 6 FCC Rcd. 4201 (1991) (granting final authority).)

8. The INTELSAT board of governors did not expressly propose that the Orion Satellite system should be favorably consulted. Instead, it reported to the assembly of parties that it had received adequate assurances from the United States and other nations interested in using the system that they would not oppose efforts by INTELSAT to respond to competition. (See International Telecommunications Satellite Org., Assembly of Parties, *Record of the Fourteenth (Extraordinary) Meeting*, AP-14-9, 12 July 1989.)

yet-to-be launched satellites using technical criteria with a lower threshold for interference than typically used by the ITU.

On the economic front, INTELSAT in past consultations objected to separate systems at the same time it reported robust traffic statistics, evidencing the need for additional satellite capacity.[9] Such a dichotomy called into question the legitimacy of the INTELSAT executive organ's analysis, particularly because it appeared to assume that INTELSAT would otherwise handle any traffic carried over a separate system. In effect, INTELSAT assumed one-for-one traffic diversion, thereby ignoring the possibility of market growth and demand stimulation by separate systems. It is quite possible that consumers' demand elasticities and risk tolerance profiles might limit the extent of direct competition between INTELSAT and separate systems. The latter might serve extremely price-sensitive users who otherwise would use air couriers and other less immediate delivery options in the absence of cheap satellite service options.

On the other hand, U.S. separate systems aiming to serve dense transoceanic routes collectively could skim the cream off the most profitable dense routes, leaving INTELSAT as the only carrier for less attractive routes, particularly to less developed nations. Even speculation about such harm had the effect of presenting foreign policy and national security concerns that the government parties to INTELSAT had to consider. The prospect of INTELSAT retaliation through procurement of non-U.S. satellites and launch vehicles made the separate system policy all the more risky.

The INTELSAT executive organ's initial findings on PanAmSat, as it had been with other separate systems, pointed to peril and adverse financial consequences. Many nations had expressly articulated their opposition to U.S. separate systems, despite the fact that they had participated in other separate satellite ventures and their international carriers regularly invest in transoceanic cables that generate intermodal competition with INTELSAT and have no consultation obligation.

The INTELSAT executive directorate used an unofficial 10% revenue diversion figure to serve as a threshold point for triggering a finding of economic harm. That figure represented a running total of all previously consulted separate systems, making it increasingly difficult for new systems to pass muster. However, both PanAmSat and Orion Satellite achieved favorable consultations with INTELSAT. The INTELSAT management recognized that the U.S. intended to authorize conditional international satellite competition regardless of its consultation findings. A confrontation over INTELSAT's right to veto compe-

---

9. See Communications Satellite Corp., 5 FCC Rcd. 753 (1990) (approving Comsat's financial participation in INTELSAT's procurement of additional satellites, costing approximately $100 million each, that would expand in the number of operational satellites in orbit to meet higher demand); see also Communications Satellite Corp., DA 92-955 (rel. 23 July 1992) (authorizing Comsat's participation in the construction and operation of a single special-purpose Ku-band satellite to operate over the Atlantic Ocean region).

tition might have threatened the viability of the cooperative, or at least the extent to which the U.S. would continue to support the organization.

The cooperative has demonstrated progressively greater flexibility in the consultation process.[10] It streamlined the consultation process for satellites with less than 30 transponders, providing fewer than 1,250 bearer circuits (later raised to 8,000)—each with 64-Kbps throughput that can be subdivided into channels with lower bit rates—that access the PSTN.[11] While initial U.S. system consultations generated much debate, lobbying, and acrimony, the politically savvy leadership at INTELSAT has recognized the futility of attempts to block competition. Instead, INTELSAT has reached agreements with governments supporting market entry by separate systems to limit provisionally the scope of competitive operations in terms of transmission capacity and types of services. Having waged a vigorous informational campaign, including advocacy before the U.S. Congress, INTELSAT managers realized that the United States was intent on seeing its separate system policy implemented, even in the face of a negative assembly of parties recommendation. INTELSAT chose to avoid a confrontation by moderating its economic harm findings. By 1997, when all U.S. competitive limitations will lapse, presumably INTELSAT will have streamlined and adapted to changing conditions.

13.2.2　Revisions to the U.S. Separate Systems Policy

Since 1984, most nations situated under the footprint of a U.S. separate system have associated with the INTELSAT consultations, meaning that they will permit international traffic to be carried via the new carrier. Rather than incur any semblance of economic harm, INTELSAT has needed to expand its inventory of satellites to satisfy robust demand. What was once characterized a threat to INTELSAT's very survival has become a relatively insignificant matter, resolved in many instances by expedited and streamlined procedures to free the cooperative to attend to more pressing matters. INTELSAT now has become less preoccupied with separate system competition and more involved in the business at

---

10. On several prior occasions, INTELSAT had identified ways to avoid findings of economic harm, primarily by deeming INTELSAT unlikely to have provided services to be offered by a particular separate system. In the case of U.S.-Canada transborder service, INTELSAT concluded that domestic satellites would augment existing terrestrial facilities. In the case of EUTELSAT, a regional system with service throughout Europe, INTELSAT reduced its preliminary finding of 10% economic harm to less than 1% on a revised view that EUTELSAT would more likely stimulate traffic migration from terrestrial microwave facilities than divert INTELSAT traffic.

11. See Permissible Services of U.S. Licensed International Communications Satellite Systems Separate From the International Telecommunications Satellite Organization (INTELSAT), 7 FCC Rcd. 2313C1992), modifying 9 FCC Rcd. 347 (1994); applied in Alpha Lyracom d/b/a part American Satellite, 9 FCC Rcd. 1282 (1994).

hand: arranging for more capacity and deploying it in ways that make it attractive to users with both fiber-optic cable and separate satellite options.

Separate System Access to the PSTN

The fragile peace between INTELSAT and separate systems was challenged in 1990 with the proposal by PanAmSat that the FCC abandon the PSTN access restriction [11]. The vast majority of comment filers favored such abandonment, but the FCC initially chose to confer with the executive branch before unilaterally acting to change the separate system policy, a product of interagency coordination. The lobbying campaigns and policy debates returned, and new concerns about national security were raised concerning the ability of the intelligence community to monitor international traffic.

However, in 1991 the Secretaries of Commerce and State announced a major revision to the executive branch's position on separate systems in a letter to the FCC chairman. The executive branch proposed that the FCC "complete[ly] eliminat[e]...the restrictions on interconnection of international satellite systems with the public switched networks by January 1997" [12]. The FCC subsequently adopted this goal, incorporated INTELSAT's "safe harbor" consultation exemption for systems with less than 1,250 bearer circuits (later raised to 8,000), and authorized immediate PSTN access for private lines [13].

It appears that INTELSAT, Comsat, and others now accept that the consultation process generates too much distraction and confrontation. By streamlining the process and seeing only a gradual relaxation of the U.S. policy on access to the PSTN, INTELSAT turned its attention to satisfying user requirements, streamlining, and becoming more competitive.

13.3 TYPES OF SEPARATE SYSTEMS

The passage of time and maturation in the manner in which INTELSAT and incumbent satellite operators view separate systems have encouraged existing and prospective separate systems to pursue ambitious business plans. The satellite services marketplace appears able to absorb separate system capacity, and a number of different models describe various approaches.

13.3.1 Provocateur/Trendsetter

The provocateur/trendsetter model involves a proactive system willing to challenge the status quo even at the risk of creating the perception that it is a threat to PTT revenues and incumbent satellite systems. For any nation's separate system policy to result in actual market entry, at least one satellite system applicant, typically the first or most entrepreneurial, must take an aggressive posture promising users new and cheaper service opportunities. In the domestic U.S. interexchange telephone market, MCI Telecommunications, Inc., fits the provocateur/trendsetter model, and in the international satellite arena, the label applies to PanAmSat.

The mere threat of competition forced incumbent carriers like INTELSAT and Inmarsat to embrace a more businesslike approach to service. But at least initially it fostered a siege mentality that stimulated thinking on ways to thwart and delay market entry. The consultation requirements, ITU coordination process, and regulatory oversight became vehicles to frustrate the provocateur/trendsetter. These administrative delays, in conjunction with the fact that the system initially would involve only one in-orbit satellite, works to limit the advantage of early entry and the potential to capture a larger market share than latecomers. On the other hand, real or perceived opposition by incumbent satellite operators and the telephone companies, which must confer operating agreements to accept and hand off traffic, fuels the perception that the provocateur/trendsetter is an iconoclast and innovator who must overcome unfair obstacles to market entry.

The provocateur/trendsetter will generate a lot of publicity in its search for a market. But such press coverage will have a mixed message. Depending on the sympathies of reporter and reader, the satellite system will provide the benefits of competition or create trouble not worth the extra effort users and carriers might have to exert for the privilege to receive and offer service. This type of system appears to savor the battle for initial market access, and later for elimination of any impediment or service barrier (e.g., services that access the PSTN).

13.3.2 Low Profiler

The low profiler model capitalizes on the market development efforts of provocateur/trendsetters while deploying additional "plain vanilla" capacity. Carriers of this sort, like Columbia Communications and Orion, seek to ride the coattails of others with an eye toward capturing market share with less expense and travail. They enter the market later, after others have fought policy battles, and secure market share when the risk of PTT retaliation or noncooperation has abated. What the low profiler loses in terms of an untapped market, it presumably gains in the ability to sell or lease capacity with few, if any, challenges to its legitimacy.

Over time, the low profiler may have more available markets to serve as the concept of separate systems becomes more palatable to governments and users. The low profiler can easily lease capacity to users when incumbents face a shortfall or lack the specific type of facility or service desired. This type of separate system may provide services accessing the PSTN without the fallout previously encountered by earlier market entrants. On the other hand, when incumbents have ample capacity and fear lost market share, even the low profiler constitutes a competitive threat. Irrespective of whether it will take away market share from incumbents, the low profiler may experience the same difficulties in securing operating agreements with PTTs as the provocateur/trendsetter. The marketplace prospects for the low profiler are handicapped by consumer perception that it has nothing new to provide without significant discounts. As a later entrant, the low profiler may have to operate from a less attractive orbital slot.

13.3.3 Innovator

The innovator model concentrates on providing greater convenience, flexibility, customer service, and access to the rest of the world. Prospective LEO satellite systems, which promise global or regional mobile telecommunication services via handheld terminals, represent the innovator model. They primarily aim to provide new services to untapped markets or to provide better service to underserved areas. Accordingly, it will emphasize the ability to generate new traffic rather than migrate traffic from existing carriers. To the extent that the innovator can serve a rich, untapped market, it can exploit inelastic demand and profit handsomely. But it must predict which market is worth entry and be prepared to front-load substantial investment before it can provide service, often despite uncertainty in terms of consumer demand, ITU spectrum allocations, and licensing by national regulatory agencies.

The innovator expects to reap a hefty profit before other carriers match its innovations. That profit accrues only if the innovator can attract the first rung of users, so-called *early adopters*, willing to invest in a new terminal and other equipment needed to access the satellite. When the innovator proposes the use of a new frequency band, it raises spectrum allocation, regulatory, licensing, and policy issues that can convert its status to provocateur/trendsetter in short order, particularly if its proposals are perceived as a bypass threat to the status quo.

13.3.4 Regional Marketer/Niche Player

The regional marketer/niche player brings new or improved services to a specific region, often simply by employing a better satellite orbital position, beam, or signal. Such a carrier can become the INTELSAT for the region or concentrate on market niches (e.g., video and service to very small rooftop dishes). A regional market player requires close cooperation of investing national carriers and regulatory authorities. Therefore, it is more likely that the regional marketer/niche player will have secured regulatory approval well in advance of launch. Such authorization may come quickly in view of the concrete benefits that flow to the region and the absence of a financial threat to incumbent carriers.

While the regional marketer/niche player may encourage others to enter the market and provide some degree of price and service competition, the more likely outcome depends on the adequacy of demand and the response of INTELSAT. In the case of Arabsat, internal management problems and unwillingness to stimulate demand by authorizing diverse video products has led to a less than stellar financial outcome. On the other hand, Palapa, the Indonesian satellite system, has thrived, because of that nation's dispersed island geography, the willingness of other nations in the Alliance of Southeast Asian Nations to lease capacity for telecommunications and video service, and the disinclination or inability of INTELSAT to offer discounts or improved satellite beam coverage for the region.

AsiaSat has managed to accommodate the diverse interests of investors from Hong Kong, China, and the United Kingdom while convincing various governments in the region that diversified television options—"television without borders"—will enhance consumer welfare without threatening political stability. Its success may also result from superior signal strength and customer service, despite the higher transaction costs many multinational consumers incur in having to do business with more than one satellite carrier.

13.3.5 Opportunist/Extortionist

The opportunist/extortionist model exploits ITU procedures for satellite orbital slot registration and the desire of developing nations to become a participant in the satellite-delivered information age. The opportunist seeks to capture some of the benefits in satellite technology for its citizens by providing a "flag of convenience" to a foreign satellite carrier. In exchange for cash or free transponder capacity, a nation might agree to serve as the "Notifying Administration" for filing orbital slot registration materials with the ITU and for initiating consultations with INTELSAT. This designation might apply to Tonga's Tongasat, since outside entrepreneurs have provided guidance on how it might participate in the satellite sweepstakes.

The extortionist goes several steps further and seeks to broker orbital slot registrations much like one might sell a New York City taxicab authorization or stock market membership. Tongasat registered a total of 31 satellites in 26 orbital slots before agreeing to narrow its Advance Publication requests with the ITU's Radio Regulations Board to 6. Its agent allegedly sought to "sell" to INTELSAT and others claims to one or more of these valuable orbital parking places. Even though an orbital slot registration does not establish a property interest, it has intrinsic value, because most nations have agreed to abide by ITU Rules and Regulations. A national commitment to avoid causing harmful radio interference in application means that most nations will respect the priority access rights of a nation who filed a registration first. However, the right of access does not translate to a right to speculate on orbital slots and to buy and sell them.

The opportunist/extortionist suffers an identity crisis. Is it a legitimate player promising regional service where it is desperately needed? Or is it a front for a foreign entrepreneur seeking to provide a thick route service or perhaps no service at all? While the possibility exists for better regional service, particularly for developing nations entitled to claim a fair portion of the shared geostationary orbital arc, it is just as likely that the opportunist/extortionist has nothing but paper satellites and orbital slot registrations to pitch. For the latter type of gambit, ITU procedures will eliminate the registration in six years, while INTELSAT and other incumbents have refrained from buying slots they normally can secure for nothing.

13.3.6 Conclusion

The diverse models of private international satellite systems demonstrate that several alternatives exist for acquiring market share and legitimacy. The choice of which model to adopt depends in part on the extent of risk a system consciously decides to embrace in terms of both regulatory delays and competitive responses by incumbents. Early market entrants and large profile-seeking systems adopt a high-risk, high-reward strategy. An equally attractive alternative involves a low profile that reduces risk, but also lowers rewards.

13.4 PRIVATIZATION OF INTELSAT AND INMARSAT

INTELSAT was established in 1964 when satellite technology was first making the transition from military and experimental uses to commercial applications. When Inmarsat was created in 1976, commercial enterprises had a firm grip on satellite technology, but uncertainty remained about whether such enterprises would serve essential but low-volume markets such as maritime communications.

Regulatory, technological, and economic changes now challenge the logic of maintaining a preference for organizations created by intergovernmental agreement, granted quasidiplomatic status and immunity from lawsuits, and partially insulated from competition. An increasing number of nations have authorized intramodal competition (i.e., market entry by other enterprises providing service via the same medium [satellites] as the cooperatives). In addition, fiber-optic submarine cables, providing broadband digital services, now provide intermodal competition.

INTELSAT and Inmarsat have achieved the goals of providing universal service and global connectivity at rates averaged over dense and sparse routes alike. It appears with greater certainty that many international telecommunication service markets can support competition and that private enterprises can and will enter the marketplace, particularly if the incumbent loses marketplace advantages conferred to it by intergovernmental agreement.

On the other hand, the reasons for creating INTELSAT and Inmarsat have not completely evaporated. Any revision to their mission or special status conferred by government should not handicap their ability to provide nondiscriminatory access to all countries of the world on an averaged cost basis. It remains likely that INTELSAT and Inmarsat will serve as the only carrier for many routes. Accordingly, the key challenge lies in pursuing facilities-based competition where feasible without jeopardizing the ability of organizations to provide services that no commercial entity would offer.

13.4.1   Privileges, Immunities, and Marketplace Advantages Available to International Cooperatives

When nations become parties to the INTELSAT and Inmarsat agreements, they in effect agree to treat these cooperatives as if they were diplomatic organizations like the ITU and the United Nations. The cooperatives are exempt from all national income, sales, excise, value-added, personal property, and real estate taxes. Officers and employees are guaranteed immunity from the legal process, except for criminal acts, when engaging in the exercise of their office. The tax exemption saves the cooperatives millions of dollars when they procure satellites and other costly facilities that otherwise might be subject to taxation.

The Headquarters Agreement executed between the cooperatives and the nation where the headquarters building is located provides for immunity from search, seizure, and law suits. This document underscores the significant immunity from law the cooperatives enjoy. In application, such immunity means that the cooperatives and their signatories are exempt from judicial oversight. A recent decision by the U.S. Court of Appeals for the Second Circuit confirms this. The court held that Comsat could not be held liable under any antitrust provision when engaging in INTELSAT signatory functions as opposed to its regulated common carrier roles.[12]

13.4.2   Reasons for Privatizing INTELSAT and Inmarsat

The executive directorates of INTELSAT and Inmarsat have recognized the need to make their cooperatives more businesslike and efficient in the face of increasing intra- and intermodal competition. In 1992, the INTELSAT board of gover-

---

12. See *Alpha Lyracom Space Communications, Inc. v. Communications Satellite Corp.*, 946 F.2d 168 (2d Cir. 1991) cert. den. 60 U.S.L.W. 3578 (24 February 1992). For a discussion of the case, see P. Brillson, "The Empire Strikes Back in the International Satellite Telecommunications Industry," *Rutgers Computer & Technology Law J.*, Vol. 18, 1992, pp. 381–402.

nors adopted recommendations of Director General Irving Goldstein, the former chief executive officer of Comsat, that the cooperative adopt "major organizational changes to reflect a new, more commercial orientation for the Company" [14]. The Inmarsat council has explored similar initiatives.

The executive directorates of INTELSAT and Inmarsat have explored and in many cases implemented ways to:

- Improve the cooperatives' ability to serve emerging markets and new profit centers;
- Price services with greater flexibility, including selective rate deaveraging for users that have a choice of carrier;
- Facilitate direct access to the cooperative without having to secure capacity at rates marked up by a national signatory;
- Streamline the organization, including reduction in staff;
- Secure funding from external sources as a way to augment self-financing and reduce the cost of capital, because the cooperatives have agreed to compensate signatories for the use of capital (not converted into capacity leases) at a rate in excess of 14%;
- Reduce the financial risks to existing signatories by securing external funding;
- Provide current owners with the opportunity to sell a portion of their investment at rates in excess of net book value;
- Allow nonsignatories an opportunity to invest in the cooperative;
- Consider creating a separate entity with a more liberal mandate and ownership structure to provide services beyond the core mission.

Collectively, these changes could achieve the virtual privatization of the cooperative while retaining the government-conferred privileges and immunities. Competitors object to this strategy as unfairly capturing the benefits of both the protected, cooperative environment and the flexible corporate world.[13]

13.4.3 Should INTELSAT and Inmarsat Privatize?

The conversion of INTELSAT and Inmarsat into for-profit, privately or publicly owned corporations would necessitate the dismantling of intergovernmental agreements creating the cooperatives. Under such circumstances, few nations would continue to confer privileges and immunities, knowing that they would

---

13. "It is time, however, to stop INTELSAT from having it both ways: if it wishes to raise private capital and compete like a commercial company, it must give up its governmental privileges and immunities." (PanAmerican Satellite Company, White Paper: *A New Private Enterprise INTELSAT*, Executive Summary, 20 April 1992.)

accord the former cooperative a marketplace advantage without a commensurate obligation to guarantee universal services at averaged rates to all users.

While it is possible that privatization would generate a market value in excess of the cooperatives' net book value, it is absolutely certain that the cooperatives would lose substantial cost-saving opportunities. Similarly, they would lose a preferential status at spectrum management forums like the ITU and at global policy making forums like the United Nations, where they have official observer status.

A more subtle adverse outcome would be the elimination of a balance the cooperatives have been able to maintain between divergent interests on issues pertaining to management, mission, and markets. It is far from certain whether a privatized entity could satisfy still essential universal service objectives while at the same time responding to shareholder imperatives such as profit maximization.

If the newly privatized entities got into a price war with other carriers, it seems quite possible that they would want to deaverage rates to meet competition, particularly on high-volume routes. Deaveraging could result in the abandonment of a tariffing structure that builds from a least common denominator utilization charge equivalent to a single voice-grade half circuit. Even if the cooperatives retained a standard utilization charge, that rate would have to rise because of instances where competitive necessity required discounts for high-volume users and those users with competitive alternatives.

Privatization would eliminate instances in which INTELSAT and Inmarsat enjoy something akin to sovereign status, subject only to general oversight by the infrequently convened assembly of parties. Because most of the world's governments have voluntarily relinquished jurisdiction, the cooperatives can operate independent of any single nation's laws. A privatized entity conceivably would be subject to the diverse and possibly conflicting laws and regulations of each country in which it operates.

Ownership of a privatized entity would probably include investors outside the existing "club." With PTTs participating in both submarine cable and satellite investment pools, they can coordinate facilities deployment and view international transmission facilities as primarily complimentary rather than as vehicles for intermodal competition. A shift in the ownership composition of the former satellite cooperatives could challenge such coordination, making it more likely that carriers would emphasize use of the currently cheapest medium on a per-unit basis, while discounting the long-term benefits of having two different modes of transmission.

The fair trading and antitrust laws of nations would apply to a privatized entity. Because INTELSAT and Inmarsat would no longer have any right to evaluate the prospective technical and economic harm from separate systems, their primary vehicle for retaining market share would be price. Privatized management might migrate from flexible pricing into the anticompetitive realm of predatory pricing.

Because most international Earth stations access INTELSAT, courts may determine that INTELSAT has market power and the ability to drive out competitors.

13.5 LOW-EARTH-ORBITING SATELLITES

Entrepreneurial visions of wireless technology that can provide telecommunication services anytime from anywhere prompted the United States and other nations to propose major new spectrum allocations at the ITU. Consumers have embraced mobile terrestrial services such as cellular radio, accessible from small, lightweight, and portable transceivers. While future terrestrial systems like personal communications networks will serve urban locales, satellites are best suited as gap fillers between systems in primarily rural areas.

GSO satellites cannot yet provide service to very small, low-powered transceivers (i.e., devices that transmit and receive signals). A number of proposals for LEO satellites have emerged to serve this market. These systems will operate only a few hundred miles above the Earth, thereby reducing the power needed to communicate. While more LEO satellites are needed to cover the globe than their GSO counterparts, they typically weigh less and have lower orbits. LEO satellites are 1/20th as expensive as GSO space stations to launch [15] and can be launched in groups of five or more.

Constellations of LEO satellites provide the prospect for ubiquitous mobile radio telephone service. Contrary to terrestrial cellular radio service, where users physically move from one fixed cell site to another, LEO satellite footprints constantly change as the satellites orbit above. Users appear stationary relative to the fast-moving satellite footprints.

The FCC recognized the need to accommodate LEO satellites with spectrum reallocations and a service classification in its preparations for WARC-92.[14] The FCC revised its domestic spectrum allocations following WARC-92, substantially expanding the spectrum available for mobile services.

The manner in which nations agree on frequencies, technical parameters, and operational rules for LEO satellite systems furnishes a timely case study of the difficult task in balancing self-interest with international comity. The United States achieved much of its spectrum objectives for LEO satellites, despite the efforts of numerous opponents with differing strategic visions, industrial policies, or uses for the targeted frequencies. The future viability of new services and spectrum allocations depends in large part on the willingness of

---

14. "We believe that the demand for mobile-satellite service [MSS] is beginning to grow. Until recently, most MSS has been limited to maritime systems, but recent years have seen a significant increase in interest in providing land and aeronautical MSS." (An Inquiry Relating to Preparation for the International Telecommunications Union World Administrative Radio Conference for Dealing With Frequency Allocations in Certain Parts of the Spectrum, GEN Docket No. 89-554, Second Notice of Inquiry, 5 FCC Rcd. 6046, 6055 (1990).)

the ITU's community of nations to reach closure. The FCC may await a global consensus or take unilateral actions on domestic allocations to expedite the availability of new services and technologies and perhaps also to affect future global allocations.[15]

13.5.1   The 1992 World Administrative Radio Conference

The 13th ITU Plenipotentiary[16] determined that a WARC should convene in 1992 to address frequency allocations in various parts of the spectrum for various services [16] including terrestrial and satellite-delivered mobile services. Accordingly, WARC-92 had a diverse agenda spanning the frequency spectrum from 3 MHz to above 20 GHz, addressing additional spectrum requirements of existing services such as shortwave and terrestrial mobile radio and considering spectrum requirements of new services such as HDTV and digital audio broadcasting. Prior to the conference, the ITU Administrative Council agreed to expand the WARC-92 agenda to include LEO satellite issues.[17]

13.5.2   The Battle of the Haves and Have Nots

The domestic administrative process to prepare U.S. positions for WARC-92 and actions at the conference exemplify substantial differences in terms of spectrum and satellite orbital arc priorities. A number of dichotomies and schisms developed between:

- Incumbent users and proponents of new services;
- LEO satellite proponents and advocates for protecting GSO satellites from potential interference and competition;
- Developed nations having the option to establish spectrum priorities between competing technologies and developing nations perceiving the need to avoid foreclosing any single option that might serve as their best, and

---

15. See, for example, Amendment of the Commission's Rules With Regard to the Establishment and Regulation of New Digital Audio Radio Services, GEN Docket No. 90-357, Notice of Inquiry, 5 FCC Rcd. 5237 (1990), Notice of Proposed Rulemaking and Further Notice of Inquiry, 7 FCC Rcd. 7776 (1992); Report and Order, FCC 95-17, 10 FCC Rcd. 2310 (1995).

16. The Plenipotentiary Conference is the supreme organ of the ITU, having both regulatory authority and the power to modify the ITU's Constitution and Convention. It has been scheduled in 8- to 10-year increments, but a 1989 amendment establishes a 5-year interval not to exceed 6 years. (*Constitution of the International Telecommunication Union*, Nice, 1989, Art. 8.) In addition to scheduling WARCs, the Plenipot determines general policies, establishes budgets, elects the major office holders of the ITU management, and selects which nations will have a representative at the ITU's executive board, known as the Administrative Council.

17. See U.S. WARC-92 Preparations: Second Notice of Inquiry, 5 FCC Rcd. at 6046, App. B.

possibly only, near-term technological solution to basic telecommunications requirements;

* Blocs of nations and cooperatives whose industrial policies and strategic visions could be furthered or frustrated by WARC-92 decisions;
* Advocates for looser, generic service definitions and proponents of retaining service-specific block allocations of spectrum.

13.5.3 U.S. Preparations

On the issue of mobile telecommunications provided over L-band frequencies (around 1.5 to 2.0 GHz), the United States has aggressively advocated a generic mobile-satellite service (MSS) allocation that would accommodate the previously separate allocations for maritime, land, aeronautical, and safety/distress services.[18] In addition to affording greater flexibility among services whose spectrum demands might grow at varying rates, a generic allocation would promote flexibility so that MSS operators could provide services for a number of different applications, not all of which might individually support a dedicated system.

The United States approached WARC-92 with the following guiding principles:

* To promote the implementation of a variety of new operational programs as rapidly as practicable so that all countries might realize the benefits and spectrum savings promised by modern telecommunication technologies;
* To provide flexibility in the international regulations to ensure that the needs of all countries could be met;
* To reduce regulatory, technical, and operational barriers so that technologies could rapidly be introduced and used to the benefit of all mankind;
* To provide up-to-date regulations that assure greater safety-of-life on land, on the sea, in the air, and in space [17].

---

18. See Amendment of Parts 2, 22, and 25 of the Commission's Rules to Allocate Spectrum for, and to Establish Other Rules and Policies Pertaining to the Use of Radio Frequencies in a Mobile Satellite Service, GEN Docket No. 84-1234, Report and Order, 2 FCC Rcd. 1825 (1986); Second Report and Order, 2 FCC Rcd. 485 (1987), on recon., 2 FCC Rcd. 6830 (1987), further recon. den., 4 FCC Rcd. 6016 (1989) (allocating 27 MHz of L-band spectrum for generic MSS); 4 FCC Rcd. 6041 (1989) (licensing a mandatory consortium of applicants), partially vacated and remanded sub nom., Aeronautical Radio, Inc. v. FCC, 928 F.2d 428 (D.C. Cir. 1991) on remand, Tentative Dec., 6 FCC Rcd. 4900 (1991); Final Dec., 7 FCC Rcd. 266 (1992); Amendment of Part 2 of the Commission's Rules for Mobile Satellite Service in the 1530–1544 MHz and 1626.5–1645.5 MHz Bands, Notice of Proposed Rulemaking, 5 FCC Rcd. 1255 (1990) (proposal to add 33 MHz to the generic MSS band), First Rep. and Ord., 8 FCC Rcd. 4246 (1993). See also Amendment of Sec. 2.106 of the Commission's Rules to Allocate Spectrum at 2GHz for Use by the Mobile Satellite Service, ET Docket No. 95–18, Notice of Proposed Rulemaking, 10 FCC Rcd. 3230 (1995) (proposing to allocate additional spectrum for MSS via GSO or LEO satellites).

Taken as a whole, the U.S. principles emphasize change and flexibility over the status quo. The fact that the U.S. position reflects a forward-thinking approach belies the confrontation and compromise occurring during the multiyear preparations for the conference. The FCC sought public participation in two ways: (1) through a formal Notice of Inquiry, and (2) through formation of an Industry Advisory Committee composed of representatives from trade associations, user groups, manufacturers, carriers, and other interested parties with a mission of either advocating spectrum reallocation to accommodate new services or protecting previously allocated spectrum from reallocation. The Interdepartment Radio Advisory Committee (IRAC), convened under the auspices of the Commerce Department's National Telecommunications and Information Administration,[19] conducted a separate, largely independent process for coordinating the preparation activities for government users.[20]

13.5.4 The U.S. Position on LEO Satellites

The United States suggested that the best way to accommodate growing demand for MSS would be "to permit flexible usage to adapt to dynamic changes in communication needs," with due consideration for priority safety services [18]. In application, the United States generated several concrete proposals that collectively would reallocate a large portion of the 1.5- to 1.6-GHz[21] L-band from service-specific frequency allocations (e.g., land, maritime, *or* aeronautical mobile services) to a generic MSS, allocate addition spectrum further up the L-band for expansion, and establish a more formal basis for interference avoidance between LEO and GSO satellites providing MSS.

---

19. "In 1978, the President delegated authority to manage the spectrum used by the federal government to the Secretary of Commerce through Executive Order 12046. The Secretary of Commerce further delegated this authority, under Department of Commerce (DOC) Organizational Order 10-10, to the Assistant Secretary for Communications and Information, who is also the head of the National Telecommunications and Information Administration (NTIA)." (U.S. Dept. of Commerce, National Telecommunications and Information Administration, *Long Range Plan for Management and Use of the Radio Spectrum by Agencies and Establishments of the Federal Government*, NTIA Spec. Pub. 89-22, June 1989, p. 1-1.) NTIA "seeks the advice of Federal Government agencies through the Interdepartment Radio Advisory Committee (IRAC). The IRAC, its related subcommittees, and ad hoc groups provide information and coordinate activities that help NTIA plan for future RF requirements, assign frequencies, and resolve conflicts." (Ibid., p. 1-2.)

20. See B. Fisher, "Preparing for WARC-92 Major Decisions for Telecommunications in the 21st Century," *Via Satellite*, Vol. 5, No. 9, September 1990, p. 49.

21. The specific frequencies targeted were the radiodetermination satellite service frequencies at 1,610–1,626.5 MHz/2,483.5–2,500 MHz, the land mobile and maritime mobile frequencies at 1,530–1,544 MHz and 1,626.5–1,645.5 MHz, and the aeronautical mobile-satellite and land MSS bands at 1,545–1,559 and 1,646.5–1,660.5 MHz.

13.5.5 WARC-92 Deliberations and Decisions

Dr. Pekka Tarjanne, secretary general of the ITU, deemed WARC-92 "the biggest, most important and difficult conference of its kind" [19]. Significant controversy arose on the issues of spectrum allocations for LEO satellites and the method for coordination to avoid interference to both LEO, GSO, and land-based systems. Many of the nations that coordinate telecommunications policies in CEPT favored terrestrial solutions to future mobile telecommunications requirements.[22] U.S. advocacy for LEO MSS may have appeared as a threat to industrial and strategic policies [20]:[23]

> The United States and Europe went head-to-head over this issue because neither country was certain that the market could sustain both ...[future terrestrial services]...and LEOs above 1 GHz. As a result, both wanted to make sure they received an allocation to get their systems off the ground. Europe, however, has had a head start, because [terrestrial services]...could begin with...[digital] cellular service, which is finishing its test phase.

Incumbent radio astronomy and navigation users of the band targeted for LEO MSS also expressed concerns about the potential for harmful interference. While Russia might possibly provide launch services for U.S. LEOs, a more immediate concern predominated: the need for technical standards and proce-

---

22. The extensive efforts by CEPT to block mobile satellite spectrum allocations prompted Ambassador Jan Baran, the U.S. delegation head to WARC-92, to allege that "an organized bloc of 32 European countries...often appeared to oppose new technologies." ("U.S. 'Big LEOs' Get Allocations at WARC Largely as Proposed, but Limits Aimed at Protecting Russian Glonass System Could Restrict IRIDIUM; CEPT Nations Get 230 MHz for Future Public Land Mobile Service; BSS-Sound Gains Worldwide Allocation at L-Band," *Telecommunications Reports*, Vol. 58, 9 March 1992, p. 12 (hereafter cited as WARC-92 Summary).) The State Department's unclassified wrap-up cable declared success, in "buck[ing] the inertia and caution of the radio community, suspicion and stonewalling by a well-organized European block of thirty-two countries, and Russian singlemindedness in protecting its GLONASS radionavigation satellite system." (Jan Baran, "U.S. Success at World Administrative Radio Conference (WARC'92): Wrap-up Cable" 3 March 1992 (hereafter cited as WARC-92 Wrap-up Cable).)

23. An unnamed source alleged that the "Europeans wanted to make sure that a terrestrial system was implemented before MSS and LEOs could serve hand-held terminals—basically, they were protecting their manufacturers, such as Ericsson and Nokia." (WARC-92 Summary.) Even the State Department wrap-up cable reported: "The well-organized and cohesive European ITU members (32 CEPT countries in all) came to the WARC generally opposed to the specific LEO proposals of the U.S. This was because of a general European lack of interest in MSSs (their priorities for WARC'92 were terrestrial mobile issues), a European desire for more time for their own industrial development of these technologies, and/or difficulties with the specific bands proposed." (WARC-92 Wrap-up Cable, para. 10.)

dures to safeguard its GLONASS global navigation system, which transmits in a portion of the frequency band targeted for LEO-delivered services.

Many observers attribute the successful outcome for LEO satellites to the eleventh hour reconsideration by a number of developing countries that initially considered LEO systems potential bypassers able to siphon traffic and revenues from the government-owned and operated telephone company [21]:

> European opposition to the allocations was ultimately overcome by the merits of the U.S. proposals and their attractiveness to many developing countries, which see LEOs as a lower cost way to provide communications in rural or other areas with an insufficient telecommunications infrastructure. Politically, the Europeans were reticent to be seen as blocking the development of new technologies, particularly with regard to developing countries with low telephone densities.

13.5.6 Little LEOs

Prior to WARC-92, the FCC proposed new frequency allocations for "little LEO" satellites in a narrow portion of the VHF and UHF bands.[24] These satellites will provide low-cost nonvoice services only, such as data messaging, paging, and position determination, which can improve the efficiency of oil exploration and transportation industries, enhance remote monitoring of the environment, and provide an emergency signaling system for motorists, hikers, and other travelers.

The United States proposed the reallocation of a small sliver of bandwidth in the VHF and UHF bands [22]. WARC-92 considered this proposal noncontroversial because of the small amount of spectrum requested for a new allocation, evidence submitted to show that little LEOs would not harm incumbent operations[25] (including weather satellites), and the shared view that the two-way data

---

24. See, for example, Amendment of Section 2.106 of the Commission's Rules to Allocate Spectrum to the Fixed-Satellite Service and the Mobile-Satellite Service for Low-Earth Orbit Satellites, ET Docket No. 91-280, Notice of Proposed Rulemaking, 6 FCC Rcd. 5932 (1991), Report and Order, 8 FCC Rcd. 1812 (1993); Amendment of the Commission's Rules to Establish Rules and Policies Pertaining to a Mobile Satellite Service in the 1610–1626.5/2483.5–2500 MHz Frequency Bands, CC Docket No. 92-166, Notice of Proposed Rulemaking, 1094 (1994), Report and Order, 9 FCC Rcd. 5936 (1994) (establishing licensing rules); see also Amendment of Sec. 2.106 of the Commission's Rules to Allocate Spectrum at 2 GHz for Use by the Mobile-Satellite Service, ET Docket No. 95-18, FCC 95-39, 1995 West Law 170701 (F.C.C.) (rel. 31 January 1995) (proposing to allocate additional spectrum for MSSs).

25. Even before nations arrived at the conference, there was a predisposition to favor this U.S. proposal: "Australia could accept an allocation for LEO MSS systems below 1 GHz provided that it can be accommodated with our current and planned use of the relevant band, and that adequate protection and coordinated procedures are agreed so that harmful interference would not be caused to existing and planned national systems." (Department of Transport and Communications, Radiocommunications Division, Canberra, *Australian Proposals for the World Admin-*

services available via low-cost pocket-sized terminals could "support economic development worldwide" [23].

WARC-92 provided "everything the U.S. wanted, plus more" [24][26] for little LEOs [25]:

> Developing countries were particularly vocal that…[such] service will be a low-cost tool to relay messages, especially in emergency situations when ground-based telephone service is not available…African nations also supported the allocation because receivers for this service will be available at one-tenth the cost of hand-held telephones for [GSO] satellite-based voice communications.

The FCC expedited the licensing process for little LEOs by encouraging applicants to negotiate technical and regulatory terms under which all could operate [26]. The FCC has legal authority to create advisory committees[27] and to encourage them to negotiate regulatory solutions.[28] "If consensus is reached, it is used as the basis for the FCC's proposal" to regulate the service [27]. The FCC believed it could use a negotiated rulemaking arrangement to develop technical and licensing rules[29] for small LEOs, because of the limited number of potentially diverse interests and a reasonable likelihood that an advisory committee could reach consensus, having considered in good faith the various viewpoints. After some wrangling, the parties did reach a consensus. The FCC granted licenses[30] and opened up a second round for additional applicants [28].

13.5.7 Big LEOs

Soon after WARC-92, the FCC issued a Notice of Proposed Rulemaking to modify the domestic Table of Frequency Allocations to incorporate spectrum allocations for MSS in the L-band [29]. The FCC also created a Negotiated Rulemaking

---

istrative Radio Conference for Dealing With Frequency Allocations in Certain Parts of the Spectrum, Malaga-Torremolinos, February 1992, 20 November 1991, p. 9).

26. In addition to allocating almost exactly what the United States sought the conference also allocated, on a secondary basis, 6 MHz at 312–315 (uplink) and 387–390 MHz (downlink) as requested by the Commonwealth of Independent States. (See WARC-92 Summary, p. 14.)

27. See Federal Advisory Committee Act, 5 U.S.C. App. 2 (1990).

28. See Negotiated Rulemaking Act of 1990, Pub. L. 101-648 (1990).

29. Some of the questions for resolution include whether to require little LEOs to offer service on a common carrier basis, what transmission modulation to require, whether different rules should apply to noncommercial systems and how to resolve frequency coordination disputes between Little LEOs and terrestrial operations.

30. See, for example, Application of Orbital Communications Corp. for Authority to Construct, Launch and Operate a Non-Voice, Non-Geostationary Mobile-Satellite System, Order and Authorization, 9 FCC Rcd. 6476 (1994).

forum for the applicants to assist it "in developing regulations that will facilitate the maximum number of MSS providers" [30]. Notwithstanding the FCC's interest in having the affected parties manage the rulemaking process, a single consensus did not evolve, thereby forcing the FCC to issue a Notice of Proposed Rulemaking with its own proposal for spectrum sharing.[31] The six applicants for big LEO licenses[32] proposed incompatible transmission systems, and may have had business plans that assumed a certain amount of channel capacity, and hence a certain amount of interference-free frequency bandwidth. Some of the applicants may not have wanted an expedited licensing process that would impose a deadline for launch and operation.

In early 1995, the FCC licensed three of the five big LEO applicants; Iridium, Globalstar, and Odyssey.[33] Since many experts view the market as supporting only two or three systems, the three U.S. licensees and Inmarsat are racing to secure financing and global support.

13.5.8 Conclusion

WARC-92, along with the domestic regulatory activities that precede and follow the conference, involves high-stakes decisions. The vision of ubiquitous communications via small handsets cannot become a reality without a global consensus on MSS frequency allocations. Likewise, the vision requires timely action by the FCC and other national regulatory authorities to incorporate the spectrum allocations, issue rules, and license applicants.

The outcome of WARC-92 demonstrates that the international process can accommodate new spectrum requirements despite the lack of a consensus going into the conference. On the other hand, the domestic regulatory process, which emphasizes procedural fairness at the expense of timeliness, may take years to run its course. The success at WARC-92 means very little if the domestic agen-

---

31. See Amendment of the Commission's Rules to Establish Rules and Policies Pertaining to a Mobile Satellite Service in the 1610–1626.5/2483.5–2500 MHz Frequency Bands, CC Docket No. 92-166, Notice of Proposed Rulemaking, FCC 94-11 (rel. 18 February 1994).

32. The FCC placed the applications of Motorola Satellite Communications, Inc. (Iridium) and Elipsat Corp. (Elipso-I) on Public Notice on 1 April 1991, No. DA-91-407. On October 24, 1991, the FCC issued a Public Notice, No. DS-1134, accepting filing applications of AMSC Subsidiary Corp. (to modify its existing MSS authorization to operate with additional bandwidth), Constellation Communications, Inc. (Aries), Ellipsat Corp. (for a second-generation system), Loral Cellular Systems, Corp. (Globalstar), and TRW, Inc. (Odyssey).

33. See, for example, Application of Motorola Satellite Communications, Inc. for Authority to Construct, Launch, and Operate a Low Earth Orbit Satellite System in the 1616–1626.5 Mhz Band, File Nos. 9-DSS-P-91(87), CSS-91-010, 43-DSS-AMEND-92, 15-SAT-LA-95, 16-SAT-AMEND-95; 1995 FCC Lexis 630 (rel. 31 January 1995). The FCC deferred action on the applications of AMSC, Elipsat, and Constellation pending the receipt of more extensive evidence of financial capability to construct, launch, and operate for one year.

cies cannot follow on a timely basis with spectrum allocations and licensing rules.

13.6 U.S. DEREGULATORY INITIATIVES AFFECTING SATELLITE TELECOMMUNICATIONS

13.6.1 Private, Noncommon Carriage

U.S. procompetitive international telecommunications initiatives include the creation of a new category of carrier that is substantially free of regulation. Private carriers have great flexibility in determining whom to serve, what prices to charge, and what services to offer.[34] They operate without conventional common carrier obligations, because they lack the ability to affect the price or supply of international transmission capacity.[35] Private carriers can customize services and can stimulate a previously static market characterized by little price competition.[36]

The creation of an alternative to the conventional common carrier systems means that users have an option to negotiate contracts rather than take service on tariffed, quasipublic terms and conditions.

---

34. For an outline of the legal and regulatory nature of private telecommunications carriers, see R. Frieden, "Can the FCC's Private Carrier Concept Gain International Acceptance?" *Telematics*, Vol. 6, No. 2, February 1989, p. 8; and S. Harris, "Regulation of 'Private' Telecommunications Networks," *Telecommunications Law J.*, Vol. 1, No. 1, 1990, p. 1.

35. In Pacific Telecommunications Cable, Inc. Request for Clarification of Policies Concerning Use of Independent Cables, 4 FCC Rcd. 4454 (1989), the FCC stated that it could use its public interest mandate to assess the economic impact of private carriers in terms of traffic diversion from conventional common carrier systems. However, it emphasized that even in the face of substantial traffic migration, a prima facie case of economic harm to conventional carriers would not arise. Furthermore, private international carriers have no restrictions on the type of capacity and services they can make available, nor do common carrier lessees of such capacity face a higher burden when applying for authority under Sec. 214 of the Communications Act to use the capacity.

36. A partial reason for the lack of price competition stems from the fact that until quite recently, all carriers invested in the same transmission facilities, meaning that they incurred roughly the same costs. The availability of facilities not deployed by the typical large consortium of carriers may create some, but not substantial, price differentials. Another FCC policy that blunted cost differentials was the requirement, until 1992, that where one carrier agreed to convey capacity in an operating cable, it must use a depreciated original cost standard for setting the price. (See Reevaluation of the Depreciated-Original Cost Standard in Setting Prices for Conveyances of Capital Interests in Overseas Communications Facilities Between or Among U.S. Carriers, CC Docket No. 87-45, 7 FCC Rcd. 4561 (1992).) The FCC previously decided not to require carriers to make such conveyances to other carriers or users. (International Communications Policies Governing Designation of Recognized Private Operating Agencies, Grants of IRUs in International Facilities and Assignment of Data Network Identification Codes, CC Docket No. 83-1230, Report and Order, 104 FCC 2d 208 (1986), on recon., 2 FCC Rcd. 7375 (1987).)

13.6.2   Private Satellite Carriers and Earth Station Operators

The FCC has extended the private carrier option to most operators of satellites and Earth stations. In 1985, the executive branch articulated and the FCC implemented a policy supporting the conditional licensing of satellite systems separate from the INTELSAT cooperative [31]. The FCC granted a number of applications for the private carrier operation of international satellite systems.[37] The FCC opted for the private carrier designation, notwithstanding the fact that no other country recognized the concept or afforded such broad regulatory flexibility. The FCC may have chosen the private carrier classification to promote competition and to facilitate substantial foreign investment without having to impose more extensive regulatory burdens.[38]

The FCC's willingness to authorize non-common-carrier operations contrasts with a long history of favoring common carrier operations, and before that, of limiting the extent to which even common carriers could operate Earth stations. At the beginning of satellite communications, the FCC in 1965 permitted Comsat alone to operate the nation's first international Earth stations [32]. The FCC revised its policy in 1966 to allow a consortium of common carriers to operate Earth stations, provided Comsat retained a 50% or greater ownership share and served as manager of the facilities [33].

Eighteen years later in 1984, reflecting an increasingly deregulatory predisposition, the FCC changed its policy to allow Earth station operations independent of the Comsat-managed consortium structure [34]. In quick succession, the FCC licensed independent common carrier Earth stations[39] and determined that certain INTELSAT receive-only Earth stations did not require a license, not-

---

37. See, for example, *Pan American Satellite, Inc.*, 101 FCC 2d 1318 (1985) (conditional authorization), 2 FCC Rcd. 7011 (1987) (final C-band authorization), 3 FCC Rcd. 677 (19988) (final Ku-band authorization); Orion Satellite Corp., 101 FCC 2d 1302 (1985) (conditional authorization), 5 FCC Rcd, 4937 (1990) (granting authority to construct, launch an operate subject to filing an expanded explanation of its financial plan), 6 FCC Rcd. 4201 (1991) (granting final authority).

38. In 1992, the FCC modified its policy that previously treated any foreign-owned or controlled carrier providing international services from the United States as "dominant" for purposes of determining the scope of regulatory oversight. (Regulation of International Common Carrier Services, CC Docket No. 91-360, Report and Order, 7 FCC Rcd. 7331 (1992).) To encourage reciprocal policy liberalization, the FCC will apply the dominant carrier burden only on those specific routes in which a foreign carrier's domestic counterpart has the ability to discriminate against unaffiliated U.S. international carriers through control of bottleneck services and facilities. (See also Market Entry and Regulation of Foreign-Affiliated Entities, IB Docket No. 95-22, Notice of Proposed Rulemaking 10 FCC Rcd. 5256 (1995) (proposing to promote competition and reciprocal market access in assessing service applications filed pursuant to Sec. 214 of the Communications Act and considering waiving foreign ownership limitations pursuant to Sec. 310 of the Communications Act).)

39. See, for example, International Relay Inc., FCC No. 84-125 (rel. 11 April 1984) discussed in TRT Telecom. Corp. v. FCC, 876 F.2d at 137–138.

withstanding language in the Communications Satellite Act that addressed carrier licensing [35].

The FCC determined that it could grant Earth station authorizations to "any applicant" and that the language in the Communications Satellite Act, which appeared to restrict who could operate Earth stations, applied only to a limited set of "large satellite terminal stations which would be built as part of the global satellite system and which would become an integral part of the terrestrial networks of the U.S. common carriers" [36]. The FCC concluded it could grant non-common-carrier licenses to operators of Earth stations not "operationally connected" with a terrestrial communications system (i.e., not intended to become an integral part of the common carrier infrastructure), even if such operators leased private lines from common carriers to connect the Earth station with customers. The reviewing court affirmed the FCC primarily on the basis of the FCC's statutory interpretation.[40]

The FCC has deregulated or lightly regulates the operation of all international satellite Earth stations, regardless of who owns them or whether they provide common carrier services. In 1986, Reuters Information Services, a subsidiary of the British firm Reuters Limited, requested the FCC to issue a declaratory ruling specifying whether the FCC had authority to license non-common-carrier operation of transmit/receive Earth stations using the INTELSAT system.[41] The FCC declared that it had regulatory authority to license such operations [37] under the Communications Act of 1934, 47 U.S.C. Secs. 151–611 (1992)[42] and the Communications Satellite Act of 1962, 47 U.S.C. Secs. 701–744 (1992).[43]

---

40. In the absence of clear legislative intent, a court may not impose its own construction of the statute if the expert administrative agency has provided a reasonable interpretation. (See Chevron USA, Inc. v. NRDC, 467 U.S. 837, 843 (1984).)

41. In 1986, the FCC also determined that customers of INTELSAT's INTELNET I service could operate small receive-only Earth stations on the customer's premises without a license. (Deregulation of Receive-Only Earth Stations Operating With the INTELSAT Global Communications Satellite System, Declaratory Ruling, RM No. 4845, FCC 86-214 (rel. 19 May 1986).) More recently, the FCC granted Comsat a waiver of the licensing requirement for international receive-only Earth stations accessing the INTELSAT K satellite for a number of video and audio services, provided such reception: (1) involves stand-alone, passive devices not an integral part of (i.e., not "operationally connected with") any common carrier terrestrial network, (2) involves direct-to-user applications, (3) involves reception of encrypted video and audio transmissions, and (4) does not infringe on any obligations owed to INTELSAT. (See Communications Satellite Corp. Request for Waiver of Sec. 25.131(J)(1) of the Commission's Rules as It Applies to Services Provided Via the INTELSAT K Satellite, 7 FCC Rcd. 6028 (Com. Car. Bur. 1992), application for review pending.

42. Section 307(a) of the Communications Act grants broad authority to the FCC to regulate radio communications, including authority to grant a radio license to "any applicant" if such a grant will serve "the public interest, convenience and necessity."

43. Section 201(c)(7) of the Communications Satellite Act of 1962 authorizes the FCC to "grant ap-

The FCC subsequently granted an application of Reuters to construct and operate an Earth station, "provided that the station is operated on a non-common-carrier basis for private use by the licensee and will not be operationally connected with the terrestrial communications system within the meaning of Sections 103(2) [defining satellite terminal station] and 207(a) of the Communications Satellite Act" [38].

13.6.3 License-Free Operation of All Receive-Only Terminals

In 1993, the FCC began considering a rule modification that would expand the extent to which it would permit license-free operation of international receive-only Earth stations.[44]

The FCC contemplated allowing license-free operation of Earth stations that receive any type of service from INTELSAT, international satellite systems separate from INTELSAT,[45] or fixed-satellite services from other countries using either domestic or foreign satellites. The FCC's proposal paralleled what it had done domestically[46] in response to marketplace and technological developments that favor the proliferation of VSAT Earth stations, typically installed on user premises.[47] The FCC determined that a license requirement was not necessary for purposes of satisfying obligations to INTELSAT, Inmarsat, or ITU Radio Regulations [39].

---

propriate authorizations for the construction and operation of each satellite terminal station, either to...[Communications Satellite C]orporation, or to one or more authorized carriers or to the corporation and one or more such carriers jointly, as will best serve the public interest convenience and necessity." (47 U.S.C. Sec. 721 (c) (7).)

44. See Amendment of Sec. 25.131 of the Commission's Rules and Regulations to Eliminate the Licensing Requirements for Certain International Receive-Only Earth Stations, CC Docket No. 93-23, Notice of Proposed Rulemaking, 8 FCC Rcd. 1720 (1993).

45. For these types of satellites the FCC will require completion of all technical and economic consultations required under Article XIV(d) of the INTELSAT Agreement before permitting license-free reception.

46. See Amendment of Part 25 of the Commission's Rules and Regulations to Reduce Alien Carrier Interference Between Fixed-Satellites at Reduced Orbital Spacing and to Revise Application Processing Procedures for Satellite Communications Services, Notice of Proposed Rulemaking, 2 FCC Rcd. 762 (1987), Report and Order, 6 FCC Rcd. 2806 (1991) (eliminating the licensing requirement for domestic receive-only Earth stations in favor of a voluntary registration program); see also Deregulation of Domestic Receive-Only Satellite Earth Stations, 104 FCC 2D 348 (1986).

47. "For instance, the high power and large coverage area of...[recent vintage] satellite[s] permit the downsizing of receive-only Earth stations to such small dimensions that it is now economically feasible to transmit international video and audio programming directly to user locations that potentially number in the thousands. Administration of a licensing program for these stations would be burdensome and possibly hinder the rapid introduction of these new services." (International Receive-Only Earth Station NPRM, 8 FCC Rcd. at para. 8.)

13.6.4 Expanded Non-Common-Carrier Licensing of Transmit/Receive Terminals

In 1993, the FCC also issued a Declaratory Ruling that qualified video transmit/receive Earth stations for licensing on a non-common-carrier basis under Title III of the Communications Act.[48] Acting on a petition for Brightstar, a British corporation operating in the United States, the FCC agreed to license, on a private carrier basis, enterprises that receive international video traffic and redistribute it via domestic satellites. The FCC used its Reuters Declaratory Ruling as the basis for determining that video distribution arrangements, which were not private in the sense that they served a number of unaffiliated customers, nevertheless did not constitute the kind of satellite terminal stations constituting an integral part of the terrestrial networks of U.S. common carriers.[49]

13.6.5 Foreign-Ownership Issues

As foreign-controlled corporations, both Reuters and Brightstar raise questions of their eligibility to receive a license from the FCC that involves the use of spectrum. Section 310 of the Communications Act of 1934 imposes restrictions on the extent to which aliens can own or participate in the ownership of stations licensed in the United States. Section 310(a) bans licensing radio facilities, including international Earth stations, to foreign governments or representatives. Section 310(b) places limits on the ownership share of broadcast and common carrier facilities, but does not bar foreign ownership of radio facilities by non-common carriers.[50]

The FCC concluded that Brightstar's video distribution service did not constitute broadcasting or common carriage [40]:

> Its business is intended to service specific commercial customers under private contract, not the public at large. Brightstar's transmissions are intended to be on a point-to-point or, at times, point-to-mul-

---

48. See Licensing Under Title III of the Communications Act of 1934, as Amended, of Non-Common Carrier Transmit/Receive Earth Stations Operating With the INTELSAT Communications Satellite System, File No. I-S-P-92-002, Declaratory Ruling, 8 FCC Rcd. 1387 (1993).

49. "We conclude that our analysis in Reuters of 'satellite terminal stations,' under Section 201(c)(7), applies equally to all noncommon carriers, whether using their facilities for private use (as was the case in Reuters), or for providing service to others. That a noncommon carrier connects a common carrier–supplied private line to its Earth station does not alone make that Earth station an integral part of a carrier's network, and hence a 'satellite terminal station.' To become an integral part of the carrier's network, the Earth station would have to be used to exchange the carrier's common carrier traffic with the global satellite system." (Title III Licensing of Private Carrier Transmit/Receive International Earth Stations, 8 FCC Rcd. at para. 10).

50. See Brightstar Declaratory Ruling, para. 20, note 6.

tipoint basis, and not over the air for reception by any individual with a properly tuned receiver.

The FCC uses the non-common-carrier classification to afford regulatory flexibility in markets where the FCC wants to stimulate competition, dismantle regulatory burdens, and allow relatively free market entry by foreign enterprises. Private carriers enjoy significant freedom to operate, notwithstanding the FCC's refusal to modify the industrial structure of the international satellite, characterized by legislatively or regulatively imposed market tiering.

The market for international telecommunication transmission facilities remains limited to a few satellite cooperatives and companies and a few cable consortia. As in most nations, the U.S. continues to support an exclusive INTELSAT signatory, Comsat, operating as a carrier's carrier. However, captive lessees of INTELSAT capacity no longer have to rely on Comsat to operate and manage Earth station facilities. While international satellite operators cannot bypass Comsat and deal directly with INTELSAT, they can take their leased capacity, interconnect it with private lines, and make service generally available to the public without incurring common carrier regulatory burdens.

References

[1] P.L. 87-624, 76 Stat. 419, 31 August 1962, codified at 47 U.S.C. Sec. 702(A).

[2] Preamble of *Agreement Establishing Interim Arrangements for a Global Commercial Satellite System (Interim Agreement)*, signed 20 August 1964, 15 U.S.T. 1705, T.I.A.S. No. 5646; see *Agreement Relating to the International Telecommunications Satellite Organization "INTELSAT" (INTELSAT Agreement)*, 20 August 1971, entered into force 12 February 1973, 23 U.S.T. 3813, T.I.A.S. No. 7532, *Operating Agreement Relating to the International Telecommunications Satellite Organization "INTELSAT" (INTELSAT Operating Agreement)*, 20 August 1971, entered into force 12 February 1973, 23 U.S.T. 3892, T.I.A.S. No. 7582.

[3] Reagan, R., Presidential Determination No. 85-2, 49 Fed. Reg. 46,987 (28 November 1984), implemented in Establishment of Satellite Systems Providing International Communications, 101 FCC 2d 1046 (1985), on recon., 61 Rad Reg. 2d (P&F) 649 (1986), on further recon., 1 FCC Rcd, 439 (1986) policy liberalized in Permissible Scope of United States Licensed International Communications Satellite Systems Separate From the International Telecommunications Satellite Organization (INTELSAT), 7 FCC Rcd. 2313 (1992).

[4] INTELSAT Agreement, Art. XIV(d).

[5] Establishment of Satellite Systems Providing International Communications, 101 FCC 2d 1046 (1985), on recon. 61 Rad. Reg. 2d (P&F) 649 (1986), on further recon., 1 FCC Rcd. 429 (1986); policy liberalized in, Permissible Scope of United States Licensed International Communications Satellite Systems Separate From the International Telecommunications Satellite Organization (INTELSAT), 7 FCC Rcd. 2313 (1992).

[6] Pan American Satellite, Inc., 101 FCC 2d 1318 (1985) (conditional authorization), 2 FCC Rcd, 7011 (1987) (final C-band authorization), 3 FCC Rcd 677 (1988) (final Ku-band authorization).

[7] Communications Satellite Act of 1962, Pub. L. No. 87-624, Sec. 101 et seq., 76 State. 419, codified at 47 U.S.C. Sec. 701(a) (1990).

[8]   Authorized User Policy, 97 FCC 2d 296 (1984), reaff'd, 99 FCC 2d 177 (1985), aff'd sub nom.,
      Western Union Int'l v. FCC, 804 F.2d 1280 (D.C. Cir. 1986).
[9]   Communications Satellite Act of 1962, as amended, Sec. 735(a) (1986).
[10] Columbia Communications Corp. Application for Auth. to Use and Offer for Lease the C-Band
      Transponder on the NASA TDRSS Satellites, 7 FCC Rcd. 122 (1992).
[11] Pan American Satellite Corp., Petition for Rulemaking to Modify Commission Policies Estab-
      lished in CC Docket No. 84-1299 Relating to Separate Satellite Systems, RM-7562 (filed 18
      July 1990).
[12] Letter from Secretary of Commerce Robert Mosbacher and Secretary of State James A. Baker
      III to FCC Chairman Alfred Sikes, 27 November 1991.
[13] Permissible Scope of United States Licensed International Communications Satellite Systems
      Separate From the International Telecommunications Satellite Organization (INTELSAT), 7
      FCC Rcd. 2313 (1992).
[14] International Telecommunications Satellite Organization, News Release, No. 92-09, 6 March
      1992.
[15] PanAmerican Satellite Company, White Paper: *A New Private Enterprise INTELSAT*, Execu-
      tive Summary, 20 April 1992, para. 9.
[16] *Final Acts of the Plenipotentiary Conference, Nice, 1989*, Resolution No. 1, Future Confer-
      ences of the Union, Sec. 1.4.
[17] U.S. Dept. of State, *United States Proposals for the World Administrative Radio Conference,
      Malaga-Torremolinos, Spain, 1992*, Washington, D.C., July 1991 (hereafter cited as U.S.
      Proposals).
[18] U.S. Proposals, p. 4.
[19] "WARC Comes to an End: Final Acts Approved," *PCN News*, Vol. 3 No. 5, 5 March 1992, p. 1.
[20] "U.S. Big LEOs Get Allocations at WARC Largely as Proposed, but Limits Aimed at Protecting
      Russian Glonass System Could Restrict IRIDIUM; CEPT Nations Get 230 MHz for Future Pub-
      lic Land Mobile Service; BSS-Sound Gains Worldwide Allocation at L-Band," *Telecommuni-
      cations Reports*, Vol. 58, 9 March 1992, p. 4 (hereafter cited as WARC-92 Summary).
[21] Baran, J., "U.S. Success at World Administrative Radio Conference (WARC'92): Wrap-up Ca-
      ble," 3 March 1992, para. 13.
[22] U.S. WARC-92 Proposals, pp. 2–3.
[23] Department of Transport and Communications, Radiocommunications Division, Canberra,
      *Australian Proposals for the World Administrative Radio Conference for Dealing With Fre-
      quency Allocations in Certain Parts of the Spectrum, Malaga-Torremolinos, February 1992*,
      20 November 1991, p. 3.
[24] WARC-92 Summary, pp. 12, 14.
[25] Marcus, D., "Delegates Bestow Mobile Mandate," *Space News*, Vol. 3, No. 9, p. 20.
[26] FCC Asks for Comments Regarding the Establishment of an Advisory Committee to Negotiate
      Proposed Regulations, CC Docket No. 92-76, Public Notice, Release No. DA 92-443, 7 FCC
      Rcd. 2370, 1992 FCC Lexis 1968 (rel. 16 April 1992); see also Alternate Dispute Resolution,
      Initial Policy Statement and Order, 6 FCC Rcd. 5669 (1991).
[27] Little LEO Negotiated Rulemaking Notice, 7 FCC Rcd. at 2370.
[28] Non-Voice, Non-Geostationary Low Earth Orbit Satellites Applications Accepted for Filing,
      Public Notice, Report No. DS-1484, 9 FCC Rcd. 7921 (1994).
[29] Amendment of Sec. 2.106 of the Commission's Rules to Allocate the 1610–1626.5 MHz and
      2483.5–2500 MHz Band for Use by the Mobile-Satellite Service, Including Non-geostationary
      Satellites, Notice of Proposed Rule Making and Tentative Decision, ET Docket No. 92-28, 7
      FCC Rcd. 6414, 1992 FCC Lexis 5094 (rel. 4 September 1992).
[30] FCC Asks for Comments Regarding the Establishment of an Advisory Committee to Negotiate
      Proposed Regulations, CC Docket No. 92-166, Public Notice, Release No. DA 92-1085, 7 FCC
      Rcd. 5241, 1992 FCC Lexis 4451 (rel. 7 August 1992).

[31] Establishment of Satellite Systems Providing International Communications, 101 FCC 2d 1046 (1985), recon. den., 61 Rad. Reg.2d (P&F) 649 (1986), further recon., 1 FCC Rcd. 439 (1989), modified 7 FCC Rcd. 2313 (1992) (reflecting changes in the INTELSAT consultation process).

[32] Proposed Global Commercial Communications Satellite System, 38 FCC 1104 (1965), recon. denied in pertinent part, 2 FCC 2d 658 (1966).

[33] Ownership and Operation of Earth Stations, Second Report and Order, 5 FCC 2d 812 (1966).

[34] Modification of Policy on Ownership and Operation of U.S. Earth Stations, Report and Order, 100 FCC 2d 250 (1984).

[35] Deregulation of Receive-Only Satellite Earth Stations Operating With the INTELSAT Global Communications Satellite System, RM No. 4845, Declaratory Ruling, FCC No. 86-214 (rel. 19 May 1986).

[36] Private Carrier Declaratory Ruling, 3 FCC 2d at 1587.

[37] Licensing Under Title III of the Communications Act of 1934, as amended, of Private Transmit/Receive Earth Stations Operating With the INTELSAT Global Communications Satellite System, Declaratory Ruling, 3 FCC Rcd. 1585 (1988), aff'd sub nom., TRT Telecommunications Corp. v. FCC, 876 F.2d 134 (D.C. Cir. 1989).

[38] Reuters Information Services, Inc., 4 FCC Rcd. 5982 (Com. Car. Bur. 1989) (granting Earth station license provided the facility is operated on a non-common carrier basis for private use by the licensee and not operationally connected with terrestrial communication systems).

[39] International Receive-Only Earth Station NPRM, 8 FCC Rcd., para. 8.

[40] Brightstar Declaratory Ruling, para. 20.

Spectrum Management

The efficient management of spectrum constitutes an essential element for effective use of a nation's telecommunications infrastructure.[1] While few countries have yet opted to treat spectrum like real estate and create a market for its sale, spectrum has substantial, if unrealized, value,[2] particularly when demand far exceeds the amount of bandwidth allocated.[3]

Some spectrum uses have the characteristic of a public good in that one person's consumption of, for example, an educational program on broadcast television does not exhaust or reduce what can be received by others. Spectrum also can constitute a "common pool" economic resource, like offshore drilling sites owned by the government, in that it is exhaustible, is subject to congestion, can be allocated for specific uses, and can be sold or leased to particular users. Technological innovations have enabled productive use of progressively higher frequencies and the ability to derive usable channels with less bandwidth. But along with innovations, which conserve spectrum and provide more throughput, are new ideas and services that generate additional spectrum requirements.

Because of increasing demand for spectrum and the costs incurred by incumbents or newcomers to conserve it, international and national agencies

---

1. The U.S. government has begun to realize the need for effective and market-oriented spectrum management. (See U.S. Dept. of Commerce, National Telecommunications and Information Administration, *U.S. Spectrum Management Policy: Agenda for the Future*, NTIA Spec. Pub. 91-23, Washington, D.C.: GPO, February 1991, p. 13 (hereafter cited as *U.S. Spectrum Management Policy*.)

2. In 1990, shipments of radiocommunication equipment generated over $55 billion. (*U.S. Spectrum Management Policy*, Executive Summary). NTIA estimated the spectrum value of cellular radio services, which consumes 50 MHz, to be over $79 billion. (See Ibid., Appendix D, "Estimating the Value of Cellular Licenses.")

3. Sales of VHF television stations in major markets can exceed $500 million, far in excess of the physical assets involved. (See H. Geller and D. Lampert, *Charging for Spectrum Use*, Washington, D.C.: Benton Foundation Project on Communications and Information Policy Options, 1989, p. 13.)

---

must conserve and manage spectrum. This endeavor involves allocating spectrum among competing uses and serving as a traffic cop of the airwaves to avoid interference and to resolve conflicts. Spectrum managers need to fashion compromises based on a number of factors, including:

- *Technology:* The duty to prevent harmful interference and to achieve efficient activation of channels. For example, in allocating spectrum for broadcast television in the VHF band, the FCC had to create large geographical spacing between stations to prevent interference. This limited the number of available stations in any locality, thereby generating demand for an additional allocation in the UHF band.
- *Regulatory policy:* Regulation may direct spectrum allocations in ways designed to serve public policies. For example, the FCC sought to promote the doctrine of localism by allotting broadcast channels for as many different localities as technologically possible. This policy reduced the number of stations available in urban localities that otherwise could have served nearby towns.
- *Commerce:* The need to conduct a comparison of spectrum requirements with an eye toward allocating spectrum to uses that will maximize social welfare primarily and individual profitability of firms secondarily. For example, the FCC reallocated portions of the UHF television band for mobile radio services when it determined that most localities could not support a full inventory of UHF television stations, but desperately needed additional spectrum for public safety and private wireless services. Commerce concerns may also include maximization of revenues for the general treasury by auctioning spectrum to the highest bidder.
- *Social welfare:* The public interest merit in allocating spectrum for a particular service in the face of other requirements that accordingly have to make do with less, different, or possibly no spectrum. For example, in allocating spectrum for new wireless mobile services like personal communications networks, the FCC forced existing microwave users like railroads and public utilities first to share the spectrum and subsequently to move to higher, less congested frequencies.
- *National security:* Compelling requirements for safety, public welfare, national defense, and emergency applications. For example, the ITU has allocated particular emergency calling frequencies.

Spectrum allocation decision making blends engineering with social sciences to provide the basis for making cost/benefit analyses. Policy makers need to determine who deserves spectrum allocations and what entitlements accrue to incumbent users. Managers must devise procedures for assessing new spectrum requirements and determining how to accommodate incumbent users and operators.

Currently, most nations generally allocate spectrum on the basis of consensus decisions reached at global or regional conferences convened under the auspices of the ITU. In the United States, the FCC uses the public interest as the basis for determining whether to implement ITU decisions by incorporating changes in the national table of spectrum allocations.[4] The FCC also typically uses the public interest standard to assign spectrum, but in 1993 it received Congressional authority[5] [1] to auction portions of the spectrum. The FCC also imposes user and licensing fees to compensate it for processing applications and granting licenses and for the cost of regulation. Demand typically will exceed supply without a market mechanism[6] for clearing out spectrum inventory to the highest bidder.[7]

---

4. See 47 C.F.R. § 2.106 (1995).

5. See also H.R. 707, "Emerging Telecommunications Technologies Act of 1993"; U.S. Congress, House Report 103-19, "Emerging Telecommunications Technologies Act of 1993," 103d Congress (24 February 1993); Policies and Rules for Licensing Fallow 800 MHz Specialized Mobile Radio Spectrum Through a Competitive Bidding Process, 7 FCC Rcd. 8590 (1992) (denying petition for competitive bidding access to spectrum for mobile radio due to lack of explicit legislative authority). The FCC's in-house "think tank" endorsed spectrum auctioning for nonbroadcast spectrum, particularly for advocates of new services who need already used spectrum. (See, for example, D. Webbink, *Frequency Spectrum Deregulation Alternatives*, FCC Office of Plans and Policy Working Paper No. 2, October 1980; E. Kwerel and A. Felker, *Using Auctions to Select FCC Licensees*, FCC Office of Plans and Policy Working Paper No. 16, May 1985.) The Commission has authorized cash payments from newly licensed applicants to incumbent users for relocating to another frequency band before spectrum reallocations subordinated their status. (See Redevelopment of Spectrum to Encourage Innovation in the Use of New Telecommunications Technologies, ET Docket No. 92-9, Notice of Proposed Rulemaking, 7 FCC Rcd. 1542 (1992) (proposing to reserve 220 MHz in the 2-GHz band for new services like personal communications networks while accommodating incumbent microwave licensees), Further Notice of Proposed Rulemaking, 7 FCC Rcd. 6100 (1992), 1st Report and Order, 7 FCC Rcd. 6886 (1992), 2d Report and Order, 8 FCC Rcd. 6495 (1993) (reallocating spectrum above 3 GHz to ease migration of incumbents in 2-GHz bands to accommodate emerging technology mobile services); 3d Report and Order and Memorandum Opinion and Order, 8 FCC Rcd. 6589 (1993) (adopting a plan for fair sharing in the 2-GHz band between incumbent microwave users and emerging technology services and for accelerated relocation by incumbents).) The FCC can also collect fees to recoup the cost of regulation. (See Implementation of Sec. 9 of the Communications Act, MD Docket No. 94-19, Report and Order, 9 FCC Rcd. 5333 (1994).)

6. Until passage of enabling legislation, the FCC could consider marketplace forces when making spectrum allocations, but could not auction off spectrum to the highest bidder. "Nothing on Sections 303(A)-(C) [of the Communications Act] suggests that the Commission is not permitted to take into account marketplace forces when exercising its spectrum allocation responsibilities under the public interest standards." (Amendment of Parts 2, 15 and 90 of the Commission's Rules and Regulations to Allocate Frequencies in the 900 MHz Reserve Band for Private Land Mobile Use, GEN Docket Nos. 84-1231, 1233, 1234, Report and Order 2 FCC Rcd. 1825, 1939 (1986).)

7. In 1959, R. H. Coase argued for a definitive spectrum ownership rights as a market-driven way to foreclose chaos and interference. (R. H. Coase, "The Federal Communications Commission," *J. Law and Economics*, Vol. 2, 1959, p. 1). In 1991, the chairman of the FCC reported that the

New Zealand leads all nations in the creation of a spectrum marketplace with passage of the Radiocommunications Act of 1989, which privatized frequencies between 44 MHz and 3.6 GHz. The law created a management right equivalent to an exclusive, transferable 20-year ownership interest. By auctioning unoccupied frequencies, New Zealand established a property-based alternative[8] to public interest-based spectrum allocation and assignment.

14.1 THE ROLE OF SCARCITY IN SPECTRUM MANAGEMENT, LICENSING, AND REGULATION

Policy makers and courts have justified regulation of spectrum usage on account of spectrum scarcity and the inability of all desiring access to reserve a license. Scarcity has different meanings, and some critics of regulation dispute that it exists at all in an age of spectrum-conserving technological innovations and closed-circuit options.[9] One can consider spectrum scarcity from a number of perspectives: technological, financial, marketplace, allocational, and resource.

14.1.1 Static and Dynamic Technologies

From the technological perspective, one can look at innovation as forestalling scarcity with ways to achieve more throughput in a digital format at faster transmission speeds carried via closed-circuit fiber-optic lines which do not use

---

spectrum requested for new services exceeded 1,200 MHz, while the FCC had only 3 MHz of unallocated spectrum available to accommodate such requests. (*Emerging Telecommunications Technology Act of 1991: Hearings on H.R. 1407 Before the Subcomm. on Telecommunications and Finance of the House of Representatives*, 102 Cong., 1st Sess. (1991) (prepared testimony of Alfred C. Sikes).)

8. Douglas Webbink identifies the following rights accruing from property-based ownership of spectrum: (1) exclusive access to and use of the property, (2) the right to generate income from the property, (3) freedom to transfer the property right to someone else, and (4) freedom from confiscation of the property without some extreme situation, such as war, nonpayment of tax, or natural catastrophe. (See D. Webbink, "Radio Licenses and Frequency Spectrum Use Property Rights," *Communications and the Law*, Vol. 9, June 1987 pp. 3–29); see also T. W. Hazlett, "The Rationality of U.S. Regulation of the Broadcast Spectrum," *J. Law and Economics*, Vol. 33, April 1990, pp. 133-175; D. Huang, *Managing the Spectrum: Win, Lose or Share*, P-93-2, Harvard University, Center for Information Policy Research, Program on Information Resources Policy, February 1993; T. W. Hazlett, *The Political Economy of Radio Spectrum Auctions*, Working Paper Series No. 1, Inst. of Governmental Affairs, University of California, Davis, June 1993.

9. See, for example, N. Negroponte, "Products and Services for Computer Networks," *Scientific American*, Vol. 265, September 1991, pp. 106–113; and G. Guilder "Into the Telecosm," *Harvard Business Review*, Vol. 60, No. 2, March/April 1991, pp. 150–161 (supporting the view that no real spectrum scarcity exists because many users of radio spectrum can migrate to closed-circuit media like fiber-optic cables).

spectrum. This dynamic view of technology considers spectrum shortages unlikely, just as innovations in agriculture prevent widespread famine despite massive increases in population. For example, compression technology provides the basis for deriving four channels or more from the bandwidth of spectrum that previously could generate only one channel. Compression provides the basis for 500-channel cable television systems and for 100-channel direct-to-home satellite broadcasting. The technology conserves spectrum by allocating processing power only to changes detected from the preceding frame of information, sound, or video.

A static view of innovation considers spectrum scarcity a very real threat as the population grows and increasing requirements for frequencies exceed spectrum-conserving innovations. For example, in many urban corridors the demand for microwave frequencies has reached a point where additional licenses cannot be issued due to the likelihood of interference. New or additional requirements typically can be served only if the applicant agrees to operate on higher frequencies. The migration to another frequency band can impose higher costs, because existing operators must procure new equipment, and often higher frequencies require more expensive equipment and closer spacing of relay towers due to poorer transmission characteristics.

The technological template of scarcity considers spectrum from either a positive or pessimistic viewpoint. For example, increases in radio modem processing speeds mean that equipment using the same amount of bandwidth can operate faster and with greater throughput. It also means that less spectrum would be needed to operate at the slower, formerly cutting-edge speeds. The dynamic view believes some users will do more with less, while the static view believes that most users will occupy the same or more bandwidth, despite the increase in throughput and productivity.

14.1.2 Financial Scarcity

The financial approach looks at the individual's ability to pay for desired spectrum. Spectrum cost varies as a function of the cost of technology needed to activate the spectrum, whether the user must pay for the right to use the spectrum and the services that can be provided.[10] Significant transaction costs create disincentives for individuals to seek spectrum, but the extent of these costs vary with the type and amount of spectrum sought. Relatively few individuals can secure the funds necessary to buy a broadcast facility or to pursue an application through the FCC regulatory process for a new major market broadcast

---

10. While broadcasters continue to secure spectrum without having to bid for it, operators of personal communication network services competed in a high-stakes auction that raised over $7.7 billion for the general treasury. (See D. Rohde, "Big Winners Thrown a Curve as the PCS Auction Closes," *Network World*, Vol. 12, No. 12, 20 March 1995, pp. 8, 85.).

license. On the other hand, virtually anyone can secure spectrum for such applications as private microwave radio networks and unlicensed services such as citizen's band radio, cordless telephones, baby monitors, model airplane controllers, and garage openers.

For high-cost, contested spectrum, the FCC assigns frequencies on an exclusive noninterference basis only to qualified individuals. For low-cost, largely unused spectrum, the FCC makes spectrum available without a screening process on a nonexclusive basis. The FCC does not even require a license for operation of low-powered devices in specific frequency bands, such as cordless telephones, baby monitors, and radio-controlled model cars and planes.

14.1.3 Marketplace-Determined Scarcity

The marketplace approach to spectrum scarcity applies economic analysis and determines where an equilibrium exists that accounts for overall supply and service-specific demand. This approach is academic, because policy makers have yet to allocate and assign most spectrum on the basis of marketplace resource allocation. It appears likely that the FCC and other national regulatory agencies will use competitive bidding to assign some unused or underused spectrum allocated for private or government use. However, major portions of the spectrum, such as broadcasting, public safety, and police bands, are not likely to be auctioned because of countervailing public interest concerns.

Currently, spectrum availability is tied to perceived demand and the comparative public interest merits of the services to be provided. The process of calibrating spectrum bandwidth allocations with need and merit is far from exact. Even as some claim spectrum scarcity now has grown acute, U.S. UHF television channel allotments for specific localities—even for major metropolitan areas—remain unclaimed. Arguably, the FCC has allocated too much spectrum for UHF television relative to what the broadcast television marketplace can support in any particular region and relative to available functional equivalent services like cable television. In 1970, the FCC reallocated UHF television channels 70 to 83 (806 to 890 MHz) to mobile radio services. UHF television overcapacity may yet reach equilibrium with the FCC's decision to allow broadcasters the opportunity to acquire a second 6-MHz channel to simulcast an HDTV signal.[11]

For other services, the claim can be credibly made that the FCC and other national regulatory agencies have not allocated sufficient spectrum relative to the bandwidth needed and the public and private benefits that could accrue. Spectrum advocates for essential requirements like public health, protection,

---

11. See Advanced Television Systems and Their Impact Upon the Existing Television Broadcast Service, MM Docket No. 87-268, Second Report and Order and Further Notice of Proposed Rulemaking, 7 FCC Rcd. 3340, 3344 (1992).

and safety demand spectrum reallocations to accommodate rising demand. To satisfy these demands, agencies must balance the interests of incumbents, who may have failed to activate all the spectrum allotted, such as UHF television broadcasters versus newcomers with compelling requirements.

14.1.4 Allocational Scarcity

Allocational scarcity results when the decision making process fails to balance marketplace imperatives and other possibly countervailing public interest considerations. A regulatory agency may reserve spectrum for particular applications on the basis of a public interest determination that does not necessarily gauge the extent of the beneficiary's actual or perspective use relative to the needs of others. For example, the FCC has reserved spectrum in the FM radio band from 88 to 92 MHz for noncommercial and educational radio, irrespective of the requirements in any particular region.

In many major broadcast markets of the United States, additional commercial broadcasters cannot operate, because the FCC has assigned all usable channels. A disenfranchised broadcaster might be willing to pay the federal government for the privilege to activate an unused noncommercial channel. The prospective broadcaster could get on the air for less than the marked-up price an operating commercial broadcaster would charge, and the number of allotted channels in the noncommercial FM band appears ample enough so that no existing or prospective noncommercial broadcaster would be foreclosed from operating. Despite what appears to be a win-win situation, public interest mandates prevent the FCC from permitting such an arrangement. Spectrum allocational decisions impose real scarcity in terms of access when the regulatory agency reserves otherwise usable and sought-after spectrum for a particular service or application that may not have takers for all of the available allocation.

14.1.5 Resource Scarcity

Spectrum constitutes an exhaustible resource like air, space on a highway, or offshore oil drilling sites. As a finite resource, spectrum must be managed, but scarcity need not be a foregone conclusion. Licensing can ration spectrum allocated on a service-specific basis. On the other hand, spectrum cannot be replenished or expanded as would be the case by planting more trees for more newspaper or widening the highway to accommodate more cars.

Resource scarcity is real and becomes acute when spectrum users have no duty or incentive to maximize efficiency. Without a financial incentive, existing spectrum users may delay replacing outmoded and inefficient technologies, and proponents of new services may face a lengthy administrative process for demonstrating that the public interest favors requiring incumbents to share spectrum, use narrower channels, or even to relocate to another frequency band.[12]

The extent and severity of spectrum scarcity depends on a number of variables, including the number of incumbent users, the scope of sunk investment, the suitability of relocating to another frequency or migrating to a closed-circuit, wireline option, and the costs that would be incurred.

14.2 INTERNATIONAL SPECTRUM ALLOCATION

Domestic spectrum management and policy making cannot operate outside the context of a parallel and interdependent international process [2]:

> Domestic allocations...generally conform to the international Table of Allocations and the Radio Regulations maintained by the I[nternational] T[elecommunication] U[nion] and revised at the W[orld] A[dministrative] R[adio] C[onference]s. Those international allocations and regulations, in turn, are the product of negotiation among many countries, each pursuing national goals.

Beginning with the introduction of wireless telegraph service, spectrum management has constituted a key function of the ITU.[13] Its intermittent WRCs (often referred to by the older acronym, WARC) define services, allocate spectrum on a global or regional basis, and promulgate binding regulations on frequency use. Nations typically implement the consensus decision reached at a WRC,[14] notwithstanding the sovereign right of domestic regulatory authorities to formulate their own rules and policies on spectrum issues.

---

12. See, for example, Redevelopment of Spectrum to Encourage Innovation in the Use of New Telecommunications Technologies, ET Docket No. 92–9, Third Report and Order and Memorandum Opinion and Order, 8 FCC Rcd. 6589 (1993), modified, (adopting a plan for sharing of 2-GHz spectrum between incumbent microwave users and new personal communication service (PCS) operators, including a mechanism by which newcomers could compensate incumbents for expedited movement to a different frequency band), Mem. Op. and Order, 9 FCC Rcd. 1943 (1994), further modification, 2d Mem. Op. and Order, 9 FCC Rcd. 7787 (1994).

13. For a comprehensive history of the ITU, see G. Codding, Jr., *The International Telecommunication Union: An Experiment in International Cooperation*, New York: Arno Press, 1972; G. Codding, Jr., and A. Rutkowski, *The International Telecommunication Union in a Changing World*, Norwood, MA: Artech House, 1982. A more concise summary is available in J. Savage, "The ITU and the Radio Frequency Spectrum: Use and Management of a Shared Universal Resource," Chap. 2 in *The Politics of International Telecommunications Regulation*, Boulder, CO: Westview Press, 1989, pp. 61–129.

14. Decisions taken at WARCs "will determine how and when new radio services will be implemented and will influence the development of new technologies and applications." (U.S. Congress, Office of Technology Assessment, *The 1992 World Administrative Radio Conference:*

Even if domestic regulatory action precedes the ITU process,[15] nations expect the ITU to establish an international or at least a regional frequency allocation based on the consensus decision that particular frequencies best meet the requirements of specific services. The "Radio Regulations that result from ITU conferences have treaty status and provide the principal guidelines for world telecommunications operations" [3]. In this way, other nations can make similar spectrum allocations and assignments, thereby coordinating spectrum usage and reducing the potential for harmful interference.

Conflict results when usage in particular nations exhausts the amount of spectrum allocated for a particular service or when users find the operational rules unduly restrictive. For example, the ITU establishes a hierarchy of access rights and entitlements to protection from interference. *Secondary* service users are subordinate to both existing and future users of the *primary* service designated for a particular frequency band. Under this system, existing and prospective users of primary service have a superior right of access to the allocated spectrum and the privilege to operate free from interfering users. Existing or future secondary service users operate on a subordinate basis and must not cause interference to primary service users.

In many instances, the spectrum allocation process involves a win-lose, zero-sum situation in that one or more authorized users will be designated to the exclusion of other users. Spectrum is earmarked in blocks for specific services with an eye toward matching its technical characteristics with particular user requirements. For example, point-to-point microwave applications need extremely high frequencies (in the gigahertz range) where radio waves propagate in a line-of-sight pattern. Long-distance broadcast applications need middle or high frequencies that are high enough to enable signals to bounce off the ionosphere. More numerous, geographically separated local broadcasters can operate on the same frequencies at lower power or at higher frequencies (e.g., in the FM radio band), which typically do not skip off the ionosphere, instead of traveling close to the ground.

Matching technical characteristics of the spectrum with user requirements requires answers to the following questions:

What application(s) should occupy this frequency band?
For any particular frequency band, certain uses appear best suited, but other commercial, social, or national security factors may prevail, particularly when

---

Issues for U.S. International Spectrum Policy—Background Paper, OTA-BP-TCT-76, November 1991 (hereafter cited as OTA Pre-WARC Study), p. 1.

15. The prospect of an approaching WARC can prompt the FCC to expedite its consideration of a spectrum allocation matter to ensure that the United States will have reached a position it can then advocate for other nations to endorse. "[I]t is necessary to make such authorizations now to clarify future U.S. domestic satellite interests at the upcoming Space WARC." (Licensing Space Stations in the Domestic-Fixed Satellite Service, 101 FCC 2d 223, 224 (1985).)

more than one application can use a frequency band. In the U.S., commercial interests supported allocation of UHF channels 14 to 82 to augment the number of television channels previously available in the VHF band. But more compelling public safety and mobile communication requirements combined with incomplete use of the band by broadcasters have eroded the UHF band so that the upper portion of the band (channels 70 to 82) was reallocated in 1970 and the lower portions (channels 14 to 20) is shared in some regions.

Who should use this allocation and how should they be selected?
The technical characteristics of the frequency band will target certain types of users for particular allocations. For example, militaries, telephone companies, and large, geographically dispersed corporations like railroads have requirements best served by microwave allocations. International broadcasters need allocations that support long-distance transmission via ionospheric skip, while local broadcasters can operate with lower power or higher frequencies. National regulatory authorities have responsibility for matching spectrum requirements with suitable users and subsequently to license qualified users.

How will the spectrum be used?
The ITU Radio Regulations establish spectrum rules of the road that include specifications on what type of transmissions are permitted and what operators must do to avoid causing harmful interference.

Where can the spectrum be used?
ITU conferences attempt to allocate spectrum on a global basis, but frequently allocations are limited to one of three regions: region 1 (Europe and Russia, west of the Urals), region 2 (North, Central, and South America), and region 3 (Asia-Pacific). Nations may make a reservation for any allocation, in effect refusing to go along with the consensus. Alternatively, they may request the ITU to attach a footnote to the allocation, contained in the ITU's Radio Regulations, specifying limitations to the consensus allocation, or a different allocation altogether. Such departures from the consensus typically are rare, because individual nations realize that deviations run the risk of causing or incurring interference from the predominant spectrum use and because of the sense that national manufacturers will have less to gain in manufacturing equipment for frequencies usable in only one nation or region.

The location of spectrum use is also important, because the first duly authorized, registered, and operating user is entitled to continue operating interference-free when other operators subsequently seek to operate. Alternative uses for the frequency typically are permissible only on a restrictive, subordinate basis whereby the newcomer must not interfere with incumbents or future operators providing the service for which the band was allocated.

Incumbent users of spectrum are quite reluctant to relinquish spectrum or to agree that the spectrum can be shared with new services, despite technological innovations that make it possible to make do with less spectrum, or to migrate to higher frequencies. For some, particularly users in developing nations, such technological innovations are too expensive, particularly where less efficient but cheaper equipment is operational and in short supply. Users in developed countries are disinclined to change frequencies or share, based on the view that they should not have to incur additional expense merely to accommodate users in nearby developed nations.

The international spectrum allocation process primarily involves decisions on *how* spectrum will be used and not *who* will use it. The process assumes that nations for the most part will register spectrum requirements as they do for satellites on an a posteriori, first-come, first-served basis. Nations should register uses with the ITU when and if they exist or will soon develop. Developed countries typically will have a head start when it comes to actual use of newly allocated or reallocated spectrum, because they will have the earliest requirements and the resources necessary to convert an allocation into a registered use.

Developing countries have voiced objections to the a posteriori system and have proposed that the ITU allocate some spectrum and satellite orbital slots on an a priori basis (i.e., that nations agree to a plan for reserving shared global resources for use by one or more specific countries in advance of actual requirements). While such a system might foster distributional equity, it could result in a glut of unused spectrum for developing countries and a scarcity for developed countries. The matter of equity versus efficiency in the spectrum allocation process, like that for satellite orbits, has on occasion led to sharp disagreements between developed and developing nations.

Spectrum management has become more controversial at both international and domestic levels because of increasing demand for what is a finite resource by both incumbent users and proponents of new services. WRCs and the domestic regulatory process must balance the duty to ensure that existing users do not experience harmful interference, with the obligation to accommodate new spectrum requirements that would serve the goals of national interest and international comity.

WRC decisions on spectrum allocation can either validate previous domestic actions or isolate nations whose previous or subsequent spectrum allocations do not jibe with the international consensus.[16] "The failure to aggressively link long-term international policy efforts with domestic needs could threaten

---

16. "The more advanced our technology becomes, and the more complicated our frequency utilization, the more apparent it is that there must be complete correlation of the national and international aspects of frequency use." (H. Felloes, Testimony at Hearings Before a Subcommittee of the Committee on Interstate and Foreign Commerce on Allocations of Radio Spectrum Between Federal Government Users and Non-Federal Government Users, 86th Cong., 1st Sess., 8–9 June 1959, p. 36, quoted in OTA Pre-WARC-92 Study, p. 9.

U.S. technological and policymaking leadership and could undermine future success in U.S. international spectrum policymaking" [4]. However, with limited exceptions, the United States concurs with the consensus decisions reached on spectrum allocation and service definitions reached at WRCs.

14.3 SPECTRUM MANAGEMENT IN THE UNITED STATES

In the United States, the Communications Act of 1934[17] authorizes the FCC to allocate and license the nongovernmental use of radio spectrum, while NTIA holds delegated authority to manage and register the federal government's allocations.[18] Some of the frequency spectrum is shared on a coordinated basis by both federal and nonfederal users.

Spectrum management in the United States largely has been reactive.[19] Because adequate unused spectrum was available, managers could respond to new requirements and proposals of entrepreneurs with new uses for spectrum, rather than plan for them. Growing commercial and governmental requirements and the proliferation of technological innovations that use spectrum require proactive planning.[20]

---

17. Pub. L. No. 73-416, 48 Stat. 1064 (1934) (codified as amended at 47 U.S.C. Secs. 151–610 (1990). For a legislative history of the Communications Act, see G. Robinson, "The Federal Communications Act: An Essay on Origins and Regulatory Purpose," in *A Legislative History of the Communications Act of 1934*, M. Paglin, ed., 1989, p. 9; see also Note, "Allocating Spectrum by Market Forces: The FCC Ultra Vires?" 37 Cath. U. L. Rev. 149 (1987).

18. Under Sec. 305 of the Communications Act of 1934, as amended, the President retains the authority to assign frequencies to all radio stations belonging to the federal government. The President has delegated this authority to the secretary of commerce, who has in turn delegated it to NTIA. (See Exec. Order No. 12,046, as amended, 3 C.F.R. Sec. 158, reprinted in 47 U.S.C. Sec. 305 app. at 127 (1990).) See also U.S. Dept. of Commerce, Department Organization Orders 10-19 and 25-7. The applicable executive order and an Office of Management and Budget Circular No. A-11 "provide NTIA with the power to assign frequencies and approve the spectrum needs of new [federal] systems." (*U.S. Spectrum Management Policy*, p. 20.)

19. Because of the need to prepare for international conferences, which address particular spectrum allocation issues, "larger policy issues have been overlooked or neglected, and insufficient consideration is being given to the long term consequences of implementing new technologies and services. The result has been an often reactive and short-sighted approach to spectrum policy." (U.S. Congress, Office of Technology Assessment, *The 1992 World Administrative Radio Conference: Technology and Policy Implications*, OTA-TCT-549, Washington, D.C.: GPO, May 1993, p. 7.)

20. NTIA, in a comprehensive study of spectrum management, proposed greater reliance on market forces and increased flexibility in management approaches. It identified four elements essential for long-term spectrum planning: (1) identifying requirements through the collection and dissemination of data concerning existing and anticipated use, (2) forecasting future requirements through both empirical methods and informed judgment, (3) publication of long-range plans, and (4) planning for unforeseen requirements. (*U.S. Spectrum Management Policy*, p. 161.)

14.3.1 The Federal Communications Commission

The FCC manages nongovernmental spectrum use in three fundamental ways:

1. *Allocating* blocks of spectrum that have propagational characteristics suitable for a particular class of user with specific service requirements. For example, the FCC allocated spectrum in the VHF and UHF bands for broadcast television.
2. *Allotting* channels within allocated blocks for use in particular localities. For example, the FCC allotted particular broadcast television channels for use in specific localities to prevent harmful interference and also to promote widespread availability of local stations. In other instances, the FCC has developed no allotment plan and grants licenses when applications are filed.
3. *Assigning* licenses to use spectrum allotments subject to conditions and procedures to achieve fair, efficient, and noninterfering use, including the comparative evaluation of applications from more than one party who vie for a single license.[21]

Until congressional amendments allowing for limited auctions were passed in 1993, the Communications Act prohibited private ownership of spectrum.[22] The Act obligated the FCC to serve the public interest, convenience, and necessity in such broad matters as spectrum allocation as well as specific classifications of radio stations, prescribing the nature of service to be provided by such stations[23] and ensuring fair distribution of media outlets in rural localities.

The FCC's overall spectrum management mission remains to allocate, assign, license, and regulate [5]:

---

21. Procedural fairness requires the FCC to consider in a single proceeding all license applications deemed mutually exclusive. (Ashbacker Radio Corp. v. FCC, 326 U.S. 327 (1945); Johnston Broadcasting v. FCC, 175 F.2d 351 (D.C. Cir. 1949) (a comprehensive comparative hearing is necessary for choosing between 2 or more mutually exclusive applicants). But compare United States v. Storer Broadcasting, 351 U.S. 192 (1955) (a hearing is not necessary for purposes of determining whether to grant a waiver of limitations on the number of licenses held by applicant).)

22. "The policy of the Act is clear that no person is to have anything in the nature of a property right as a result of the granting of a license." (FCC v. Sanders Bros. Radio Station, 309 U.S. 470, 475 (1940); Red Lion Broadcasting Co. v. FCC, 395 U.S. 367 (1969) (approving regulations on the basis of scarcity and the view that broadcasters are public trustees); Office of Communication of the United Church of Christ v. FCC, 425 F.2d 543 (D.C. Cir. 1969) (broadcasters are temporary permittees of a great public resource).) Both the FCC and commentators have begun to question whether a marketplace approach would better serve the public interest in lieu of the view that spectrum scarcity justifies pervasive regulation. (See, for example, M. Fowler and D. Brenner, "A Market Approach to Broadcast Regulation," *Texas Law Review*, Vol. 60, 1982, p. 207).

23. See Communications Act of 1934, as amended, 47 U.S.C. Secs. 302(a), 303, 307(a), 308, 309, and 316 (1990).

[S]o as to make...a rapid, efficient, Nation-wide, and world-wide wire and radio communication service with adequate facilities at reasonable charges, for the purpose of the national defense, [and] for the purpose of promoting safety of life and property.

This process remains one largely within the discretion of the FCC as tempered by procedural fairness requirements imposed by the Administrative Procedure Act [6], international spectrum allocational efforts at the ITU, and the actual scarcity that occurs when consumer and government demand for spectrum exceeds supply.

Courts typically defer to the FCC's expert determination of the public interest,[24] provided the FCC has satisfied procedural due process requirements and the duty to provide a "fair, efficient and equitable distribution of radio service" [7]. The FCC typically uses a rulemaking process to allocate spectrum or to modify an allocation. In application the rulemaking process affords interested parties, including proponents of differing uses for a particular frequency band, a forum to articulate the public interest merits of a particular use.

Once it allocates spectrum, the FCC either assigns frequencies according to an allotment plan for localities or issues licenses on a first-come, first-served basis to qualified applicants, with some rules on sharing, spacing of transmitters, and proof of adequate channel loading before assigning additional spectrum. However, in the event more applicants file for licenses than can be assigned, which regularly occurs in broadcasting and cellular radio, the FCC conducts a comparative hearing, lottery, or other proceeding to determine who shall receive an assignment.

For instances where many applicants vie for a single spectrum assignment, the licensing process involves a comparative assessment of each applicant's financial, legal, and technical qualifications and a determination of which applicant would best serve the public interest, convenience, and necessity. When awarding licenses in a comparative proceeding, the FCC must adhere to rules of evidence and procedure to ensure a fair hearing for each applicant and a complete record for possible review by an appellate court.

The FCC primarily selects among competing applicants by means of a "paper hearing" where staff reviews responsive pleadings of interested parties, possibly augmented by live testimony before an administrative law judge, the FCC's Review Board, or the full FCC, depending on the level of review. The comparative hearing process imposes substantial costs and does not guarantee that it will select the applicant most likely to serve the public interest because of the difficulty in articulating and evaluating comparative criteria. In recognition of these problems, Congress authorized the FCC to award some, but not all, li-

---

24. See, for example, FCC v. WNCN Listeners Guild, 450 U.S. 582 (1981).

censes by random selection (e.g., a lottery [8]), and in 1993 authorized limited use of spectrum auctions.

The FCC has considered liberalizing its rules for some frequencies to permit use by more kinds of licensees.[25] The FCC also has expanded the type of services certain licensees can provide [9]. Recent FCC spectrum allocation decisions have involved such new services as LEO satellite-delivered portable mobile telephone service, position-indicating and low-speed data services from small LEO satellites [10], digital audio broadcasting [11], HDTV,[26] and personal communication networks [12].

The FCC assigns spectrum by issuing licenses that require periodic review to assess whether operators have complied with service requirements including technical regulations and standards (e.g., type of signal emission, signal strength, bandwidth). Most FCC allocations are made on a nationwide basis, although the FCC has granted narrow exceptions, such as shared use of UHF television channels 17 to 20 by land mobile service providers in major metropolitan areas and spectrum use near an international border. The FCC's spectrum allocations process may reserve additional bandwidth to satisfy future demand. The FCC generally does not consider weighing future or alternative uses for already allocated spectrum. When it does reallocate spectrum from one service to another, the FCC, on its own initiative or in response to a rulemaking petition, must evaluate existing or prospective demand in the context of which would better serve the public interest.

14.3.2 The National Telecommunications and Information Administration

NTIA's spectrum management objective lies in guiding federal radio users and ensuring that the federal agencies "make effective, efficient, and prudent use of the radio spectrum in the best interest of the Nation, with care to conserve it for uses where other means of communication are not available or feasible" [13]. In application, this means that NTIA's management functions are [14]:

---

25. See, for example, Amendment of Parts 1, 21, 22, 74, and 94 of the Commission's Rules to Establish Service and Technical Rules for Government and Non-Government Fixed Service Usage of the Frequency Bands 932–935 and 941–944 MHz, Gen Docket No. 82-243, First Report and Order, 50 Fed. Reg. 4650 (1985), Second Report and Order, 4 FCC Rcd. 2021 (1989); Mem. Op. & Order, 5 FCC Rcd. 1624 (1990); Amendment of Part 90 of the Commission's Rules to Make Additional Channels Available for Private Carrier Paging Operations in the 929–930 MHz Band, PR Docket No. 85-102, Report and Order, 58 Rad. Reg.2d (P&F) 1290 (1985) (ordering sharing between commercial, common carrier, and private use of paging frequencies).

26. See, for example, Advanced Television Systems and Their Impact Upon the Existing Television Broadcast Service, MM Docket No. 87-268, Mem. Op. and Order, Third Report and Order, and Third Further Notice of Proposed Rulemaking, 7 FCC Rcd. 6924 (1992).

(1) to provide frequency assignments to federal government agencies; (2) to develop plans and policies for the effective and efficient use of the spectrum in coordination with the FCC; (3) to undertake substantive preparation for ITU radio conferences on behalf of the Executive Branch including pertinent foreign counterpart consultation, presentation of coordinated Executive Branch views on telecommunications matters, and promoting international cooperation regarding telecommunications-related policy matters.

Both NTIA and the FCC must coordinate the spectrum management function, not only because of divided jurisdiction,[27] but also because some spectrum is allocated for shared private and government use.[28] The two agencies maintain separate databases of spectrum use authorizations with the FCC compiling a record of private license grants and NTIA maintaining the Government Master File, containing information on stations using government spectrum.

The FCC and NTIA coordinate spectrum allocations that are incorporated by the FCC in a consolidated Table of Frequency Allocations, with nongovernmental allocations codified in Part 2 of the FCC's Rules and Regulations [15]. The coordination process takes into consideration the nature and number of users involved, propagational characteristics of target frequencies, the cost to deploy services, and the possibility of sharing between services and between government and nongovernment users.

Governmental frequency assignments are coordinated by the Interdepartment Radio Advisory Committee, convened under the auspices of NTIA and attended by representatives of most government agencies with significant spectrum requirements. IRAC's Frequency Assignment Subcommittee resolves spectrum differences between government agencies.

---

27. Section 305(a) of the Communications Act, as amended, provides that "Radio stations belonging to and operated by the United States shall not be subject to the provisions of sections 301 and 303 of this Act [that broadly state the scope of the FCC's regulatory mission]. All such Government stations shall use such frequencies as shall be assigned to each or to each class by the President." (47 U.S.C. Sec. 305(a) (1990).) "Federal Government spectrum is delegated by The President under Executive Order 12046 to the Department of Commerce, NTIA, which is aided by other federal departments and agencies through an advisory group, the Interdepartment Radio Advisory Committee (IRAC)." (U.S. Dept. of Commerce, National Telecommunications and Information Administration, Appendix A in *Telecom 2000—Charting the Course for a New Century*, Washington, D.C.: GPO, 1988, p. 657.)

28. Sixty-three percent of the spectrum below 30 GHz is shared. (*U.S. Spectrum Management Policy*, p. 17.)

14.4 BLOCK ALLOCATIONS

WRCs and, in turn, the FCC allocate spectrum in blocks of frequency bandwidth earmarked for a particular service. Politics and nontechnological factors may dominate the process, and the FCC may not fully articulate the criteria used to determine the relative merits of one service versus others. The process places a premium on incumbency, with the result of creating an expectation that once allocated spectrum, a service and its users will never be ousted or forced to share the allocated spectrum. New services, technological innovations, or user constituencies with expanded spectrum requirements must vie for spectrum with incumbents. Newcomers typically receive a spectrum allocation on a *co-primary* basis, meaning that they must share the spectrum with incumbent, primary users, but qualify for interference protection from subsequent users proposing to use the spectrum for a primary service, as well as secondary service uses. A secondary allocation would subordinate the newcomer, not only to existing primary service users, but also to subsequent ones.

The block allocation process awards bandwidth on the basis of the technologies, services, and user requirements effectively advocated at the time of decision. For example, satellite services have been divided as a function of transmitter and receiver location. There are maritime, land mobile, fixed, and aeronautical services with separate allocations. Discrete service definitions and spectrum allocations made sense when users could not easily move terminals, but now they can easily operate the same portable transceiver in fixed, mobile, land, maritime, and aeronautical locations.

The FCC acknowledged the flexibility afforded by technological innovation and proposed at a 1987 and 1992 WARC that a generic MSS spectrum allocation incorporate previously discrete maritime, land, and aeronautical mobile allocations. When the international consensus persisted in maintaining separate, geographically specific allocations, the United States "took a reservation... with respect to these allocations, indicating its continuing desire to implement MSS in an appropriate manner to satisfy U.S. requirements" [16].

Frequency spectrum users and commentators have begun to recognize the financial stakes in the spectrum planning and allocation process [17]:

> The United States is at a crucial turning point in the history of spectrum use and management. Technological, economic, and political forces are converging to radically alter the context within which domestic and international spectrum decisions and policies are made.

Figure 5.2 contains a summary of the block allocations created at WARC-92.

References

[1]   103d Cong., 1st Sess., Omnibus Budget Reconciliation Act of 1993, Title VI, Communications Licensing and Spectrum Allocation Provisions, PL 103-66 (HR 2264), 107 Stat. 312, 379 et seq. (1993).

[2]   U.S. Congress, Office of Technology Assessment, *The 1992 World Administrative Radio Conference: Issues for U.S. International Spectrum Policy—Background Paper*, OTA-BP-TCT-76, November 1991, p. 9 (hereafter cited as OTA Pre-WARC Study).

[3]   U.S. Dept. of Commerce, National Telecommunications and Information Administration, Appendix A, *Telecom 2000—Charting the Course for a New Century*, Washington, D.C.: GPO, 1988, p. 655 (hereafter cited as Telecom 2000).

[4]   OTA Pre-WARC Study, p. 9.

[5]   Communications Act of 1934, as amended, 47 U.S.C. Sec. 151 (1990).

[6]   Administrative Procedure Act, 5 U.S.C. Secs. 551–559, 701–706 (1990) (requiring federal agencies, inter alia, to solicit public comment before issuing rules and regulations).

[7]   Communications Act of 1934, as amended, 47 U.S.C. Sec. 307(b).

[8]   47 U.S.C. 309(1); see Communications Amendments of 1982, Conference Report No. 97-765, 97th Cong. 2d Sess., 19 August 1982.

[9]   Amendment of Parts 2 and 22 of the Commission's Rules to Permit Liberalization of Technology and Auxiliary Service Offerings in the Domestic Public Cellular Radio Telecommunications Service, Gen Docket No. 87-390, Report and Order, 5 FCC Rcd. 1138 (1990).

[10]   Amendment of Section 2.106 of the Commission's Rules To Allocate Spectrum to the Fixed-Satellite Service and the Mobile-Satellite Service for Low-Earth Orbit Satellites, ET Docket No. 91-280, Notice of Proposed Rulemaking, 6 FCC Rcd. 5932 (1991), Report and Order, 8 FCC Rcd. 1812 (1993); Amendment of the Commission's Rules to Establish Rules and Policies Pertaining to a Mobile Satellite Service in the 1610–1626.5/2483.5–2500 MHz Frequency Bands, CC Docket No. 92-166, Notice of Proposed Rulemaking. 1094 (1994), Report and Order, 9 FCC Rcd. 5936 (1994) (establishing licensing rules); see also Amendment of Sec. 2.106 of the Commission's Rules to Allocate Spectrum at 2 GHz for Use by the Mobile-Satellite Service, ET Docket No. 95-18, 10 FCC Rcd. 3230 (1995), (proposing to allocate additional spectrum for mobile satellite services).

[11]   Amendment of the Commission's Rules to Establish New Digital Audio Radio Services, Gen. Docket No. 90-357, Notice of Inquiry, 5 FCC Rcd. 5237 (1990), Notice of Proposed Rulemaking and Further Notice of Inquiry, 7 FCC Rcd. 7776 (1992), Report and Order, FCC 95-17, 1995 FCC Lexis 329, 76 Rad. Reg. 2d (Pike & Fischer) 1477 (rel. 12 January 1995).

[12]   Amendment of the Commission's Rules to Establish New Personal Communications Services, GEN Docket No. 90-314, Notice of Inquiry, 5 FCC Rcd. 3995 (1990), Policy Statement and Order, 6 FCC Rcd. 6601 (1991), Notice of Proposed Rule Making and Tentative Decision, adding ET Docket No. 92-100, 7 FCC Rcd. 5676 (1992) (tentatively granting 3 Pioneer's Preferences for wideband PCS), erratum, 7 FCC Rcd. 5779 (1992), 7 FCC Rcd. 7794 First Report and Order, 8 FCC Rcd. 7162 (1993) (establishing rules for narrowband PCS in the 900-MHz band and awarding 1 Pioneer's Preference), erratum, 9 FCC Rcd. 6388 (1994); Second Report and Order, 8 FCC Rcd. 7700 (1993), Third Report and Order, 9 FCC Rcd. 1337 (1994) (granting pioneer's preferences to American Personal Communications, Cox Enterprises, Inc., and Omnipoint Communications, Inc., for wideband PCS) recon. den., 9 FCC Rcd. 7805 (1994).

[13]   U.S. Department of Commerce, National Telecommunications and Information Administration, *Manual of Regulations and Procedures for Federal Radio Frequency Management*, Sec. 2-1, May 1990; see also 47 C.F.R. Sec. 300.1 (1990).

[14]   Telecom 2000, p. 659.

[15]   47 C.F.R., Part 2 (1992).

[16] An Inquiry Relating to Preparation for the International Telecommunication Union World Administrative Radio Conference for Dealing With Frequency Allocations in Certain Parts of the Spectrum, GEN Docket No. 89-554, Second Notice of Inquiry at 26, para. 56, citing Notice of Proposed Rulemaking, GEN Docket No. 89-103, 4 FCC Rcd. 4173, 4178 (1989); see also Report and Order, FCC 91-188 (rel. 20 June 1991).

[17] OTA Pre-WARC Study, p. 2.

Introducing Private Enterprise in Telecommunications 15

Privatization in telecommunications involves a change in the legal status of a nation's government-owned telecommunications carrier or equipment manufacturer to permit private investment. Privatized companies may retain a monopoly service franchise, but in some instances, nations may permit limited competition coupled with liberalization of regulatory requirements imposed on the incumbent carrier. At the same time as they privatize the PTT, governments typically streamline and reduce oversight to allow the privatized incumbent flexibility to respond to changing conditions, including the need to compete in some markets with one or more new facilities-based carriers and numerous resellers.

The onset of privatization represents an explicit recognition by governments that [1]:

> [T]elecoms has become too important to be left to engineers exchanging incomprehensible acronyms. A rich country with outmoded laws that blocks innovation will set its companies at a huge disadvantage against foreign competitors; a poor country will find it immeasurably easier to attract foreign investors if it ensures that they have access to telephones and fax machines that work. So the thrust of the debate is everywhere the same: how to encourage the fullest possible modernisation of domestic networks?

Simply put, "much of the recent enthusiasm for privatization is the rather simplistic notion that government ownership results in economic inefficiency" [2]. Notwithstanding this view, some government-owned entities have performed quite well, either because they have achieved a sufficient degree of independence or because government has targeted telecommunications for subsidization from tax receipts. For example, the government-owned carriers in France, Germany, and Singapore have successfully invested abroad and have developed a global marketing presence through joint ventures even as they up-grade domestic facilities. France Telecom, the Deutsche Telekom (DT), and Sprint have formed a joint venture aiming to provide global one-stop shopping to large, multinational enterprises. France Telecom holds a 20.4% ownership share in the Mexican telephone company, TelMex, and Singapore Telecom has invested in U.S. paging (Mobile Telecommunication Technologies, Inc.) and mobile satellite (American Mobile Satellite Corp.) ventures.

Raymond Duch suggests the likelihood that a privatization initiative will occur depends on two critical dimensions [3]:

1. The ease with which challenges to established public policy can be mounted;
2. The institutional resources available to threatened interests that resist any changes in the policy status quo.

Using these two variables, one can predict that pluralist nations like the United Kingdom and the United States will promote greater access to the political process and provide limited protection to incumbent institutions like telephone companies, despite their political clout and large base of employees. Nations like France, which are keen on stability and maintenance of political order, limit access to the political process, but likewise provide limited protection of entrenched interests if they cannot help maintain the status quo. "Corporatist" nations like Germany, concentrate on expanding the wealth and welfare of the populace and have done so by limiting access to the political process and by supporting successfully operating government-owned institutions like the PTT.

15.1 WHY DID GOVERNMENT PROVIDE TELECOMMUNICATION SERVICES IN THE FIRST PLACE?

Ironically, economic analysis has been used to support both state ownership of telecommunications and privatization. The justification for government ownership lies in the view that telecommunications constitutes a natural monopoly. Under this paradigm, competition would increase risk, prevent the monopoly from achieving economies of scale necessary to operate efficiently, and jeopardize the prospects for achieving public policy goals like universal service at affordable rates. A regulated monopoly can extend and subsidize services to accrue social benefits like national cohesiveness achieved when most citizens have access to telephones. Regulation mandates or supports carrier pricing decisions that set basic service rates at levels that would be unsustainable if business considerations, such as return on investment, were considered exclusively. The incumbent may claim that it alone can be trusted to serve all areas, lucrative or not. It may also raise national security concerns about the strategic nature of telecommunications, particularly in times of war or civil unrest.

France Telecom provides a case study for assessing how continued government ownership can deliver many of the benefits touted by advocates for privatization. France used to have one of Europe's worst telecommunications infrastructures. With legislated financial support, France Telecom accelerated system upgrades and speedily achieved near-ubiquitous digital service capability. It also distributed at no charge Minitel terminals for use by telephone subscribers to access databases, including online directories, and to communicate via electronic mail. At least in the case of France Telecom, "enlightened paternalism" achieved widespread deployment of cutting-edge technologies to the hinterland, an outcome that might not result if a private company operated without close government scrutiny.

Proponents for privatization claim that monopolies seek to perpetuate the status quo, even in the face of technological innovations that reduce barriers to market entry and undermine justifications for a single service provider. Having invested in expensive technology, subject to regulatory limitations on how quickly the investment may be recouped, the incumbent may not innovate or make timely additional investments. But to foster goodwill among the electorate, the incumbent, perhaps in collaboration with elected officials, may pursue political and popular objectives, regardless of whether they might adversely impact economic efficiency. In application this means that the incumbent may scrimp on infrastructure improvements while intentionally underpricing domestic POTS with subsidies from overpriced international and corporate services.[1]

Reformers claim that carriers and equipment manufacturers must respond to elasticities of supply and demand (i.e., whether a particular class of consumer has alternative ways to communicate its intensity of need for and price sensitivity to services and equipment).[2] Pricing too out of line with the available

---

1. "Often these policies confer benefits on a very diffuse electorate, the costs are borne by a small segment of society, and they negatively affect the efficiency of the public enterprise. One example is the subsidization of residential telephone rates with revenues from business and long-distance telephone tariffs. The costs of subsidizing a very diffuse segment are borne by a more concentrated constituency. Although such pricing policies tend to command widespread political support, they are inefficient from a strictly economic perspective because the tariffs have little relation to the cost of providing the services." (R. Duch, *Privatizing the Economy—Telecommunications Policy in Comparative Perspective*, Ann Arbor: University of Michigan Press, 1991, p. 30.)

2. Adverse financial consequences to society, known as deadweight social losses, may increase when governments refuse to authorize competition or mandate privatization: "Small changes in the cost of supplying products and services can now have a very significant impact on the quantities of goods made available on the market. Similarly, consumers have become increasingly sensitive to changes in the pricing of telecommunications goods and services. As a result, the impact of subsidies on overall industry demand and growth have become increasingly mag-

options encourages users to devise methods for bypassing overpriced services, and at the extreme to relocate personnel and facilities to nations with better and cost-based telecommunication rates. Privatization, or at least action that affords greater operational flexibility for the PTT, results when government and the public believe that less governmental oversight and regulation can:

- Better respond to consumer demand;
- Incorporate new technologies faster than a government bureaucracy could;
- Achieve greater efficiencies than the government-owned PTT would;
- Survive in an increasingly competitive and volatile environment intolerant of inefficiency and poor productivity;
- Comply with preexisting public service mandates that include the duty to offer uneconomical but socially desirable services.

15.2 GROWING INTEREST IN AND METHODS FOR PRIVATIZATION

While some nations may want to maintain restrictions on market entry and access by foreign carriers and equipment manufacturers, the philosophical and political justifications have lost much of their appeal. Nations have permitted foreign investment to augment what the government incumbent can internally generate or secure through loans. Typically, nations limit such investment to a minority position, thereby promoting the formation of strategic alliances that involve:

- Investing in a privatized incumbent (e.g., the privatizations in Argentina, Chile, Hungary, Japan, Mexico, New Zealand, the United Kingdom, and Venezuela);[3]
- Creating new ventures (or adding investments) in companies seeking to serve markets made accessible by deregulation and liberalization (e.g., Sprint's joint venture with Cable & Wireless (C&W) in the PTAT-1 transat-

---

nified." (Ibid., p. 102.) Duch speculates that adversely affected consumers will become more inclined to seek remedies with the legislature and regulatory authorities. Increased elasticities of supply and demand provide "a signal and an incentive for consumers and taxpayers to press for the reduction of entry barriers. Such a change in supply and demand elasticities has been occurring in the telecommunications industry over the last twenty years. As a result of dramatic technological advances in the telecommunications industry, capital investment costs have fallen, thereby increasing the elasticity of supply and dramatically reducing the cost of services. All of this has, in turn, increased the elasticity of demand." (Ibid.)

3. Many privatizations have involved a series of public stock offerings that prevent a single company from acquiring an ownership majority. Shares in British Telecom, Cable & Wireless, Telefonica, Telecom New Zealand, Chile Telecom, Tele Denmark, Portugal Telecom, and Telefonos de Mexico are traded on the New York Stock Exchange.

lantic submarine cable that links the U.S. with Bermuda, Ireland, and Britain);

- Teaming with a PTT to fund infrastructure development in a third country; for example, a number of ventures that build, own, and operate cellular radio carriers, overlay microwave systems, and satellite Earth stations in the Commonwealth of Independent States (CIS) and Eastern Europe (e.g., IDB Communications Earth stations in Moscow and Lithuania); Sprint International data services throughout the CIS; AT&T-led modernization of Kazakhstan's telecommunications services and infrastructure; and U S West's plans for a trans-Siberian fiber-optic cable;
- Entering into ventures with foreign enterprises to capture larger market shares and to develop new markets (e.g., AT&T's joint venture with the Ukraine State Committee of Communications and PTT Telecom of the Netherlands to modernize the domestic network of the Ukraine and to install an international telecommunications infrastructure, thereby eliminating the need to route traffic via Moscow).

The pace of privatization opportunities has increased, particularly because some nations (e.g., the United Kingdom, Mexico, New Zealand, Australia, Japan, and even Germany) have relaxed or reinterpreted foreign ownership and licensing restrictions to permit foreign investment in all but core services reserved for the incumbent carrier. Domestic restrictions imposed by judicial rulings (e.g., the former prohibition on local exchange carrier provision of information and cable television services)[4] and legislation (e.g., prohibition on telephone company provision of cable television services in the same location)[5] have created incentives for the regional Bell operating companies (RBOCs) to invest abroad. RBOC investments in foreign cable television ventures typically can be made at a lower per-subscriber rate than in the United States, particu-

---

4. The Modification of Final Judgment in the AT&T divestiture case prohibited the spun-off BOCs from (1) providing broadly defined information services, (2) providing interexchange, long-distance services, and (3) manufacturing telecommunication equipment. (United States v. AT&T, 552 F. Supp. 131 (D.D.C. 1982), aff'd sub nom., Maryland v. United States, 460 U.S. 1001 (1983).) As part of its triennial review and determination of whether line-of-business restrictions should continue, the District Court narrowed the information service restrictions. (United States v. Western Electric Co., 714 F. Supp. 1 (D.D.C. 1988); 767 F. Supp. 308 (D. D.C. 1991).) On appeal, the Circuit Court of Appeals determined that the narrowing was inadequate and used improper evaluative criteria. (United States v. Western Electric Co., 900 F.2d 283 (D.C. Cir.) cert. den. sub nom., MCI Telecommunications Corp. v. United States, 111 S.Ct. 283 (1990).) The Circuit Court of Appeals subsequently lifted the ban on BOC provision of information services. (United States v. Western Electric Co., 1991-1 Trade Cases (CCH), 69,610 (D.C. Cir. 1991).)

5. *Cable Television Consumer Protection and Competition Act of 1992*, Pub. L. No. 102-385, 106 Stat. 1460 (1993) codified at 47 U.S.C. Secs. 534–535 (1993); see also 102d Congress, 2d Session, House of Representative Report No. 102-862, Conference Report.

larly when localities award initial franchises. But even without these enticements, some investments might still occur to provide the RBOCs with opportunities to test broadband integrated networks providing both cable television and telephone services.

Similarly, the RBOCs can operate outside the United States substantially without line-of-business restrictions imposed by the Modification of Final Judgment that applies domestically. NYNEX, which is prohibited from providing interexchange long-distance services in the United States, plans on constructing and operating a $1 billion fiber-optic submarine cable venture, the Fiber Optic Link Across the Globe (FLAG). Previously it failed to receive a line-of-business waiver to permit an investment in the PTAT-1 cable because that cable provided interexchange telecommunications and made a U.S. landfall. While the AT&T divestiture court subsequently allowed minor investments in companies owning such facilities (e.g., Pacific Telesis' 10% investment in the Japanese carrier International Digital Communications, whose assets include the North Pacific Cable), that do make U.S. landfalls, the RBOCs ironically find safety in making off-shore investments.

AT&T, other major telephone companies, and nearly every U.S. telecommunications equipment manufacturer have made foreign investments. Proximity to users, profit potential, market growth opportunities, and the need for operating experience using foreign standards generate incentives for such investments. Additionally, American companies may perceive the need to develop business alliances with foreign companies in lieu of a unilateral attempt. A consortium or joint venture of multinational companies that have done business together may have improved odds for winning the bidding sweepstakes for a privatizing PTT or for a new service franchise.

As a result, the proliferation of multinational corporations, private lines and enterprise networks, leaky PBXs, shared use and resale, IVANs, accounting rate bypass schemes, and increased user sophistication have profoundly changed the international telecommunications marketplace. PTOs can no longer act cavalierly—users can vote with their feet and currency. Users and some carriers have achieved some downward pressure on rates through hubbing options (just like the airlines). Networking based on functionality has little concern for geographical boundaries. Governments have listened to users and have responded by allowing market entry, particularly in niche markets; so long as access is affordable and achievable, the resale carrier can diversify and expand. Governments have endorsed accelerated deployment of mobile and overlay microwave and data networks, particularly in less developed nations.

PTOs have belatedly acknowledged the need to manage deregulation proactively: anticipating new competitors' likely strategies (e.g., creamskimming, cherry picking routes and customers, seeking regulations that favor the newcomer and underprice access to the "first and last" mile, exploiting the incumbent's weakness and consumer dissatisfaction with the incumbent, and implementing a more realistic corporate culture that embraces rather than resents or blocks change.

15.3 STATISTICAL INDEXES SHOW THE NEED FOR LARGER INVESTMENT IN MANY NATIONS

Telecommunications is expected to remain a high-growth industry in many nations whose number of telephones per 100 inhabitants[6] ranges from single digits to 25. Some developing nations have considered privatization as a way to accelerate improvements in telephone penetration and the overall telecommunications infrastructure. Privatization tenders may require bidders to agree that they will achieve certain benchmarks specifying minimum levels of investment commitments and network upgrades (e.g., the number of new telephone lines that must be installed per year).

The trend toward privatization of PTTs has gathered momentum, presenting outsiders with possibly the most substantial investment opportunity: the prospect for financial participation in a telecommunications administration with an exclusive or nearly exclusive franchise in reserved services not subject to any facilities-based competition.

The concept of privatizing PTTs has become more attractive in large part due to deregulatory initiatives in the United Kingdom and the United States which have been perceived as favorably stimulating the national economy and the telecommunications sector. But at the time of such groundbreaking initiatives, most observers expressed concern and skepticism. In 1984, Britain's telecommunications monopoly became publicly traded and forced to share a domestic duopoly with Mercury, a C&W subsidiary. The boldness of Britain's initiative resulted in part from the shared philosophical views of then Prime Minister Margaret Thatcher and President Ronald Reagan, and the belief that AT&T's divestiture and other U.S. deregulatory initiatives had stimulated innovations, efficiency, and beneficial competition. Notwithstanding actions taken in these two key nations, it would take several more years to achieve a growing consensus for reform.[7]

---

6. "Telephone penetration and related measures are important barometers...[and] their utility with respect to international comparisons...seems to center on their use as measures of comparative development." (U.S. Department of Commerce, National Telecommunications and Information Administration, *Telecommunications in the Age of Information*, NTIA Special Publication 91-216, Washington, D.C.: GPO, October 1991, p. 193.)

7. In 1988, the European Commission issued a telecommunications directive setting a timetable for member countries to liberalize the terminal equipment market. The commission used its authority established in Article 90 of the Treaty of Rome that grants it power to ensure that "public undertakings and undertakings to which member states grant special or exclusive rights" do not thwart free competition, but operate in the general interest of the EU. The gov-

15.4   PRIVATIZATION OPPORTUNITIES REQUIRE PUBLIC AND POLICY-MAKER SUPPORT FOR ENDING THE STATUS QUO

For privatization to occur, nations need to reject long-standing economic arguments that efficient telecommunications require a single service provider that can spread risk over a large captive user population and achieve economies of scale and scope. National legislators and regulators require evidence that a more efficient telecommunications industry can evolve through privatization and market entry, while not handicapping the ability of the incumbent carrier to recoup its investment in infrastructure development.

A marketplace orientation supports entry by new operators installing competing transmission facilities that are neither a waste of resources nor a threat to the incumbent's ability to provide essential services. In view of the expanded applications demanded by sophisticated users involved in key sectors of the international economy, many nations now consider the single incumbent unable or unwilling to provide all product and service solutions to user requirements.

For a marketplace philosophy to be predominate, policy makers must decide that the PTT should no longer have a mandate to operate as one system with complete responsibility for all telecommunication requirements. Once the government decides to foster competition, the PTT must abandon its expectation that all revenue streams will remain free from encroachment by newcomers. In exchange for accepting the prospect of some competition, the incumbent typically receives significant relaxation of government oversight, including the migration from rate-of-return regulation, which prescribes the maximum level of profitability, to a price cap regime that creates incentives and financial rewards for efficiency.

15.5   PHILOSOPHICAL AND PRACTICAL REASONS FOR PRIVATIZATION

A number of philosophical and pragmatic elements factor in the decision of whether to privatize, the manner in which privatization occurs, and the corporate structure that results. The primary basis for acting falls into several different categories.

---

ernment of France challenged this action in the European Court. In middle 1989, the threat of another Article 90 directive on value-added network services prompted the EU ministers to order incremental deregulation from 1990 onward. "This prospect [of increasing deregulation in telecommunications] amounts to a minor revolution in all member states except Britain." (N. Colchester and D. Buchan, *Europower—The Essential Guide to Europe's Economic Transformation in 1992*, London: Economist Books, 1990, p. 158.)

15.5.1 Real or Perceived Need to Stimulate Efficiency and Change

Some of the largest privatizations have occurred on the basis that a more commercial approach should replace a public utility service orientation [4]:

> [T]hat transfer from the public sector to the private sector will conduce to greater efficiency and superior customer service by injecting entrepreneurial energy and freeing independent managers to formulate their strategies in accordance with more commercial motives.

The privatizations in the United Kingdom, Japan, Australia, and New Zealand fit this model and place telecommunications in the context of a larger marketplace exit by government. U.S. companies that have participated in this type of privatization include BellSouth, with its participation in forming the second Australian carrier Optus, and Bell Atlantic and Ameritech, with their participation in the acquisition of New Zealand Telecom.

15.5.2 The Need or Desire to Secure Outside Capital

Many privatizations occur when national governments seek to encourage private financing of telecommunications ventures. This category includes privatizations of carriers with up-to-date telecommunication systems and high telephone line penetration rates, as well as those with obsolete equipment and low penetration rates. In the former case, adequate capital might have been available from government sources, albeit in competition with other public works. Privatizations in Portugal, Chile, and Malaysia provide examples here. In the latter case, capital is needed to expedite development of a functional infrastructure, even if all or portions of privatization franchise fees flow into the general treasury, as typically has been the case for PTT revenues in developing nations. Argentina, the Czech Republic, Hungary, and Mexico provide examples of this category.

Differences in a PTT's conversion to partial or complete private ownership often depend on the scope of permissible foreign ownership and retained government control. Many nations, including the United States, have laws limiting the percentage of foreign investment. Often governments retain a "golden share" representing a degree of ongoing oversight and perhaps veto power if they consider a business decision inconsistent with the national interest. These limitations can adversely affect the number of bidders and the amount they offer.

15.5.3 A Change in Government Regime or Economic Policy

A broader government commitment to market-driven economies also promotes privatization. In the case of Eastern Europe and the CIS, telecommunications becomes one of numerous previously state-controlled industries targeted for privatization, often occurring in fits and starts. Germany also fits in this category. Once a stalwart for government control, Germany has divided the PTT into three quasiprivate enterprises: banking, postal operations, and telecommunications. Germany has impressed observers with its change in regulatory philosophy,[8] even as it confronts the expensive task of achieving resource and infrastructure parity between two formerly divided nations.

15.5.4 Other Pragmatic Reasons

The newfound attractiveness of telecommunications privatization probably also results from recent imperatives and incentives. Telecommunications has substantially become less a public utility undertaking, like the provision of electricity or water, and more a competitive undertaking where minutes of network use become indistinguishable (i.e., fungible) between different companies and customers can migrate to carriers providing the best deals.

Carriers have belatedly recognized consumer sovereignty and the need to satisfy user requirements or risk traffic and revenue losses as consumers migrate to other carriers and service providers. While carriers still match half circuits, they see real financial advantages in achieving market share outside their domestic markets. Accordingly, telecommunication service providers, particularly new carriers, no longer consider off-limits the customers of other carriers.

As never before, these carriers vie for high-volume customers, particularly ones with multinational traffic streams. Very much like airline carriers, telecommunication administrations have to devise innovative ways to confer discounts to frequent travelers on their networks to retain loyalty, particularly where carriers can capture traffic volumes by erecting a hub for routing traffic throughout a region.

This newfound pragmatism encourages aggressiveness to the point of poaching the customers of other carriers. So far the incentive to generate greater traffic volumes has only resulted in selective price cutting. Carriers have recognized the profit potential in providing greater customer service, network functionality, and general flexibility so that customers will design or reroute network to traverse the innovative carrier's regional hub. Customer networks increasingly serve specialized customer requirements, making functionality the key factor with geography and political boundaries deemed insignificant or an increasingly avoidable impediment. Sophisticated users design intracorporate networks that can be reconstituted to route around outages and to exploit new arrangements with carriers.

---

8. See H. Riche, "Germany's TELEKOM: A New Way of Doing Business in a Liberalized Market," *Telecommunication Journal*, Vol. 58, X-1991, p. 711. Note, however, that "TELEKOM will remain the only provider of the telephone network in Germany." (Ibid., p. 712.)

Telecommunications carriers have to demonstrate greater flexibility to accommodate customer requirements, because the twin impact of technological innovation and policy liberalization all but mandates it. If somewhere within the region, resellers of leased lines and users have opportunities to exploit such innovations, then incumbents face the potential for lost traffic and revenues if they cannot or refuse to innovate.

The ability of private branch exchanges to access the PSTN, the permeability of regional private line networks, transborder satellite footprints, resale of leased lines, accounting rate evasion, and a host of other factors result in tipping the balance decidedly in favor of the customer. The prudent telecommunications administration heeds the call for one-stop shopping, heightened responsiveness to user requirements, and upgraded networks. A satisfied customer is less likely to exercise the freedom to lease or buy terminal equipment from new vendors, to shift traffic to a different carrier's facilities, or to relocate all or some operations to take advantage of upgraded networks elsewhere.

The smart telecommunications administration will also seek strategic alliances with other carriers and equipment manufacturers if the alliance consolidates needed skills and technologies, promotes competitiveness, and enhances the possibility for replacing domestic or international traffic volumes lost to competitors. This new found receptiveness to collaboration with outsiders who can inject new thinking parallels quickening efforts by policy makers and regulators to spur innovation through privatization, deregulation, and a global or regional market orientation.

15.6 HANDICAPS AND ADVANTAGES IN THE PRIVATIZATION SWEEPSTAKES

Privatization opportunities trigger significant interest and competitive bidding. In certain instances, some companies may face handicaps regardless of having readily available investment capital and expertise. Such handicaps include:

- Lack of a shared cultural heritage like that benefiting Telefonica of Spain in its successful efforts to invest in Chilean, Argentinean, and Puerto Rican carriers;
- Alien ownership restrictions that limit foreign investment, but may have loopholes for investors in the region; for example, efforts to achieve more coordinated Europewide policies might result in broader use of a common currency and relaxation of foreign ownership restrictions on enterprises incorporated in a member nation of the EU;
- Familiarity with and use of equipment operating on standards different from the privatizing carrier;

- Incomplete vertical integration, that is, the integration of telecommunication equipment manufacturing and telephony operating experience, that would enable the acquiring company to recoup its investment partially through an affiliate's equipment sales;
- The lack of government foreign aid in telecommunications that would enable the acquiring company to raise its bid on the expectation that it could secure some financial support from its government.

On the other hand, some prospective investors can achieve significant advantages in terms of access to commercial capital, operational and management skills, administrative expertise including tariffing and working in both regulated and deregulated environments, and technological leadership.

15.7 PRIVATIZATION OCCURS IN A NUMBER OF WAYS

Nations increasingly recognize that privatization "makes it possible in practice to get rid of certain bureaucratic procedures and other inhibitions on the efficacy of the businesses concerned" [5]. Most countries readily accept that telecommunications has a significant impact on a nation's economic viability by providing an infrastructure for transacting commerce.

Nations pursuing a privatization campaign have a variety of options and models to consider. To assess fully the privatization opportunities available, a review of the types of privatizations may prove helpful. Set out below is an outline of several investment opportunities available to companies and individuals.

15.7.1 Public Stock Offerings

Many privatizations result in government issuance of stock reflecting ownership shares in all or a portion of the PTT. This sale may first involve one or more transactions involving a large portion of the available shares, a portion of which the purchaser may have to resell to meet foreign ownership limitations. Privatizations in Argentina, Chile, Denmark, Guyana, Hungary, Italy, Japan, Mexico, New Zealand, the Philippines, Portugal, Spain, and the United Kingdom involved public stock offerings. Each case has its own particularities in terms of what percentage, if any, the government retained; the percentage of stock held by single investors; whether labor unions, pension funds, or subscribers received stock ownership opportunities; and the extent to which government retained a "golden share" capable of vetoing any decision deemed contrary to the national interest.

15.7.2 Investment Opportunity in a New Facilities-Based Competitor

Some privatization observers are surprised when they learn that relatively few nations have authorized facilities-based competition. Many nations have spun off the PTT into a private or quasiprivate enterprise or have authorized niche market competition (e.g., cellular radio). Few nations have opened markets to facilities-based competition. These include Australia (Optus, whose investors include BellSouth and C&W), Canada (Unitel, whose investors include AT&T), Chile, Japan (International Digital Communications and International Telecommunications of Japan, whose investors constitute a consortium of major trading companies and manufacturers), Korea (Dacom), New Zealand (Clear), Sweden (Tele2), the United Kingdom (Mercury, with opportunities for additional facilities competition after 1997), and the United States (AT&T, MCI, Sprint, World-Com, ALC Communications, etc.).

Opportunities to acquire the former PTT or to invest in a new facilities-based competitor present substantial financial rewards over time, but do carry significant risk. Accordingly, the typical bidder in these sweepstakes is a major telecommunications company interested in developing a multinational presence.

15.7.3 Competition in Niche Markets

A number of nations, keen on accelerating facilities construction and attracting foreign investors to build networks, have devised alternatives to basic telephone service competition. The frequently used mechanism involves granting a franchise for niche markets like mobile services (e.g., cellular radio, business teleports), which provide reliable international satellite access, and microwave voice and data networks that overlay the existing, often unreliable wireline plant. Franchise terms often include a timetable for building a particular network and may include a variety of benchmark service obligations. Nations can improve the attractiveness of a franchise by granting other financial incentives, such as easier repatriation of funds and favorable tax treatment.

Three primary models are used for niche market franchise opportunities:

1. *Build, Own, and Operate:* This model allows new domestic and foreign enterprises to invest, own, and operate a franchise. Examples include a variety of cellular radio franchises, such as GTE and U S West in Hungary and Ameritech in Poland; private overlay microwave radio networks, such as Bell Atlantic/U S West in the Czech and Slovak Republics; and digital data overlay networks, such as IBM's partnership with the Czech and Slovak PTTs and Tesla, an equipment manufacturer;

2. *Build, Operate, and Transfer:* This model allows an outsider to generate profits from niche or even core reserved services, but only for a fixed time period after which the franchise lapses and ownership of the plant transfers to the government franchisor. This option provides developing nations with the opportunity to acquire cutting-edge technology that they could not afford to buy, and previously acquired the technology only after granting a long-term or permanent franchise. Foreign enterprises participate in such tenders because they present new opportunities to establish a presence in new markets. The franchise terms dampen short-term returns, but the participating foreign carriers and manufacturers consider longer term opportunities, such as establishing an early presence that may lead to other franchise awards and market share for equipment. Examples of build, operate, and transfer arrangements can be found in Thailand and Indonesia.[9]

3. *Build, Transfer, and Operate:* This model allows an outsider to sell equipment and operational services (e.g., a VSAT network) to a nation desiring outside investment but leery of permitting foreigners to own and operate facilities. The foreign enterprise installs needed facilities, confers title to the government, and receives either installment payments or a portion of the operating revenues as compensation for both the equipment and facilities management.

15.7.4 Joint Ventures Including Government Participants

Joint ventures are another vehicle for public and private enterprises to achieve a critical mass of finances, political pull, and operational skills. This structure also enables a government to participate in entrepreneurial telecommunication ventures without even privatizing the PTT.

EUTELSAT represents a cooperative approach enabling nations to sponsor satellite competition, including direct broadcast satellite television, without expressly changing national broadcasting laws and regulations. Another example is the AsiaSat satellite venture with government owners—the China International Trade and Investment Company—and private investors—C&W of Britain and Hutchison Whampoa of Hong Kong.

15.7.5 Limitations on Foreign Investment

Privatization and its foundation in law, policy, and deregulation favor foreign investment opportunities, as do regional and multinational trade agreements. However, privatization does not mean that governments, which have closed

---

9. See P. Smith and G. Staple, *Telecommunications Sector Reform in Asia—Toward a New Pragmatism*, World Bank Discussion Paper No. 232, Washington, D.C.: World Bank, 1994, p. 55.

markets to foreign investment or restricted procurements to domestic sources, suddenly have become free marketeers. Most nations impose a variety of limitations on foreign participation, including the outright prohibition of or percentage limitation on foreign investment in telecommunications. Many nations limit competitive opportunities to niche markets, and even where competition is allowed, the incumbent facilities-based carrier retains plenty of opportunities to stifle competition (e.g., providing inferior, more expensive and inflexible interconnection to the PSTN, predatory pricing, unfair trade practices, and cross-subsidies).

Other factors may dampen the attractiveness of investment opportunities. Governments may place an unrealistically high price tag on the privatized carrier, in view of political factors restricting the level of permissible profitability and the scope of infrastructure investments and benchmark service commitments that the winning bidder must make. In the best of circumstances, it is difficult to reduce to present value a revenue stream generated by a nation's telephone company, particularly where substantial upfront investment is needed to upgrade the infrastructure and the possibility that government might subsequently limit the scope of reserved services or allow additional market entry.

Some nations view privatization as a one-time opportunity to acquire hard currencies and to replace an obsolete physical plant. The daunting task of building largely from scratch a nation's telecommunication infrastructure should lower the franchise cost, but the promise of newly accessible growth markets abroad in conjunction with limited further growth in mature domestic markets often triggers a bidding war. Carriers like AT&T, British Telecom, C&W, Telefonica of Spain, and France Telecom intend to develop a global marketing presence primarily to serve multinational telecommunications companies. Ownership of a geographically diverse set of telephone companies supports their participation in a global network. Such investments also reflect the need to find new markets, because domestic policies have opened previously captive markets to competition.

Part of the heightened interest in foreign telecommunication service opportunities may have resulted from a defensive strategy. Some companies believe that market access restrictions will grow worse over time, particularly after the implementation of initiatives aiming primarily to improve intraregion trade and global competitiveness. This "fortress" view stimulates current investment on grounds that an incountry operation will fare better in the face of ongoing or worsening discriminatory practices in equipment standardization, carrier procurements, testing and certification, and licensing of service providers.

15.8 MULTINATIONAL STRATEGIC ALLIANCES

The pace of technological innovation, privatization, experimentation with competition, and globalization provide unprecedented opportunities and risks. By far, strategic alliances have become the most common vehicle for such investments. Incumbents with declining control over domestic consumers find in alliances the opportunity to serve new markets and also to acquire resources and expertise, perhaps unavailable within the organization. If an incumbent faces what it considers the worst possible policy initiative, loss of monopoly control, then an alliance with a large multinational enterprise is one often successful way to bolster skills to retain a dominant market share or to acquire shares in new markets. For example, the incumbent Canadian carrier Stentor, Inc., has developed a strategic alliance with MCI, in part to achieve access to MCI's cutting-edge technological leadership in such services as virtual private lines and networks. Ironically, AT&T has invested in the newly authorized long-distance service challenger, Unitel.

Increasingly, incumbents cannot rely on cartel management of new technologies, facilities, and services. Previously, PTTs held significant ownership interests and voting power in global satellite cooperatives like INTELSAT and submarine cable consortia. They were able to control the deployment of additional capacity by creating a quasiofficial consultative process, ostensibly to forecast demand, plan for new transmission facilities, and allocate investment shares. PTTs were a dominant force in regional policy making, primarily because operational and regulatory matters were controlled by single entities in most countries. PTT coalitions such as CEPT bolstered individual PTT powers by promulgating policy suggestions readily accepted by the few independent regulators.

In the last few years a number of strategic ventures have responded to the deteriorating prospects for marketplace control by preexisting organizations. In the satellite marketplace, private systems, separate from the cooperative ownership structure, have been established to serve growing consumer demand for "television without borders" (e.g., new networks for program distribution directly to home satellite antennas, satellite master antennas, and cable television systems). These new enterprises include program packagers, like Sky Television in Europe and Star TV in Asia, and new carriers like Astra in Europe and Asia Pacific Telecommunications. Even incumbent carriers like C&W (AsiaSat) and Stet of Italy (Orion) have invested in separate systems, particularly as such carriers acquire greater consumer acceptance and fewer service restrictions.

In the submarine cable marketplace, governments have accepted private ventures separate from the conventional PTT-AT&T consortium approach. The consultative process cannot prevent market entry by operators of private submarine cables. The FCC has decided that private carrier capacity should not even be considered when forecasting demand, and the FCC has authorized new submarine cable projects on the basis of promoting competition at private investor risk rather than on conventional grounds based primarily on traffic forecasts. Examples include PTAT-1 in the North Atlantic, a joint venture of Sprint and

C&W, and North Pacific Cable, a consortium that includes C&W, Pacific Telecom of the U.S., and International Digital Communications of Japan. Planned projects include a trans-Siberian cable, in which U S West has played a key role and the Fiber Optic Link Across the Globe that NYNEX has organized.

Thus, both incumbents and newcomers have determined that global or regional alliances blend diverse skills and knowledge of the region to achieve success unachievable by a single carrier. But there have been several notable failures, such as cable television ventures in Hong Kong, IBM-MCI-Rolm, BT-Mitel, and AT&T-Olivetti.

The traditional models are consortia, such as TAT 1-13, and cooperatives, such as INTELSAT. In the late 1980s, cooperative members cut individual deals (e.g., global maritime and aeronautical service via BT, Norwegian Telecom, and Singapore Telecom).

Incumbents have pursued a number of recent alliances, including:

- *Outsourcing/one-stop shopping:* AT&T-organized World Partners; the Atlas venture of Deutsch Telekom, France Telecom, and Sprint; the Concert venture of British Telecom and MCI; and the Unisource alliance of Telia AB (Sweden), Swiss Telecom, PTT Telecom (the Netherlands), and Telefonica (Spain).
- *New or existing carrier franchises:* Australia (Optus-C&W, BellSouth); Venezuela (AT&T, GTE); Puerto Rico (Telefonia of Spain), Japan (ITC and IDC Japan, second and third KDDs with most major trading companies and manufacturers participating).
- *Private submarine cables* (e.g., PTAT, NPC).
- *IVANs.*

15.8.1 Types of Alliances

With changing rules of the road, additional and more diverse opportunities exist for market entry by strategic alliances. Set out below is an outline of the various types of alliance opportunities.

15.8.1.1 New Facilities

The bilateral and multilateral nature of international transmission facilities promote joint ventures. The PTAT-1 submarine cable in the Atlantic Ocean provides an example. C&W envisioned the development of a global digital highway to link the world's trading centers and the various telephone companies it owned or managed. C&W understood that most nations, including the United States, impose restrictions on alien ownership of transmission facilities. The Submarine Cable Landing Act of 1921 requires reciprocity in the landing of cables on U.S. soil. Accordingly, C&W sought strategic partners in the United States to own and operate the western half of the PTAT-1 cable and the eastern half of the North Pacific Cable. Through investment in a Pacific and Atlantic submarine cable network, C&W can provide a seamless link between Hong Kong, Japan, and other cities in the Asia-Pacific region with the U.S. and Europe.

15.8.1.2   Government-Created New Market Niches

Governments create investment opportunities when they negotiate bilateral agreements authorizing new types of services and carriers. An agreement between the United States and the United Kingdom legitimized the market for IVANs by distinguishing between networks that provide enhancements to leased lines and networks that simply resell basic services (e.g., long-distance telephone calling via leased lines). The U.S. and Britain recognized the former as an exception to the general prohibition on leased-line resale. Other nations have joined this view, making the U.S.-U.K bilateral agreement a basis for market growth, investment, and future revisions to ITU Recommendations. AT&T has taken advantage of this opportunity by acquiring the British IVAN, Istel. In turn, British Telecom has acquired the U.S. IVANs, Tymshare and Dialcomm.

15.8.1.3   New Carrier and Franchise Opportunities

The largest number of strategic opportunities results when governments license new carriers, franchises, and new technology trials. These opportunities require less capital than a PTT privatization, but can still involve quite large initial investments and provide lucrative opportunities.

Additional Local and Long-Distance Companies

A few nations, led by the United Kingdom and New Zealand, have abandoned the natural monopoly view for core basic services like local and long-distance telephone service. Other nations, which include Australia, Canada, Korea, and Sweden, have created a duopoly (i.e., a monopoly shared by two competitors). For example, in the U.K., a C&W subsidiary, Mercury, competes with the incumbent carrier British Telecom. Cable television companies provide additional local exchange service competition, and in 1997 additional facilities-based carriers will be authorized. In Sweden, Tele2, a joint venture of C&W and Kinnevik, provides across-the-board competition with Sweden's incumbent carrier Telia AB (known as Televerket until a reorganization in 1993).

Special-Purpose Networks

Nations disinclined to authorize market entry for core local and interexchange services can still stimulate competition in niche markets. For example, AT&T, IBM, Westinghouse, Itohchu of Japan, and Deutsche Aerospace have teamed with the former Soviet Union's Aeronavigation Research and Development Institution to modernize the CIS's air traffic control system by the year 2005. EDS provides turnkey technology integration and systems support for Polish, Czech, and Slovak cellular systems.

Overlay Data, Mobile Service, and Satellite Networks

Many nations with developed or developing economies have authorized two or more competitive mobile radio operators. Nations with obsolete, unreliable, and overtaxed wireline facilities have permitted foreign ventures to install microwave and fiber-optic cable networks for business applications. Competition in this sector is the result because decision makers believe the market can support multiple carriers or because the incumbent carrier has shown an unwillingness or inability to provide reliable service.

For example, Hong Kong has four cellular radio carriers, some of the lowest rates, and one of the highest market penetrations for service in the world. Ameritech and France Telecom operate the Polish cellular radio systems. San Francisco-Moscow Teleport and the IDB Communications Group have teamed with the Russian Inmarsat signatory to create Russian Satellite Communications Co. This company operates a satellite teleport that provides convenient and reliable international telecommunication services. A joint venture between Sprint and RosTel, a new organization formed by the Central Telegraph of the Russian Ministry of Communications, provides new high-speed data networks.

Cable Television

Many nations award cable television franchises instead of allowing the incumbent telephone company to provide the service. In authorizing a separate cable television wire into the home, decision makers have created the potential for facilities-based telecommunication service competition. Developments in digital, fiber-optic communications present the potential for cable television operators to provide local telephone service in addition to video services. A variety of foreign enterprises have entered this market. For example, U S West, Time Warner, Kinnevik of Sweden, Tele-Communications, Inc., and United International Holdings are partners in cable television ventures in the United Kingdom, Sweden, Norway, and Hungary.

Manufacturer-Carrier

Manufacturers pursue joint ventures with carriers to develop markets and to test equipment. For example, Motorola teamed with C&W to develop personal communications networks in the United Kingdom. Andrew Corp., a U.S. manufacturer of microwave and satellite and local-area network equipment has teamed with the Moscow Metro Subway to develop a 162-mile high-speed fiber-optic network.

Multiple-Carrier Alliances

Perhaps belatedly, the world's telecommunications carriers have acknowledged consumer sovereignty and the need to become more responsive. Carriers have expanded the array of available service options and have convinced increasing numbers of users to trade internal telecommunications management for *outsourcing*: relying on the expertise of carriers and systems integrators to provide and manage all necessary design, negotiation, procurement, coordination, and project management functions. These enterprises deliver a turnkey network (ready to use).

The combination of new multiple-carrier global alliances and success in convincing users to rely on outsiders for telecommunications management presents numerous investment opportunities for enterprises. In fact, the prevailing view seems to be that if a carrier does not become a member of one or more alliances, it will lose market share and perhaps much more.[10] Many global alliances manage international private networks used by multinational businesses or provide some sort of quasipublic networking option, commonly known as *virtual private networks* or *software-defined networks*. This market is estimated to have a value of approximately $3 billion and is growing at a 15% to 20% annual rate.[11]

15.8.2   Why Alliances Fail

While press releases herald the onset of yet another strategic alliance, little is reported when parties abandon a venture. Failed alliances demonstrate that not

10. "Speaking in London at the Third Economist Conference on Telecommunications..., DBT's Director General of International Affairs, Klaus Grewlich, stated boldly that 'to survive, operators must become global players,' adding that those who do not master globalisation 'may become candidates for not surviving, even in their own markets.'" ("Diversification: safeguard or suicide?" *Public Network Europe*, Vol. 1, No. 10, October 1991, p. 60.)

11. See J. Williamson, "Navigating Through the Global Managed Network Maze—Outsourcing: Hit or Myth?" *Telephony*, 4 May 1992, p. 9. A European publication reports that by 1994 this market will have a value of $6 billion. ("BT's Global Push for Outsourcing," *Public Network Europe*, Vol. 1, No. 10, October 1991, p. 13.)

all investment opportunities result in profits and successful blending of corporate cultures and objectives.

15.8.2.1 Failure to Stimulate Demand

Telecommunications alliances generally result when two or more enterprises perceive a strategic opportunity for higher revenues and profits from collaborating rather than individually marketing. Ventures fail when they do not stimulate demand, capture additional market share, or result in customer migration from one set of services to new ones. If the venture is forced to price its new services at a discount to rates otherwise available, the venture's success "cannabalizes" and adversely impacts participating carriers' existing revenue streams. Some global one-stop shopping ventures and virtual private networks may generate user churn (i.e., approximately equal numbers of new customers and existing customers canceling service) and migration between services instead of growth, but users increasingly demand such discounted service packages.

15.8.2.2 Countervailing Cultures and Objectives

Joint ventures and alliances in telecommunications require extensive coordination. While PTTs used to share similar objectives and philosophies, varying degrees of privatization, liberalization, market entrants, and public policy initiatives have created significant differences in approach that may not be fully understood by joint venturing parties. The alliance may stimulate vastly different strategies among investors, and the operating climate and strategy for success may vary by country, particularly because regulatory policies vary.

In Europe, for example, the United Kingdom stands at the vanguard for promoting marketplace competition and deregulation. Countries like Belgium, Germany, the Netherlands, Sweden, and Switzerland stand midway: retaining complete or partial government ownership for the time being, but permitting limited competition by resellers and prohibiting or substantially limiting the extent of facilities-based competition. Greece, France, and Italy lag behind, primarily because of the view that telecommunications continues to be a natural monopoly in most markets, or that "enlightened paternalism" through centralized planning best serves the national interest.

Blending the national telecommunications policy-making posture and a company's individual business plan presents an even greater diversity of cultures and objectives. For example, some telecommunication companies have learned how difficult it is to develop a successful venture that combines expertise in telecommunication equipment and services. AT&T's joint venture with the Italian manufacturer Olivetti, British Telecom's acquisition of the Canadian telecommunication equipment manufacturer Mitel, and IBM's attempt to vertically integrate with investments in MCI and Rolm failed to achieve desirable synergies. Other negative impacts on the likelihood for alliance success result from changing corporate strategies or the need for restructuring due to poor financial performance. An unexpected change in regulatory policy can also have an adverse impact (e.g., permitting more quickly installed master antenna television systems to compete with cable television in Hong Kong).

15.8.3 Prescriptions for Successful Alliances

The most successful alliances occur when each participating enterprise can make a contribution and concludes that the investment of time, money, and effort fits with overall corporate objectives. The parties must understand the markets they will attempt to serve, including the climate and policies of the nations where they will operate. Similarly, the parties must achieve a synergy that creates concrete disincentives for any participant to exit, on the assumption that it could perform all roles and therefore should capture all financial benefits. If the venture seeks to establish a global presence, it should have a geographically diverse composition, including representatives operating in all key markets.

The parties must commit to the long term, something that many corporations and business environments cannot support. Likewise, they must conduct preliminary research to determine whether corporate cultures, regulatory climates, and trade factors promote a positive collaboration.

A successful alliance may also result from less obvious and external factors than expertise, financial resources, access to markets, proximity to customers, and reputation. Likewise, the opportunity to erect an alliance may also depend on personal relationships, a company's desire to evidence its global orientations, labor and tax considerations, and the desire to operate in an environment using different technical standards.

The ability to contribute something meaningful to the venture may also depend on whether a country's technology transfer policies permit sale of computers, fiber-optic cables, and other high-technology innovations, despite national security concerns. For example, before relaxation of technology transfer restrictions, U.S. manufacturers complained that they could not propose state-of-the-art fiber-optic cable transmission capabilities for projects in the CIS.

Additionally, a venture may have to satisfy public policy goals of regulatory administrations in the host nation (e.g., contributing skills training, committing to local employment, and providing venture capital).

15.8.4 Structuring Ventures to Satisfy the Requirements of Multinational Enterprises

Over the last 10 to 15 years, the global economy has grown more integrated and interdependent. For example, consumers have grown to expect that merchants everywhere will accept credit cards. A complex global network must exist to process credit card verification orders across borders, time zones, bank systems, and different carrier networks in a few seconds. Newspapers like the *Wall Street Journal*, the *New York Times*, the *Washington Post* (through a joint venture that publishes the *International Herald Tribune*), and *USA Today* seek a global presence to acquire and disseminate news. "Cross-border business has been driven forward by three main things: falling regulatory barriers to overseas investment; tumbling telecommunications and transport costs; and freer domestic and international capital markets in which companies can be bought, and currency and other risks can be controlled" [6]. Additionally, foreign investment no longer automatically triggers fears of lost sovereignty[12] and control in most nations.

A key technological characteristic of telecommunications is its ability to make data flows and broadcasts of one nation accessible in other nations. Some, primarily developing, nations still inveigh against "transborder data flows" and cultural imperialism. However, most nations acknowledge that transborder operations expand the utility derived from a telecommunications investment, help spread costs over a larger set of users, and enable enterprises to achieve scale economies.

Globalization exploits fundamental technological characteristics of telecommunications. For example, satellite footprints traverse national boundaries, and three geostationary orbiting satellites can illuminate and thereby provide service to all populated regions of the world. Globalization results when corporate enterprises perceive strategic, tactical, or defensive reasons for investing in a foreign country. It can be more easily achieved when foreign governments treat firms on a relatively equal basis regardless of their nationality, and where goods, services, transactions, capital, and labor can move freely.

Foreign investment also occurs simply because of the need for closer proximity to customers. Companies that "think globally and act locally," can reduce costs (e.g., transportation) and risk (e.g., currency volatility). Additionally, they can minimize the potential for harm from discriminatory policies on imports through duties, taxes, and subsidies. Telecommunications enterprises also have globalized in part simply to keep pace with their multinational customers.

15.9 GLOBALIZATION STRATEGIES

Telecommunication equipment and service providers have to make strategic decisions on how to maximize profits and market share in a global economy. They can choose to:

---

12. "[I]n the political domain, the power of satellite telecommunications to transmit text and pictures cheaply has done more to weaken the grip of governments than any scheming capitalist giant." ("Multinationals Back in Fashion," in "A Survey of Multinationals," *The Economist*, Vol. 326, No. 7804, 27 March 1993, pp. 6–7.)

- Compete from afar by not investing to create a foreign presence;
- Compete incountry where allowed;
- Generate market share incountry through acquisitions;
- Cooperate through joint ventures.

15.9.1 Competition

Until quite recently, historical, political, regulatory, and economic forces made competition unlikely. Try as they might, equipment manufacturers succeeded, if at all, in capturing limited market share in a region or cluster of nations outside the home market. Manufacturers might compete in open procurements, but the combination of political and financial pacts, foreign aid tied to procurement of the donor nation's equipment, cultural affinity, and industrial policy tended to favor a national hero.

In the services arena, carriers matched half circuits and entered into correspondent relationships that respected national sovereignty and positioned carriers in a cooperative posture. Carriers still match half circuits, invest collectively in consortia and cooperatives, agree on toll division arrangements, and split responsibility for the maintenance of transmission facilities.

Governments and users, not telecommunications incumbents, have stimulated the urge to compete. By eliminating absolute monopolies through authorized market entry by resellers and even facilities-based carriers, governments have forced incumbent carriers and manufacturers to consider how to make up for lost market share. Users have grown more sophisticated in articulating their service requirements and adept at pursuing routing opportunities that reduce cost and best satisfy their needs. Often the potential for demand stimulation at home, even if the incumbent becomes a more efficient operator, pales in comparison to the opportunities abroad.

The greatest opportunity for a company to globalize may lie in pursuing infrastructure improvements in nations that lack a reliable network and have telephone line penetration below 25 per 100 inhabitants, particularly if per capita income will so increase that the majority of the population can afford what heretofore constituted a luxury. Many of these nations recognize the need to allow more qualified outsiders to assume the logistical and financial responsibility for speedy development of an advanced telecommunications infrastructure. New independent nations (e.g., the split-up of the Soviet Union) and nations no longer subject to central planning in a political bloc (e.g., Eastern Europe) can now pursue long overdue infrastructure improvements without first receiving outside authorization.

The next best opportunities lie where national trade policies favor some degree of market access and foreign investment. Several nations with vast populations and single-digit line penetration per 100 inhabitants have accelerated efforts to remove telecommunication bottlenecks and outages, recognizing their drag on the economy. Such nations include China, India, Malaysia, and Indonesia.

15.9.2 Buying Market Share

Telecommunications enterprises can acquire foreign market share overnight by investing in incumbent operators, such as British Telecom's acquisition of a 20% share in MCI and efforts by France Telecom and Deutsche Telekom to invest in Sprint. A captive customer base, the absence of a requirement to flow through revenues to the national treasury, and perhaps also relatively ineffectual rate regulation make it possible for many incumbents to hold substantial cash reserves. Foreign investment occurs when an enterprise thinks it can achieve operational synergy with the acquired company while building on the company's existing market share.

Many globalizing telecommunications enterprises prefer to pursue joint ventures and secure global or regional alliances with counterparts of equal capitalization (e.g., AT&T's World Partners alliance with such carriers as KDD of Japan, Singapore Telecom, and the Swiss and Dutch PTTs). Limitations on foreign ownership support this alternative, as does the recognition that ventures marketing global services need geographically diverse participants.

15.9.3 Cooperation

Many incumbent players face new domestic competition and substantial threats to the status quo. Global and regional alliances represent a recognition of an individual firm's limitations and the real or perceived benefits accruing from an affiliation with similarly situated enterprises. A company with a long operational history in a particular region is better equipped to know the particular features, limitations, customer requirements, and other features specific to the area.

Telecommunication ventures typically involve substantial investments that a number of players are better able to share. Similarly, the risk attendant to any particular venture can be spread across a number of investors. If a venture proposes to operate in a large geographical area and to assume end-to-end responsibilities, then a strategic or tactical alliance of geographically diffused players is essential. It remains to be seen whether these ventures will have staying power and constitute the basis for a new industrial structure or are simply short-term efforts to pool investment and reduce risks.

15.9.4 Conventional Types of Global Alliances

Set out below is a list of ways enterprises agree to structure a global alliance:

- *Cooperative equity arrangements:* Carriers pool investments in costly satellites and submarine cables through global cooperatives such as INTELSAT, regional cooperatives such as EUTELSAT, and submarine cable consortia.
- *Cooperative bilateral agreements:* Carriers negotiate "accounting rates" and other contractual terms and conditions for the acceptance and routing of traffic and the division of toll revenues.
- *Memoranda of understanding:* Companies agree to exchange personnel and cross-license technology; for example, MCI's investment in the dominant Canadian carrier Stentor included an agreement to cross-license proprietary virtual private networking technology that uses software for speedy creation and disassembly of private networks.
- *Joint ventures:* Companies negotiate short- or long-term contracts to bid on projects involving, for example, the installation and maintenance of equipment and services on a regional or global basis by the AT&T-organized World Partners venture with KDD of Japan, Singapore Telecom, Unitel of Canada, Telefonica de Espana of Spain, Telecom New Zealand, Telstra of Australia, and the Unisource venture composed of Telia of Sweden, PTT Switzerland, PTT Holland, and others.

15.9.5 New Ventures

More recent alliances include:

- *Consortia of foreign telephone companies and local partners:* Privatizing the incumbent PTT or creating license opportunities for a second facilities-based carrier. The winning bidding consortium for Australia's second carrier, Optus, include Britain's Cable & Wireless, BellSouth of the U.S., and Australian partners AMP, an insurance company, and Mayne Nickless, a transport company. Most winning consortia integrate foreign players that have money to invest and telecommunications expertise with local partners that emphasize national citizenship and political ties.
- *Consortia of foreign telephone companies, foreign equipment manufacturers, and local partners:* An extension of the above model adds a manufacturer who provides expertise in ventures requiring new types of equipment. Numerous cellular radio and personal communication network ventures provide an example.
- *Consortia that include large user groups:* Sophisticated, high-volume telecommunications users have joined in projects that should better serve their requirements. Examples include the consortia of C&W, Itohchu, a Japanese trading company, Toyota, Pacific Telesis, Merrill Lynch, Singapore Telecom, and US Sprint, which developed the ASPAC trans-Pacific fiber-optic submarine cable.

- *Joint ventures targeting niche markets:* A number of ventures target profitable niche markets rather than go after the telephone company franchise. Examples include value-added networks, outsourcing, cellular radio, satellite teleport, and microwave networks that overlay the obsolete wireline network.

15.9.6 Challenges Facing Alliances

Alliances require partners to share objectives and to identify likely long-term benefits. The partners must devise a structure that maximizes the contributions each brings to the venture and allow the venture to evolve as conditions change. The parties must share a vision for the venture and agree to make it financially difficult for anyone to exit.

Telecommunication alliances face industry-specific challenges. The alliance may contain investors subject to quite different degrees of government oversight and domestic competition. The venture may have difficulty defining a single, consistent corporate culture, because its investors may have inconsistent perspectives based on the extent to which they have privatized and abandoned prosocial obligations. Complex regulatory duties like universal domestic service may challenge the attention and financial wherewithal of investors to concentrate on making the alliance work.

The nature of the venture's mission vis-a-vis other activities of the investors presents an additional challenge. Strategic alliances may contain strange bedfellows who compete vigorously for business in other markets. For the alliance to work, the parties must carve out markets where they agree to collaborate rather than compete, no matter how attractive the solitary option. The benefit of synergy must persist. Otherwise, individuals will perceive a financial advantage in exiting the alliance.

15.9.7 Future Outcomes for Global and Regional Alliances

National carriers and equipment manufacturers have rushed to join alliances, often out of concern that the failure to do so will result in lost market opportunities, particularly in the face of trade actions that do more to bolster a region's marketing prowess than to make the region accessible to outsiders. The proliferation of regional trading pacts (e.g., the Single Europe Act and NAFTA) creates concern that the marketplace will concentrate into an oligopoly of successful alliances that have a robust marketing presence in each of the three key markets of the world: (1) Europe, (2) the Americas, and (3) Asia/Pacific.

It remains to be seen whether alliances are a productive response by corporate managers to their individual firm's limitations and the inability to operate efficiently everywhere, or a passing fad abandoned when investors become disillusioned by mounting financial losses and covenants not to make individual market entry. For the time being the following justifications for tactical alliances seem strategic enough:

- Spreading risks;
- Sharing in high-cost ventures;
- Improving the information flow between equipment manufacturer and service provider;
- Promoting economies of scale and scope;
- Reducing the time from innovation to market rollout;
- Sharing research and development costs;
- Finding an expedient way to access an otherwise closed market.

References

[1]   "Politics on the Line," a survey of telecommunications, *The Economist*, 5 October 1991.

[2]   Duch, R., *Privatizing the Economy—Telecommunications Policy in Comparative Perspective*, Ann Arbor: University of Michigan Press, 1991, p. 3.

[3]   Ibid., p. 6.

[4]   Oxman, S., "Privatization: A Strategic and Financial Perspective," *Proc. 6th World Telecommunication Forum, Part 3 Regulatory Symposium: Competition and Cooperation in a Changing Environment*, Geneva: ITU, 9–11 October 1991, p. 211.

[5]   Carlsberg, B., "Regulator's Perspective in the Role and Purpose of Regulation," *Proc. 6th World Telecommunication Forum*, Part 3, pp. 7, 10.

[6]   "Multinationals Back in Fashion," in a survey of multinationals, *The Economist* Vol. 326, No. 7804, 27 March 1993, p. 6.

Case Studies in Telecommunications Deregulation

16

Many nations throughout the world have considered revamping the industrial structure and regulatory scope in telecommunications. Recent actions have fostered more change than what occurred in the 100-plus years since the start of telegraph service. Four factors contribute to growing attention to telecommunications and ways to improve performance and consumer welfare.

1. Technological innovations reduce or eliminate justifications that a natural monopoly can achieve optimal economies of scale and scope.
2. Telecommunications can stimulate national economies and market entry can increase consumer and national dividends.
3. Regulatory and industrial initiatives can achieve efficiency gains and promote the acquisition of international market share.
4. Initiatives can harmonize operational terms and conditions within a region to make its constituent nations more competitive on a global basis.

Few nations can afford to ignore the new imperatives presented above. But the manner in which any nation responds reflects individual or regional social, political, and economic factors. Therefore, outcomes will vary substantially in how a nation responds to changed circumstances. Set out below is a brief outline of what has occurred in particular nations and regions.

16.1 EUROPE

Many nations in Europe have completed or are considering significant changes to the PTT monopoly model. The changes run on a continuum from complete privatization (e.g., the United Kingdom) to partial privatization and some market entry opportunities (e.g., Denmark, Portugal, Italy (public stock offering), the Netherlands (new corporate charter)) to no privatization, but the introduction of competition in some telecommunications markets (e.g., Germany (cellular radio), Sweden (second facilities-based carrier)) and finally to little if any change in PTT ownership and scope of competition (e.g., France and Greece).[1] This consideration results not only from factors affecting the provision of telecommunications services, but also as part of a larger effort to unify Europe. Harmonized telecommunications standards, operational terms and conditions, regulations, and policies constitute part of the goals contemplated by the Single Europe Act [1]:[2]

> Most countries are interested in having competition in the supply of consumer premises equipment and in the supply of value-added or enhanced services. Several are interested in competition in providing network facilities for mobile communications. Few have shown any interest in competition in the provision of facilities for basic fixed link networks. Interest is rather limited in regulation that is independent of ministers.

Currently, a number of nations in Europe seem committed to a cautious migration to limited competition in selected markets. While the United States deregulated the provision of customer premises equipment in the early 1970s, most European nations adopted this policy in the 1980s or later, and some still authorize the PTT to lease the first instrument at each premises. Many nations also see the benefit in a robust value-added services marketplace, because such services can stimulate the national economy and, as data and enhanced serv-

---

1. See, for example, H. Ungerer and N. Costello, *Telecommunications in Europe: Free Choice for the User in Europe's 1992 Market*, Brussels: Commission of the European Communities, 1988; Fontheim, "EC 1992 and Telecommunications: Fortress Europe or Market Liberalization?" *Telematics*, Vol. 6, No. 4, April 1989, pp. 1–5; F. Cate, *The European Broadcasting Directive*, American Bar Ass'n, Sec. of Int'l Law and Practice, Communications Committee Monograph Series, April 1990; D. Elixmann and K. H. Neumann, eds., *Communications Policy in Europe*, Berlin: Springer-Verlag, 1990.

2. The Treaty of Rome, formally called the Treaty Establishing the European Economic Community, 298 U.N.T.S. 11, entered into force at Rome, Italy (effective 25 March 1957) calls upon member states to agree in principle to accept wide-ranging limitations on their individual national sovereignty to create a single market. The treaty envisaged the ultimate goal of an integrated market for the free movement of goods, services, capital, and people. (G. C. Hufbauer, "An Overview," in *Europe 1992: An American Perspective*, G. C. Hufbauer, ed., Washington, D.C.: Brookings Institution, 1990, p. 1). The Single European Act amended the Treaty of Rome (effective 1 July 1987) and created a 31 December 1992 deadline for completion of a single market. See G. C. Hufbauer, ed., Appendix 1-5, "Key Provisions of the Single European Act," in *Europe 1992: An American Perspective*, p. 58. See also Commission of the European Communities, Completing the Internal Market, COM(85) 310 Final (1985) (White Paper containing 300 proposals for legislation affecting a wide variety of economic activities).

ices, they tend to generate additional usage of incumbent carrier facilities rather than encourage migration to specialized competitor networks.[3]

16.1.1 Europe 1992

Because of the excitement and apprehension over the plans of 12 European nations[4] to unify their markets, these plans have the potential to replace ISDN as the most talked about telecommunications concept for the future. "Europe 1992," like ISDN, has different meanings depending on one's perspective and frame of reference. In the abstract, it represents the attempt by European nations to remove barriers to commerce, thereby converting the European Community (EC) into the EU.

The stakes of this campaign are incredibly high. In 1993, the then 12-nation EC (see Figure 16.1) represented a market of over 320 million people, with a gross national product in excess of that generated by either the United States or Japan. Fragmented and uncoordinated into separate markets,[5] the nations of Europe cannot establish a fortresslike collective barrier to market entry. Integrated and cohesive, these same nations can achieve a more efficient, productive, and consolidated market, and as some fear, also a more competitive regional trading bloc.

16.1.2 Efforts Toward a Single Market

In June 1985, the European Commission, which serves as the executive organ for the EEC,[6] published a White paper entitled *Completing the International*

---

3. See W. Garrison, *Case Studies of Structural Alternation in the Telecommunications Industry—The United Kingdom, the Federal Republic of Germany, the United States of America,* Washington, D.C.: Annenberg Washington Program, 1988.

4. The EC consists of 12 member states: Belgium, Denmark, France, Germany, Great Britain, Greece, Ireland, Italy, Luxembourg, the Netherlands, Portugal, and Spain. (See Treaty Between the Member States of the European Communities and the Kingdom of Spain and the Portuguese Republic Concerning the Accession of the Kingdom of Spain and the Portuguese Republic to the European Economic Community and to the European Atomic Energy Community, 1985 O.J. (L 302) 9.) Austria, Finland, Iceland, Liechtenstein, Norway, Sweden, and Switzerland subjected themselves to most EEC regulations when they joined with the EEC to form the European Economic Area. (See V. Robinson, "Recent Developments in the Law of the European Communities," *Duke J. Comparative and International Law,* Vol. 2, 1992, pp. 25–26.) More recently, Austria, Finland, and Sweden joined the EU, which is the umbrella organization over the EEC. (See "Three Nations Agree to Join EU," *Wall Street J.,* 2 March 1994, p. A6.)

5. Internal estimates of the financial loss occasioned by nonintegration have been estimated at between 5% and 7% of gross domestic product. (*Research on the "Costs of Non-Europe,"* INSEAD, December 1987 (the Cecchini Report).)

6. "The main tasks of the Commission are: (1) ensuring that the founding EC treaties and their amendments are carried out; (2) drawing up the budget; (3) exercising its sole power to initiate

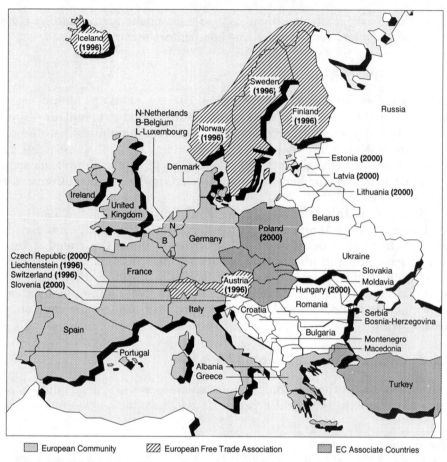

| ■ European Community | ▨ European Free Trade Association | ■ EC Associate Countries |

(1996), (2000) – Indicate projected dates countries may join the European Community

Figure 16.1 Current and projected European Community membership. (*Source:* Office of Technology Assessment, 1993.)

Market.[7] The commission generated this document to serve as a "blueprint for dismantling existing technical, fiscal, and legal barriers among member states of

Community legislation (directives and regulations) and power to amend legislation at any stage; and (4) implementing decisions reached by the Council of Ministers" [the EC's decision-making body]. (G. C. Hufbauer, ed., Appendix 1-2, "Institutions of the European Community," *Europe 1992: An American Perspective*, p. 53.

7. Commission of the European Communities, *Completing the Internal Market*, COM No. 310, June 1985.

the EC. It also contained 300 proposals for legislation affecting a wide variety of economic activities[8]...[with a view toward] *harmonization* of various laws in member states, and *mutual recognition* of authorizations and licenses granted by a member state" [2]. Through such initiatives, the European Commission sought to remove barriers to a single market, defined as any legislation, standard, procedure, or checkpoint that impedes the free flow of goods, persons, services, and/or capital among the member states.

In July 1987 the member states agreed to the Single European Act, thereby amending the Treaty of Rome. The act set out procedures for implementing the White Paper goals by December 31, 1992, and authorized the European Commission's Legislative Council to take decisions affecting unification by majority vote rather than unanimity.

16.1.3 European Telecommunications Initiatives

The European Commission's blueprint for telecommunications was articulated in a June 1987 Green Paper.[9] This document called for the lowering of barriers among member states and more uniform network operations. In a nutshell, the Green Paper, and the more concrete European Commission Report, which sought further consideration of the Green Paper within the EC governance structure [3], called for "the full liberalization of the terminal equipment market, with a reasonable period of transition," along with fair type approval procedures to ensure unrestricted provision and interconnection of CPE; "the liberalization of value-added services...to competitive provision"; "the separation of regulatory and operational responsibilities of the Telecommunications Administrations"; the implementation of tariffs that "follow overall cost trends"; the harmonization of standards to "maintain or create Community-wide and world-

---

8.   The proposals affect such diverse sectors of the European economy as financial services, investment flows, currency exchange controls, securities, insurance, intellectual property, and government procurement.

9.   Commission of the European Communities, *Towards a Dynamic European Economy: Green Paper on the Development of the Common Market for Telecommunications Services and Equipment*, COM(87) 290, 30 June 1987; see also Commission of the European Communities, H. Ungerer and N. Costello, *Telecommunications in Europe*, 1988; B. Delcourt, "EC Decisions and Directives on Information Technology and Telecommunications," *Telecommunications Policy*, Vol. 15, February 1991, pp. 15–20; and W. Garrison, Jr., *The European Telecommunications Directives: Provisions Requiring Regulatory Restructuring*, American Bar Assn., Sec. of Int'l Law and Practice, Communications Committee Monograph Series 1990/2, August 1990. "In the parlance of the EC, a 'green paper' is a consultative document released as a statement of policy and roadmap for future legislation. Although green papers have no legal force, they can have tremendous policy influence, and can be as heavily debated and revised as the resulting legislation." (G. Oberst, Jr., *EC Satellite Regulation: Steps Toward Liberalization*, American Bar Assn., Sec. of Int'l Law and Practice, Communications Committee Monograph Series 1992/1, July 1992, note 3.)

wide interoperability, while safeguarding the capability for innovation"; opening the market for satellite receive-only antennas that do not access the public switched telephone network; establishing the ETSI to promulgate EC-wide network operating standards, particularly for ISDN and "Open Network Provision," the EC counterpart to the FCC's *Third Computer Inquiry*'s "Open Network Architecture" plan for nondiscriminatory access to local and interexchange facilities by enhanced services providers; and "ensuring fair conditions of competition," that is, no cross-subsidies on PTT provision of services subject to competition, and open public procurement of telecommunications equipment.

The ambitiousness of the Green Paper and subsequent documents assumed that the goals would not be reached by 1992, with some likely to trigger substantial opposition [4]:

> While the attitude to the single market is largely positive in Europe, the difficulties lie in the practical details. For this reason, the goal will probably only have been partially reached in 1992, and the process of harmonization will continue beyond 1992.[10]

Nevertheless, the EU has begun what appears to be an irreversible course toward liberalizing telecommunications[11] in three major sectors:

1. *Terminal equipment:* Separating equipment from services and allowing a competitive marketplace for most types of equipment, including telephones, modems, telex terminals,[12] and receive-only satellite Earth stations not connected to the PSTN;[13]

---

10. A progress report on EC'92 conducted National Association of Manufacturers notes that: "Much has changed for American companies doing business in Europe...[The] major significance of EC-92 is not that there will be a 'big bang' on December 31 and that the nature of the market will change overnight. In actual practice, the Single Internal Market may take years to 'complete,' in all its ramifications." (S. Cooney, *The Europe of 1992: An American Business Perspective*, the Fourth NAM Report on Developments in the European Community's Internal Market Program and the Effects on U.S. Manufacturers, National Assn. of Manufacturers, May 1992, p. 1.)

11. For a discussion on some of the legal challenges to EC procompetitive initiatives, see J. Darnton and D. Wuersch, "The European Commission's Progress Toward a New Approach for Competition in Telecommunications," *International Lawyer*, Vol. 26, No. 1, Spring 1992, pp. 111-124; and S. Lando, "The European Community's Road to Telecommunications Deregulation," *Fordham Law Review*, Vol. 62, 1994, p. 2159.

12. See Commission of the European Communities, Directive 88/301 of 16 May 1988 on Competition in the Markets in Telecommunications Terminal Equipment, 1988 O.J. (L. 131) 73, aff'd in part, reversed in part sub nom., France v. Commission, Case No. 202/88, slip op. (Court of Justice, 19 March 1991) (affirming that Article 90 of the EEC treaty authorizes the EC Commission and not the Council in its legislative function under Article 100 to enact a directive abolishing special and exclusive rights of importation commercialization and linkup of terminal equipment). See also Council Directive 86/361 of 24 July 1986 on the Initial Stage of the Mutual Rec-

2. *Telecommunication services:* Promoting competition in telecommunication services, other than voice telephony, by withdrawing many of the special and exclusive rights vested in PTTs;[14]

3. *Open networks:* Harmonizing conditions for access to telecommunication facilities operated by incumbents to stimulate competition in non-reserved services[15] and access to incumbent facilities, such as the local exchange on fair terms and conditions.[16]

---

ognition of Type Approval for Telecommunications Terminal Equipment, 1991 O.J. (L. 217) 21 and Council Directive 91/263 of 29 April 1991 on the Approximation of the Laws of the Member States Concerning Telecommunications Terminal Equipment (establishing procedures for mutual recognition of technical compatibility approval procedures for telecommunications equipment).

13. See Commission Directive of 16 May 1988 on competition in the markets in telecommunications terminal equipment, 88/301/EEC, O.J. 131/73, 27 June 1988, aff'd sub nom. France v. Commission [1991] ECR I-1223. See also K. Platteau, "Article 90 EEC Treaty After the Court Judgment in the Telecommunications Terminal Equipment Case," *European Competition Law Reporter,* Vol. 3, 1991, p. 105; J. Naftel, "The Natural Death of a Monopoly: Competition in EC Telecommunications Terminals Judgment," *Emory International Law Review,* Vol. 6, Fall 1992, p. 449; see also A. Lin, "Telecommunications Competition in the European Union After France v. Commission—the Terminal Equipment Case," *Connecticut J. International Law,* Vol. 9, 1994, p. 355 (examining Case 202/88, France v. EC Commission (Telecom), [1991] 5 CMLR 552, [1993] 1 CEC (CCH) 748 (rejecting France's challenge to the EC Directive on Telecommunications Terminal Equipment requiring withdrawal of special or exclusive rights to import, market, interconnect, and maintain terminal equipment like telephones).

14. See Commission Directive 90/388 of 28 June 1990 on Competition in the Markets for Telecommunications Services, 1990 O.J. (L. 192), aff'd in part, annulled in part sub nom., Spain, Belgium and Italy v. Commission, Joined Cases C-271/90, C-281/90 and C-289/90 slip op. Court of Justice (17 November 1992) (affirming right of the European Commission, pursuant to Article 90, to authorize private companies to provide telecommunication and information services, except for basic voice communications, but annulling provisions prohibiting the conferral of special rights on a single entity for lack of specificity on what an EC government cannot confer and mandating application of the directive even to long-term supply contracts).

15. Because voice telephony represents 90% of EC telephone company business, it will be subject to competition on a much slower, incremental track. The European Commission initially proposed to apply open network requirements for basic voice telephony. (Proposal for a Council Directive on the Application of Open Network Provision (ONP) to Voice Telephony, COM(92) 247 final.) Its proposal generated widespread opposition, prompting it subsequently to propose only to introduce voice telephone competition only for intra-EC calls. It appears quite likely that there will be further delays in any directive regarding voice competition. (See also C. Zepos, "Liberalizing the Sacred Cows: Telecommunications and Postal Services in the EC," *Duke J. Comparative and International Law,* Vol. 3, Fall 1992, p. 203.) For background on how the EC and the United States regulate enhanced services, see Note, "A Comparative Study of the Regulatory Treatment of Enhanced Services in the United States and the European Community," *Northwestern J. International Law and Business,* Vol. 9, 1988, pp. 415-443.

16. See Council Directive 90/387 of 28 June 1990 on the Establishment of the Internal Market for Telecommunication Services Through Implementation of Open Network Provision, 1990 O.J. (L. 192); Council Directive 92/44 of 5 June 1991 on the Application of Open Network Provision to Leased Lines, 1992 O.J. (L 165) (mandating the right of individuals to lease private lines

The EU has begun the task of establishing "Community-wide network integrity...that works on the principle of full interconnectivity" [5]. Such integration requires uniform standards and operating protocols, but perhaps more importantly, it requires a shared view of the scope of competition and the manner by which the telecommunications infrastructure will evolve to furnish a pan-European network of terrestrial, satellite, and mobile highways[17] using broadband, digital technologies.[18] A shared vision will require consensus among substantially divergent points of view within the EU on the role of government in telecommunications and the extent to which competition in the sector can generate consumer benefits and enhance regional viability without handicapping the attainment of social goals like universal service. The manner by which the EU governs can often lead to delay and incrementalism in the implementation of needed reforms.

In view of opposition to its blueprint for the future, the European Commission opted for a slower, more measured timetable for fostering competition. For example, the commission issued a new Green Paper on satellite services in 1990 that specified in greater detail the steps necessary to achieve the kind of procompetitive marketplace the commission deemed essential. By returning to the Green Paper blueprint, the commission could articulate all the steps it considered key, even though it could have little confidence that the council would follow up with concrete proposals toward making the blueprint a reality.

The commission proposed [6]:

1. Full liberalization of the Earth segment (i.e., deregulation of all Earth stations);

---

across the EC on open, tariffed, and nondiscriminatory terms) and Council Recommendation 92/383 of 5 June 1992 on the Provision of Harmonized Integrated Services Digital Network Provision (ONP) Principles, 1992 O.J. (L 200) (recommending standards, usage conditions, tariff principles, etc., relating to packet-switched public data networks). See also R. Frieden, "Open Telecommunications Policies for Europe: Promises and Pitfalls," *Jurimetrics*, Vol. 31, No. 3, Spring 1991, pp. 319-327.

17. See, for example, Council Resolution of 22 January 1990 Concerning Trans-European Networks, 1990 O.J. (C 27) 8.

18. See Council Recommendation 86/659 of 22 December 1986 on the Co-ordinated Introduction of the Integrated Services Digital Network in the European Community, 1986 O.J. (L 382) 36; Council Resolution of 18 July 1989 on the Strengthening of the Co-ordination for the Introduction of the Integrated Services Digital Network in the European Community up to 1992 1989 O.J. (C 196) 4; Council Resolution of 5 June 1992 on the development of the Integrated Services Digital Network in the Community as an Europe-wide telecommunications infrastructure for 1989 and beyond 1992 O.J. (C 158) 1; and Council Recommendation 92/383 of 5 June 1992 on the Provision of Harmonized Integrated Services Digital Network Access Arrangements and a Minimum set of Integrated Services Digital Network Offerings in Accordance with Open Network Provision (ONP) Principles 1992 O.J. (L 200) 10.

2. Unrestricted access to space segment capacity, including the right of users to acquire satellite capacity from a foreign carrier;
3. Full commercial freedom of space segment providers (i.e., the ability of an EU satellite carrier to lease capacity to users in all EU nations);
4. Harmonization measures to facilitate pan-European satellite services, including license-free operation of receive-only Earth stations, minimal regulation of transmit/receive Earth stations not interconnected with the PSTN, and nondiscriminatory licensing of Earth stations that do access the PSTN, subject to compliance with any exclusive or special rights retained by incumbent carriers.

Additionally, the European Commission in 1992 proposed a directive[19] aiming to establish a procedure for EU-wide recognition of licenses, thereby allowing any licensed enterprise to provide telecommunication services anywhere in the region [7]. By the end of the century, the commission expects all nations of the EU to permit facilities-based competition even for basic local and long-distance services.

16.1.4 Will Europe 1992 Benefit Outside Telecommunications Players?

Europe 1992 will have a mixed impact on other countries.[20] In telecommunications, no single EU nation "accounts for more than 6 percent of the world's telecommunications market, whereas the United States represents 35 percent and Japan 11 percent. Yet, taken as a whole, the...[EU] has a 20 percent world market share. By 1992, the...[EU] is expected to be the largest single market" [8]. Such concentration of economic power and marketing opportunities have sent a

---

19. Directives are important vehicles for converting a policy initiative, which may be articulated in a Green Paper, into legislative instruments addressed to and binding upon the member states. Either the European Commission or the European Council may issue a directive that typically requires implementation by member nations through national legislation. See Treaty Establishing the European Economic Community, 25 March 1957, art. 189, reprinted in Treaties Establishing the European Community 207, 388, Office for Official Publications of the European Communities, Luxembourg, 1987.

20. For example, an EU Directive ostensibly designed to promote "television without frontiers" requires that broadcasts originating in a member country contain at least 50% domestically produced programming. (Council Directive 89/552 of 3 October 1989, on the Coordination of Certain Provisions Laid Down by Law, Regulation or Administrative Action in Member States concerning the Pursuit of Television Broadcasting Activities, 1989 O.J. (L298), art 4.) For analysis of the EU's "television without frontiers" directive, see J. Shelden, "Television Without Frontiers: A Case Study of Turner Broadcasting's New Channel in the Community—Does It Violate the Directive?" *Transnational Lawyer*, Vol. 7, Fall 1994, p. 523; and L. Kaplan, "The European Community's 'Television Without Frontiers' Directive: Stimulating Europe to Regulate Culture," *Emory International Law Review*, Vol. 8, Spring 1994, p.255.

number of other nations' manufacturers and service providers scurrying for EU partners, in the event a "fortress Europe" scenario occurs, triggering more pronounced market access barriers in the future.

Non-European companies seek to establish a domestic presence for a number of pragmatic reasons:

- Nationalism, culture, and geographic proximity tend to favor domestic suppliers, particularly for high-price technical products with long usable lives and extensive maintenance requirements;
- Governments, unions, trade associations, and other organizations may continue to favor "national heroes in procurement";
- The EU will not adopt policies that absolutely mirror the deregulatory and market access initiatives of other nations.

A single European market in telecommunications should create new business opportunities, because a single coherent set of standards and somewhat more liberal policies and regulations can make it easier to sell products and services in Europe. The growth forecast for the telecommunications sector should provide enough demand for non-European players. On the other hand, Mr. Robert Priddle, former head of telecommunications for the U.K. Department of Trade and Industry who sought to convince a U.S. audience in 1989 that there is nothing to fear about a single European market, acknowledged that it was "mighty lonely in Europe two years ago as a proponent of competition" [9].

European nations will address the same technological, regulatory, and policy issues as other nations. Collectively and individually their resources may parallel but will not replicate what other nations have decided. For example, the Open Network Provision Directive on access to carrier facilities will parallel but not match the FCC's open network architecture blueprint for enhanced-services provider access to local and interexchange networks. Still, it is unlikely that, even if it wanted to, the EU could erect strong access barriers to the largest telecommunications market in the world. The number of "illegal" television receive-only satellite dishes in Eastern Europe corroborates this point: consumers will secure access to hardware and software, despite government attempts to block or financially handicap imports.

If anything, Europe 1992 represents a somewhat belated recognition on the part of the EU that a dozen or more nations working in tandem can achieve synergy and mutual benefits far in excess of what any individual nation could hope to capture for itself. That kind of thinking, whether for telecommunications or tariffs in general, is what prompted formation of the EC in the first place.

16.2 DEVELOPMENTS IN INDIVIDUAL EUROPEAN COUNTRIES

16.2.1 The United Kingdom

The United Kingdom most closely approximates the deregulatory atmosphere achieved in the United States, and in some respects exceeds all nations in ground-breaking policy initiatives. With the election of the Margaret Thatcher administration in 1979, five distinct government initiatives ensued [10]:

> (1) a formal separation of telecommunications from the Post Office and establishment of British Telecom (BT) as an independent but regulated entity, (2) establishment of competition in services by permitting rival carriers and value added services, (3) privatization of the public network by selling a majority of British Telecom, (4) liberalization of the market for peripheral equipment, and (5) establishment of the [independent] regulatory body, Oftel [the Office of Telecommunications].

The United Kingdom started the liberalization of telecommunications policies with the British Telecommunications Act of 1981 that separated telecommunications from postal operations and authorized the Secretary of State to license new facilities-based network operators.[21] The act reconstituted BT as a public corporation, permitted individuals and companies to install their own customer premises equipment, except for the first telephone, and authorized both facilities-based and limited resale competition. In 1982, Mercury was authorized to compete with British Telecom, thereby creating a facilities-based carrier duopoly. Other nations, which include Australia, Canada, and Korea, have created a similar duopoly.

The United Kingdom linked deregulation, by creating market entry opportunities, with liberalization, reducing or eliminating rules that prevented the incumbent carrier from competing and operating effectively. BT was afforded greater operational flexibility and the opportunity to keep profits generated by greater operational efficiency. Oftel devised a price cap formula instead of conventional rate-of-return regulation. Rather than apply public utility rate setting to determine a fair rate of return and the revenues BT could reasonably accrue, Oftel established a new social contract. Rates were capped at current levels with permissible annual increases linked to the Retail Producers Index, an overall measure of inflation and living costs *minus* a productivity factor. Technological innovations have made telecommunications a declining cost industry. There-

---

21. For an extensive review of British telecommunication laws, see P. Strivens and V. Sinden, "Telecommunications Law in the United Kingdom," in *Telecommunications Laws in Europe*, J. Scherer, ed., 1993, pp. 129, 129–130).

fore, Oftel wanted to create new incentives for BT to become more efficient by allowing it to capture any profits remaining after the productivity offset, rather than have to flow them automatically to ratepayers through refunds or further rate reductions.

The British government subsequently provided for even more competitive stimulation by expanding the scope of permissible resale opportunities to include basic switched long-distance calling. In the domestic satellite marketplace, it granted seven new licenses, including one to a U.S. enterprise, EDS. The United Kingdom also supported mobile telecommunications competition by licensing multiple cellular radio operators, cordless telephone services, and personal communication networks. In 1991, the government proposed modifications to the Telecommunications Act of 1984[22] that would eliminate the BT/Mercury duopoly in local and interexchange telephone service and over time would open virtually all telecommunications markets to competition.[23]

The U.K. experience provides a solid model for incremental deregulation, market entry, and liberalization. Now that sufficient time has passed to weigh the benefits of such initiatives, most observers agree that the resulting changes have fostered a more streamlined, efficient, and business-minded environment. In addition, Oftel has succeeded in erecting incentive regulations that reward efficiency. Even with increases in the productivity factor to approximately 7%, BT has succeeded in generating larger profits and sufficient cash flow to support investments abroad, including the acquisition of a 20% share in MCI.

16.2.2 Germany

Until 1989, the German government owned and operated a conventional PTT.[24] The Deutsche Bundepost had a telecommunications monopoly, and no independent regulatory authority existed. Lacking competition, the German telecommunications environment developed a reputation for having Europe's highest rates and most rigid network operating rules.[25] The Poststrukturgesetz

---

22. United Kingdom, Office of Telecommunications, Telecommunications Act of 1984, Proposed Modifications of the Conditions of License of British Telecommunications plc and Mercury Communications Limited; see also United Kingdom, Dept. of Trade and Industry, *Competition and Choice: Telecommunications Policy for the 1990's*, London: Her Majesty's Stationary Office, March 1991.

23. See Oftel Statement from Professor Sir Bryan Carsberg, Director General of Telecommunications, License Modification Proposals To Implement Duopoly Review Conclusions (rel. 24 July 1991).

24. For an extensive report on telecommunications reform in Germany, see A. Thimm, *America's Stake in European Telecommunication Policies*, Westport, CT: Quorum Books, 1992, pp. 11-86; and S. Schmidt, "Taking the Long Road to Liberalization," *Telecommunications Policy*, Vol. 15, June 1991, pp. 209–222.

25. See A. A. Morris, "Germany's New Telecommunication Law," *Syracuse J. International Law*

(Postal Constitution Act) law separated telecommunications regulation from service provision within the Deutsche Bundepost, vesting regulatory functions with the Ministry of Posts.

The entrepreneurial functions of the Bundepost were divided into three separate entities serving telecommunications, banking, and postal operations. While retaining a monopoly in basic telecommunications services, the DBT entity was directed to develop cost-based tariffs and provide nondiscriminatory access to network facilities. This provision is important, because the law authorized cellular radio, satellite, and data transmission service competition, all of which require access to the bottleneck public switched network facilities to originate and terminate service. The law also opened terminal equipment markets to competition, provided the equipment received type approval by a new Office of Telecommunications Approvals to ensure compliance with operational and safety standards, including evidence that the device will not cause personal or technical harm to the DBT network.

The Poststrukturgesetz retains a basic voice communication service monopoly for DBT, but contemplates some facilities-based competition. For example, the law goes beyond the scope of liberalization in the United States on the issue of foreign ownership of the cellular radio competitors to DBT, although the first winning applicant only had minority foreign participation, including Pacific Telesis [11]. Reunification of Germany may delay progressive efforts to liberalize and privatize DBT,[26] and to achieve a global marketing presence.[27] However, the government remains committed to a future privatization.

Germany's initiatives stem from a recognition that it must take affirmative steps to improve network reliability, flexibility, global access, and pricing to retain marketplace leadership. With increasing opportunities to serve regional markets, DBT has undertaken a dual-track campaign: (1) to pursue foreign investment opportunities and to establish marketing offices abroad, including New York, Chicago, and San Francisco; and (2) to eliminate structural and regulatory barriers to developing a comparative advantage vis-a-vis other carriers in Europe. German legislators recognized that centralized management and control by government often resulted in handicapping competitiveness.

In 1995, DBT became a public stock corporation in preparation for a near-term privatization. The company changed its name to Deutsche Telekom and a variety of strategic alliances and joint ventures have formed in expectation that market entry opportunities will occur soon after privatization of Deutsche Telekom.

---

and Commerce, Vol. 16, 1989, p. 65.

26. But see "German Government Approves Finance Ministry's Privatization Proposal for DBP Telekom," *Telecommunications Reports*, Vol. 58, No. 32, 3 August 1992, p.16.

27. See, for example, "DBP Telekom of German Joins AT&T, PT Telecom in Venture With Ukraine's Ministry," *Telecommunications Reports*, Vol. 58, No. 25, 6 July 1992, p.27.

16.2.3 France

Changing political party control has dampened the pace at which France pursues deregulatory initiatives. However, through different administrations the French government has evidenced appreciation for the importance of telecommunications. Having previously allowed the telecommunications infrastructure to deteriorate, the government targeted telecommunications for development in the 1980s. It accelerated the pace of improvements through subsidies, rather than requiring carriers to improve service through internally generated funds and loans.

France has achieved progress while retaining government ownership of the PTT monopoly, France Telecom. With a growing budget, France Telecom has simultaneously undertaken an extensive domestic network upgrade while developing a global presence through foreign investments (e.g., in Mexico and Argentina and with marketing offices in Britain, China, Germany, Indonesia, Japan, Singapore, the United States, and Venezuela). Its ISDN is as advanced as that from any carrier.

France has reserved most telecommunications markets for France Telecom. However, a competitive marketplace exists for value-added networks and customer premises equipment. In 1989, France separated telecommunications regulation from the business aspects of France Telecom. In 1991, legislation revised France Telecom's charter to make it a more corporate and business-oriented enterprise. Nevertheless, the French government retains a tight tether on the domestic telecommunications market structure and France Telecom's strategic plan.

16.2.4 Other Nations in Europe

Other carriers in Europe, particularly ones in Belgium, Denmark, Italy, the Netherlands, Spain, Sweden, and Switzerland have displayed a heightened global business orientation.[28] Telefonica of Spain, a private company with significant state investment, has launched a number of international strategic alliances, including a domestic telephone network upgrade program with Bell Atlantic and an equity investment in the privatization of the PTTs in Chile, Argentina, Puerto Rico, and Venezuela. It appears likely that Spain will follow other nations in Europe and support a competitive customer premises equip-

---

28. See, for example, O. Stehmann, "Liberalizing the Intra-EC Long Distance Market," *Telecommunications Policy*, Vol. 15, April 1991, pp. 129-136; E. Bohlin and O. Granstrand, "Strategic Options for National Monopolies in Transition—The Case of Swedish Telecom," *Telecommunications Policy*, Vol. 15, October 1991, pp. 453–476; N. Garnham and G. Mulgan, "Broadband and the Barriers to Convergence in the European Community," *Telecommunications Policy*, Vol. 15, June 1991, pp. 182–194; and O. Stehmann, "Facility-Based Competition in Europe," *Telecommunications Policy*, Vol. 16, March 1992, pp. 135–146.

ment and value-added network marketplace. The 1987 law that organized the telecommunications sector did little to spur deregulation. The law did separate operational and regulatory functions, but expressly maintained a basic telecommunications monopoly for Telefonica.

The Netherlands government has committed to a progressive campaign to introduce some competition in telecommunications and restructuring the PTT. It provides a representative blueprint [12]:

> While the exact terms of liberalization have yet to be developed, the general scheme for the future structure is now evident...Monopoly provision of basic services, whether local or long distance, is to be maintained and the operations of the underlying infrastructure network required for the provision of these services will be accomplished under exclusive concession to the PTT. The resale of leased capacity, the offering of "enhanced" services and the supply of CPE will be opened to competitive provision under regulatory supervision.

The Italian government recently consolidated three separate companies, SIP, Italcable, and Telespazio with an eye toward making the country more competitive and its domestic service more efficient. Telecom Italia begins operations with a comparatively inefficient and overused infrastructure, but like France, the government considers state ownership the best vehicle to expedite service improvements. At the risk of failing to implement EU directives, Italy has only fully liberalized the value-added service market.

At this point, one can conclude that Europe has begun the process of disengaging government from the ownership and management of the telecommunications sector. One optimistic commentator concludes that [13]:

> The road ahead, however, remains long and complicated. Though some countries have approached their goal of full deregulation, others are more entrenched in their traditional, monopolistic, telecommunications organizations. Despite these difficulties, recent developments have indicated that all European telecommunications markets will be deregulated within the decade.

16.3 DEREGULATORY INITIATIVES IN THE PACIFIC BASIN

For several years now, the Pacific Basin has generated the highest growth rate in international telecommunications traffic, although its volume remains smaller than the Atlantic and Indian Ocean regions. The newly industrialized countries in the Pacific Basin, most notably the "four dragons" of Hong Kong, Singapore, Taiwan, and Korea, have experienced extraordinary growth and a concomitant explosion in international telecommunications traffic and the need for an expanded and improved infrastructure.

Internationalization of the banking and financial services industry also has a substantial impact on the Pacific Basin, since stock markets and financial institutions in Tokyo, Singapore, and Hong Kong interconnect with counterparts in New York, Toronto, London, and Paris on a real-time basis. Financial organizations now have the option of 24-hour trading achieved through digital technology and increasingly through the proliferation of high-capacity satellites and fiber-optic submarine cables that have been deployed in the region.

In a larger sense, the Pacific Rim, and perhaps Eastern Europe, promises to become the area where the need for new and higher capacity telecommunications facilities triggers a willingness by government to permit private initiatives. While many nations have only recently explored liberalization and privatization options, the trend began in 1985 when Japan began its deregulatory campaign. Other nations in the region have followed suit, fast making some Pacific Rim nations as progressive as other more developed countries.

16.3.1 Japan

Japan's telecommunications business law provided the foundation for competition in virtually all telecommunications services and equipment markets.[29] That competition has been less than robust in some of these markets perhaps has less to do with the viability of competition than with how markets are structured and managed in Japan.

Nevertheless, Japan has achieved a vibrantly competitive leased-line value-added network marketplace with dozens of market entrants, known as Type II carriers, including a number of large U.S. enterprises that have established Japan subsidiaries or have forged strategic alliances with Japanese companies. These companies include AT&T, IBM, Sprint, and General Electric Information Systems. Japan also has permitted a number of intercity microwave and fiber-optic facilities-based Type I interexchange carriers to compete with NTT, the world's largest corporation. NTT has largely converted from a wholly government-owned and controlled monopoly to a mostly private, publicly traded enterprise.

Two new international carriers, International Telecom Japan, Inc., and International Digital Communications, Inc. (IDC), have received authorization to compete with KDD for international facilities-based services. Pacific Telecom, a U.S. electric and telephone holding company, and Cable & Wireless of the

---

29. See H. Sato and R. Stevenson, "Telecommunications in Japan: After Privatization and Liberalization," *Columbia J. World Business*, Vol. 24, No. 1, Spring 1989, p. 31; M. Kojo and H. Janisch, "Japanese Telecommunications After the 1985 Reforms," *Media and Communications Review*, Vol. 1, 1991, p. 308.

United Kingdom have an ownership interest in IDC which owns the western half of the North Pacific Cable. The MPT also authorized two domestic satellite operators, Japan Communications Satellite Co. (JC Sat) and Space Communications Corp. (SCC).

Despite the proliferation of service providers in some markets, many foreign equipment manufacturers and carriers still experience less than ideal opportunities for market access. The MPT retains its licensing power for most services and can control the speed with which competition arrives and the degree to which non-Japanese equipment will be certified as safe for use on Japanese networks. For example, Motorola has objected to its exclusion from a large portion of the Japanese cellular radio market on less than clear and justifiable grounds. In 1994, President Clinton and U.S. Trade Representative Mickey Kantor used this issue to exemplify the frustration experienced from decades of trade negotiations that have achieved relatively minor reductions in market access barriers. Notwithstanding market access problems, the Japanese telecommunications regulatory climate has changed significantly, and barring unforeseen negative impact on the consumer, the change appears permanent and conducive to foreign investment in some market segments.

16.3.2 Australia

Australia has legislated a duopoly in basic telecommunications until 1997, but permits full resale of basic and enhanced services. The Telecommunications Act of 1989 reshaped the Australian regulatory framework and articulated with greater clarity the scope and need for government involvement.[30] The law created the Australian Telecommunications Authority (Austel) with responsibility for economic and technical regulation of the industry and a direct reporting line to the minister for transport and communications. It also defined what services fall within a monopoly definition to be reserved for exclusive provision by Australia's two facilities-based carriers. The law established a licensing and permit system for creation of newly competitive value-added services, private network services, customer premises equipment, and inside wiring markets.

With enactment of the Telecommunications Act, a major restructuring of the Australian telecommunications marketplace had to occur [14]. Australian Telecommunications Corp. (Telecom Australia), responsible for most domestic services, merged with Overseas Telecommunications Corp. (OTC), responsible for international services, and became Telstra. In late 1991, the Australian government created a telecommunications duopoly by authorizing a second facilities-based carrier to provide both domestic and international services. Optus Communications Pty. Ltd., which includes investment by BellSouth Corpora-

---

30. See K. C. Beazley, *Micro-Economic Reform: Progress Telecommunications*, Ministry of Transport and Communications, November 1990.

tion and Cable & Wireless, agreed to pay $800 million (Australian) for the franchise and will invest $3.1 billion (U.S.) within the first 5 years of operation [15]. As part of its franchise, Optus had to take over operations of Australia's debt ridden domestic satellite carrier, Aussat.

16.3.3 New Zealand

New Zealand has embarked on a deregulatory plan unmatched by other nations in scope. This nation considers privatization as a key public policy initiative to jump-start its economy. It has all but eliminated government intervention in whether and how a commercial enterprise can enter the nation's telecommunications markets. In April 1987, the New Zealand Post Office, a government department with both commercial and regulatory responsibilities, was restructured into three separate state-owned corporations. Telecom Corporation of New Zealand was privatized and divested of its government-sanctioned monopoly. Within 2 years, the government removed all restrictions on customer premises equipment, including private branch exchanges, inside wiring, value-added services, and finally, all network services, including international telecommunications.

With the enactment of the Telecommunications Act of 1988, effective 1 April 1989, New Zealand permitted competition in the provision of all telecommunications services. Because the law contains no restrictions on foreign ownership, other than applying the general law pertaining to foreign investment, opportunities for widespread foreign participation exist. Bell Atlantic and Ameritech have ownership interests in Telecom New Zealand, and MCI has become a facilities-based telecommunications service provider in the nation. New Zealand now has the most progressive and deregulatory regime in the world, and the actual results of its policies on rates and consumer welfare can be expected to have a significant impact on whether and how other nations proceed in the future.

16.4 DEVELOPMENTS IN LATIN AMERICA

Privatization has not been limited to the developed world with mature markets. The governments of Mexico, Argentina, Chile, and Venezuela have all but abandoned ownership of the PTT. However, for most developing countries, the PTT model remains firmly entrenched.

The decision to privatize in these newly industrialized countries typically results from broader macroeconomic policies established in response to national and international pressures for fiscal reform. Users have grown weary of waiting for the incumbent PTT to deliver reliable services and to achieve telephone penetration above single digits per 100 inhabitants. Some international development and financial aid organizations share the frustration over slow progress and have considered alternatives to blind support for the status quo.[31]

16.4.1 Argentina

Privatization in Argentina took place despite record inflation from 1975 to 1989. Privatization and easing foreign investment constituted two of the several strategies undertaken by the government to reform the economy. As part of the 1989 State Reform Law, ENTEL, the state-owned telecommunication company was privatized and divided into two franchises: a subsidiary of Telefonica of Spain received up to a 10-year exclusive franchise to operate in the southern area, including 57% of Buenos Aires. A consortium composed primarily of STET of Italy, France Telecom, and JP Morgan received the northern franchise.

The two private franchisees acquired a mostly obsolete electromechanical network prone to outages and overload. As part of the franchise agreement, the new companies agreed to install 100,000 new lines per year and to reduce rates by 2% per year starting in the third year of operation. With a line penetration of 10.9 lines per 100 inhabitants in 1991, Argentina needed a dramatic infusion of capital and expertise to upgrade its telecommunications infrastructure. The nation opted for privatization coupled with demanding franchise commitments to help jump-start its economy and phone system.

16.4.2 Chile

Chile's commitment to privatization resulted principally from a volatile, not necessarily hyperinflationary economic performance. With a dictatorship eliminated, the country pursued privatization early and comprehensively with an eye toward fostering long-term efficiency and widespread public stock ownership held by individuals and employee pension funds. The domestic telephone company, CTC, which had been owned by International Telephone and Telegraph (ITT) before its nationalization during a military domination of the government, was reprivatized with the 1978 passage of the National Telecommunications Policies Act. Alan Bond, an Australian entrepreneur, acquired a 50.2% ownership interest and subsequently sold it to Telefonica of Spain in 1988. Telefonica also holds a 20% share in the international carrier ENTEL. Privatization has increased line penetration by approximately 100% since 1988 to about 10 per 100 inhabitants. Currently both CTC and ENTEL are sparring over each other's markets and in cellular radio.

---

31. See, for example, I. Vogelsang et al., *Welfare Consequences of Selling Public Enterprises—Case Studies from Chile, Malaysia, Mexico, and the U.K.*, The World Bank Country Economic Dept., Public Sector Management and Private Sector Development Div., Washington, D.C.: World Bank, 1992.

16.4.3 Mexico

Mexico's debt crisis in 1982 prompted the government to consider drastic vehicles for economic development. The secretariat of communications withdrew from directly providing telecommunication equipment and services, and in 1990, Telmex, the national telephone company, was privatized. The winning franchisee, which includes France Telecom and Southwestern Bell, received a temporary monopoly in exchange for major service commitments including:

- Annual line growth of 12%;
- Providing electronic switching to all towns with more than 5,000 inhabitants by 1994;
- Reducing phone installation delays to 1 month by the year 2000;
- Providing basic telephone services to all towns of 500 or more people by 1994.

In the face of substantial infrastructure improvements, it appears that developing nations like Mexico can stimulate private investments despite episodic debt and currency crisis as occurred in early 1995. A demanding franchise agreement and the prospect for facilities-based competition by the year 2000 call for substantial onsite management to achieve performance benchmarks and efficiency gains. This motivates timely improvements and all but forecloses, for many years, repatriation of revenues to absentee owners.

16.4.4 Venezuela

The oil bust of the 1980s fractured Venezuela's wealth and stability, causing severe dislocation for a nation highly reliant on oil production. Despite the lack of uniform support for privatization, the government pursued it as a vehicle to achieve the needed modernization it could no longer afford. Venezuela has a telephone line density of less than 10 lines per 100 inhabitants, while developed nations average 40 to 60 lines per 100 inhabitants.

Compania National de Telefonos de Venezuela (CANTV) was partially privatized in 1991 with 49% retained by the government, 11% held by an employee trust, and 40% acquired by a consortium composed of GTE, with 51%, Telefonica of Spain, with 16%, the Caracas electric company holding 16%, a Venezuelan investment group holding 12%, and AT&T holding 5%.

The CATV franchise covers virtually all telecommunication services and runs for 35 years with a 20-year renewal period. It provides a 9-year monopoly for switched services in exchange for meeting service benchmarks. Rates have been tied to inflation.

16.5 BURGEONING OVERSEAS MARKETS

New receptiveness to foreign investment in telecommunications, regulatory liberalization, the need for massive infrastructure investments, shifting trade patterns, and political change combine to create unprecedented foreign market opportunities. These opportunities persist, despite a variety of trade and market access barriers. Companies that have not already established foreign subsidiaries vigorously seek global strategic alliances in recognition that market access often requires an ally with knowledge of local business customs, a positive and long-standing reputation, a client base, rapport with the PTT, and a working relationship with national equipment testing bodies. Mergers and acquisitions often constitute the quickest and sometimes cheapest vehicle for market access.

Achieving strategic alliances mitigates to a degree pervasive brand loyalties and the variety of visible and hidden trade barriers that make the competitive playing field tilted in favor of national heroes. Similarly, the product testing and certification process and the issue of equipment standards in general can blunt market access and marketplace success.

The Report of the Advisory Committee for Trade Policy and Negotiations to the United States Trade Representative succinctly sums up the marching orders for U.S. companies [16]:

- Be *early* in stating a position on proposals under consideration.
- Be *flexible* in helping to influence constructive compromises.
- Be *selective* in choosing battles to avoid dissipating resources and credibility.
- Be *supportive* of the process of change in other countries.
- Be *persistent* in pushing the interests of U.S. business.

16.5.1 Television Without Frontiers

Cable television has begun to proliferate throughout the world. Most nations retained government ownership of a single channel or limited group of broadcast channels. A few nations permitted limited commercial options, but restricted advertisement minutes and the amount of foreign program content. Fewer nations still permitted commercial cable television enterprises to import foreign signals that could threaten cultural homogeneity and political stability.

The proliferation of television receive-only (TVRO) Earth stations, whether legal or not, and the debut of direct broadcast satellite (DBS) services evidenced the demand for greater diversity in video programming regardless of concerns about indigenous cultures and the national origin of programming. Some commentators have alleged that the upheaval in Eastern Europe was sparked in part by the ability to receive off-air television signals broadcast from nearby non-Communist nations and by the ingenuity of citizens to erect home brew TVROs to receive satellite transmissions.

Given a whetted appetite for programming alternatives to the state's broadcasts, foreign governments have reconsidered prohibitions or restrictions on video programming. Some nations now permit private cable television ventures rather than reserve the market for the PTT or national broadcast authority. In the United Kingdom and France, for example, several of the RBOCs and U.S. cable television multiple-system operators have become equity partners. Despite some restrictions on the number of foreign channels and threats to impose quotas on the amount of foreign program content, video programmers and movie producers have experienced much success in having their fare included in foreign cable service tiers.

The RBOCs have been attracted to foreign cable television markets in view of domestic restrictions on ownership of cable television systems in urban areas, former restrictions imposed by the AT&T divestiture decree on the provision of information services, and the opportunity to provide both cable television and other telecommunication services via a single wire in some nations. The run-up of cable television system prices in the U.S. provided another incentive as foreign nations provide a ground floor opportunity for the construction of new cable television facilities. Because licenses or franchises typically are awarded by cities rather than national administrations, outside firms have a better shot at the more numerous opportunities.

SBC, previously known as the Southwestern Bell Telephone Corporation has acquired an equity interest in the company that received the franchise to build the cable television system in the Liverpool region of England, and also has acquired an interest in the Israeli company that serves Jerusalem and Tel Aviv. US West has heavily invested in European and Asian cable television with a stake in 13 areas of France, including Paris, Hong Kong, and several U.K. locales, including South London and Birmingham.

16.5.2 Satellites and Very-Small-Aperture Terminals

The deregulatory initiatives of some nations, pressure from large corporate telecommunications users, and advocacy by telecommunications equipment suppliers have created the impetus for liberalized satellite and Earth station policies.[32] Currently, most satellite services are provided by the global coopera-

---

32. The EC released a Green Paper specifically addressing satellite communications. (See Commission on the European Community, *Toward Europe-Wide Systems and Services—Green Paper on a Common Approach in the Field of Satellite Communications in the European Community*, COM 490, 20 November 1990; see also S. Mosteshar and P. A. Galleberg, *Regulation of Satellite Communications in the European Community*, American Bar Ass'n, Sec. of Int'l Law and Practice, Monograph Series 1991/2, April 1991.

tive INTELSAT. A few separate systems have developed and achieved success primarily in video program delivery.

An increasing number of nations permit intracorporate VSAT networks and residential use of one-way TVRO satellite dishes. Nations disinclined to deregulate and liberalize in this sector consider VSAT operations part of the basic services reserved for the PTT or satellite monopoly. Some nations limit or prohibit TVROs because of concerns for political instability, or that, given options, viewers will ignore the domestic operator for more provocative and diverse programming.

The U.K. leads most nations in liberalization of satellite policies. It has authorized seven "specialized satellite service operators," one of which is the U.S. company EDS. The systems cannot access the PSTN, and most lease the transponder capacity of cooperatives like INTELSAT and EUTELSAT from BT. But they can uplink signals for broadcast throughout Europe, the reception of which of course depends on the policies of the receiving nation.

The proliferation of TVRO terminals and DBS creates the impetus for further liberalization of the VSAT and terminal market for data services. While it does not appear that many additional satellite operators will enter the marketplace, the prospects are bright for terminal manufacturers and systems integrators.

16.5.3 Cellular Radio, Mobile Telecommunications, and Data Services

Nations throughout the world have launched extensive programs to develop state-of-the-art cellular radio, paging, and personal communications systems. The campaign for mobile telecommunications stems in part from the fact that many nations have yet to develop a fully operational and ubiquitous telecommunications infrastructure. In cities like Buenos Aires, citizens have erected their own "tie-lines" between office buildings out of impatience with local telephone company delays and inefficiency. Cellular radio is viewed as the means to secure reliable, albeit expensive, telephone service in quick order. Likewise, the one-way Telepoint services and two-way personal communication network services provide much needed additional capacity and alternatives to both overtaxed and unreliable wireline networks, as well as new cellular systems.

The value of ground floor mobile telecommunications and data service opportunities runs into the billions of dollars. Many nations realize now that a modern and efficient telecommunications infrastructure is a sine qua non of healthy economic development. Foreign company technical expertise is welcomed as foreign nations liberalize or make exceptions to alien ownership restrictions.

Value-added network services also present a growth and investment opportunity. Because such services tend to stimulate usage of the PTT network, rather than merely prompt migration from one type of service to the other, nations are more inclined to permit market entry and looser operational rules. Annual growth in the data transmission marketplace has run at a rate of from 9% to 20%. The electronic information services, which often use packet-switched value-added networks, have experienced an annual growth rate of between 23% and 33%, which exceeds the cable television rate of 17% to 27% and mobile communications rate of between 16% and 19% [17]. Such growth rates appear both sustainable and ripe for market entry, "countries realize the benefits of competition and [hopefully further] relax trade barriers" [18].

References

[1] Carsberg, B., Director General Office U.K. Office of Telecommunications, "The Future of Global Telecommunications From a U.K. Perspective," in Center For Strategic & International Studies, *U.K. Perspectives on International Telecommunications*, Washington, D.C., 21 March 1989, p. 5.

[2] "1992 and the United States," *The Business Lawyer Update*, Vol. 9, No. 5, American Bar Assn., Sec. of Business Law, May/June 1989 (emphasis added).

[3] Commission of the European Communities, *Towards a Competitive Community-Wide Telecommunications Market in 1992, Implementing the Green Paper on the Development of the Common Market for Telecommunications Services and Equipment*, COM(88) 48, 9 February 1988.

[4] H. Baur, "Telecommunications and the Unified European Market," *Telecommunications*, Vol. 24, No. 1, January 1990, p. 33.

[5] Council Resolution on the Development of the Common Market for Telecommunications Services and Equipment up to 1992, 1988 O.J. (C 257) 2.

[6] *Towards Europe-Wide Systems and Services—Green Paper on a Common Approach in the Field of Satellite Communications in the European Community*, COM(90) 490, 57, 20 November 1990, approved. EC Council Resolution of December 19, 1991 on the Development of the Common Market for Satellite Communications Services and Equipment, O.J. C 8/01, 14 January 1992. See also S. Mosteshar and P.A. Galleberg, *Regulation of Satellite Communications in the European Community*, American Bar Assn., Sec. of Int'l Law and Practice, Communications Committee Monograph Series 1991/2, April 1991.

[7] *Commission Proposal for a Council Directive on the Mutual Recognition of Licenses and Other National Authorizations to Operate Telecommunications Services, Including the Establishment of a Single Community Telecommunications License and the Setting Up of a Community Telecommunications Committee (CTC)*, COM(92) 254 final SYN 438.

[8] European Telecommunications Research, "Telecommunications in Europe—The Challenge of Change," *Europe 1992*, April 1988, p. 2.

[9] Priddle, R., "U.K. Telecommunications Policy and Europe 1992," in *U.K. Perspectives on International Telecommunications*, CSIS Lecture Series on Global Communication, p. 18.

[10] Noam, E., "International Telecommunications in Transition," in *Changing the Rules: Technological Change, International Competition, and Regulation in Communications*, R. Crandall and K. Flamm, eds., Washington, D.C.: The Brookings Institute, 1989, p. 265–66; For a description of Oftel's function, see Gist, "The Role of Oftel," *Telecommunications Policy*, Vol. 13, February 1990, pp. 26–51.

[11] "Mannsman Mobilfunk Begins GSM Digital Cellular Phone Service in Germany," *Telecommunications Reports*, Vol. 58, No. 27, 23 July 1992, p. 25.

[12] Garrison, W., *Case Studies of Structural Alternation in the Telecommunications Industry—The United Kingdom, The Federal Republic of Germany, The United States of America,* Washington, D.C.: Annenberg Washington Program, 1988, p. 61.

[13] Lando, S. D., "The European Community's Road to Telecommunications Deregulation," *Fordham Law Review*, Vol. 62, 1994, p. 2159.

[14] Evan, G., QC, Minister for Transport and Communications, *Australian Telecommunications Services: A New Framework*, Canberra, 25 May 1988.

[15] "Australian Government Picks Optus Group to Acquire Domestic Satellite Carrier Aussat, Competitive Licenses," *Telecommunications Reports*, Vol. 57, No. 47, 25 November 1991, pp. 18–19.

[16] Report of the Advisory Committee for Trade Policy and Negotiations, *Europe 1992*, November 1989.

[17] Sims, "Baby Bells in Europe," *New York Times*, Business Section, 10 December 1989, pp. 1, 8–9; Roussel, "Baby Bells Sprout International Wings," *Communications Week Preview Issue*, 1988, pp. 30–31.

[18] Spievack, E. B., "Telecommunications Trade: Evolving Markets and Opportunities," *Telecommunications*, Vol. 24, No. 1, January 1990, pp. 36, 38.

Telecommunications and Trade Policy Players in the United States

17

International telecommunications policy making in the United States involves more agencies and greater complexity than in most nations. The U.S. does not have an MPT with a near exclusive portfolio in telecommunications policy making and regulation. Instead, a number of agencies in the executive branch share the portfolio with an independent regulatory agency, the FCC. With a variety of players, each having different expertise, constituencies, perspectives, and agendas, the international telecommunications policy making process in the U.S. is complicated and confusing.

While subject to congressional oversight hearings and budget authorizations, which sometimes result in "regulation by lifted eyebrow," U.S. policy-making agencies generally have significant discretion in assessing what rules and regulations will serve the public interest. On a day-to-day basis, the agencies have ample discretion to determine, with the assistance of public participation, what initiatives, rules, and regulations will flesh out the general language contained in the Communications Act of 1934 and its amendments.

Such decisions affect industry structure and have an impact on individual company profitability. For example, the FCC used to monitor and regulate the manner in which international carriers activated new submarine cable and satellite circuits. The commission believed that forcing carriers to activate more expensive satellite circuits would serve the public interest by promoting routing and facility diversity. This policy also promoted the financial well-being of Comsat and INTELSAT and may have artificially reduced demand for submarine cables. The FCC also used to prevent AT&T from providing nonvoice data services, in part to promote the financial viability of a small group of IRCs and to keep AT&T from dominating another market sector.

The judiciary performs a review function to determine whether Congress, the FCC, and the executive branch agencies complied with applicable laws on both procedural and substantive grounds. Occasionally, a federal court may issue an activist, results-oriented decision, such as ordering the FCC to authorize MCI to provide switched long-distance service in addition to unswitched private line service on the grounds that the FCC did not expressly state that long-distance competition should be limited to private line service. Typically, courts limit their role to reviewing administrative agency actions to ensure that decisions are rational and that the agency followed proper procedures in fact-finding and seeking public participation.

17.1 GOVERNMENT ROLES

17.1.1 The Legislature

The Commerce Clause of the U.S. Constitution, Article I, Section 8, empowers the Congress to legislate on matters affecting interstate and foreign commerce. This provision affords Congress exclusive and preemptive power over commercial matters that impact more than one state or involve relations with other nations. Congress has ceded day-to-day regulatory authority to the FCC through the powers established in the Communications Act of 1934 and the Communications Satellite Act of 1962. However, congressional committees with a budgetary or oversight role keep close track of FCC actions.

The power of the purse and the ability to convene oversight hearings on any topic confer substantial congressional power. This means that legislators can stimulate regulatory activity and influence the decision-making process through the questions posed in a hearing via correspondence with the FCC and by "unofficial" telephone calls. Congress cannot operate outside laws it created requiring government to operate "in the sunshine," that is, in public forums that allow all interested parties to observe, participate in, and know the positions of other parties. Similarly, it cannot engage in "ex parte communications," that is, advocacy on matters before the FCC but outside the conventional process where all parties know of the communication and have an opportunity to comment on it. Nevertheless, Congress has broad powers to affect the FCC decision-making process, both officially and informally.

Primary Telecommunications Legislation

The Communications Act of 1934, 47 U.S.C. Secs. 151–699 and the Communications Satellite Act of 1962 47 U.S.C. Secs. 701–757 provide the basis for the structure and broad policies in international telecommunications. The Communications Act created the FCC to serve as an expert regulatory agency with a mission to serve the public interest, promote the widespread use and availability of radio, and ensure that the enterprises it regulates likewise operate in the public interest. In international telecommunications, the Communications Act contemplates service by common carriers, who operate as public utilities providing nondiscriminatory service on tariffed terms and conditions subject to FCC review.

The Communications Satellite Act extends the common carrier model to satellite service, and goes further by specifying that one carrier shall serve as the U.S. participant in the INTELSAT global cooperative. The Satellite Act affects the market composition in international telecommunications by creating a three-tiered satellite services marketplace: Comsat as exclusive vendor of INTELSAT capacity, a handful of international carriers authorized by the FCC to deal with Comsat directly, and other service providers, resellers, and users.

Telecommunication Laws Affecting Trade

Aside from limitations on foreign ownership in telecommunications, few laws in the United States impose absolute market access reciprocity or the ability to impose financial and access penalties in the face of unequal access. One law that does is the Submarine Cable License Act, enacted in 1921.[1] It allows foreign enterprises to land an international cable on U.S. soil if and only if U.S. carriers have such a right of access in the carrier's home country. Because national security concerns have prompted all other nations to prohibit the landfall of foreign-owned international cable facilities, the United States imposes a similar prohibition. While a reciprocity requirement may exacerbate trade barriers and access restrictions, it may enable nations to negotiate compromise positions without resorting to further market access restrictions.

The United States officially refrains from more closely scrutinizing foreign manufacturers and carrier activities. It has committed to the trade concept of "national treatment," meaning that foreign enterprises will receive the same rights and responsibilities as conferred on domestic companies. Trade commitments notwithstanding, the FCC has considered imposing more burdensome reporting requirements on foreign-dominated companies doing business in the U.S. While such records could assist the U. S. government in trade negotiations and provide empirical evidence of unequal market access, the FCC abandoned such a strategy in deference to executive branch trade policies. The FCC recently linked a finding of equivalent market access opportunities with application of deregulatory and investment opportunities for foreign enterprises. The FCC has proposed to examine whether U.S. enterprises have equivalent market access opportunities, particularly companies providing international services via leased private lines [1]. For nations providing equivalent market access op-

---

1. "An Act relating to the landing and operation of submarine cable in the United States," codified at 47 U.S.C. Secs. 34–39 (1990); see also Executive Order 10530 (10 May 1954) (delegating to the FCC certain presidential functions relating to submarine cable landing licenses); and Tel-Optik, Ltd., Cable Landing License, File Nos. I-S-C-L 84-002, I-S-C-L- 84-003, Mimeo No. 4618 (rel. 17 May 1985).

portunities to U.S. carriers, the FCC may, on a case-by-case basis, grant waivers to the 20% to 25% foreign ownership cap as permitted by Section 310 of the Communications. The FCC also has expressed the desire to use Section 214 of the Communications, which requires carriers to apply for its authorization before providing service, as a basis for gauging whether a foreign carrier's home country permits U.S. carriers to provide the very service the foreign carrier seeks authority to provide in the U.S.

17.1.2 The Judiciary

Courts interpret rather than make law. Their primary role in telecommunications policy and regulation is to resolve disputes by assessing whether the FCC has acted in compliance with applicable laws. In particular, courts determine whether the FCC has complied with the terms of the Communications Act, the Satellite Act, and the Administrative Procedure Act (APA).[2] The substantive laws affecting telecommunications establish the FCC's powers and mission with broad and flexible terms to account for changed circumstances and technological innovation. The APA establishes procedural standards that agencies like the FCC must satisfy to ensure that the public has a full and fair opportunity to participate in the decision-making process and that the agencies reach reasoned decisions based on the record of evidence and findings generated. The FCC cannot act arbitrarily or capriciously. It cannot abuse the discretion conferred on it by Congress, nor can it otherwise act outside the scope of authority vested with it by law.

17.1.3 The Executive Branch

The President and executive branch agencies have powers created by the Constitutional separation of powers (e.g., foreign relations and operation of militaries). In addition, legislation can create additional executive branch responsibilities, such as participation in shaping international satellite policy and monitoring the conduct of Comsat in its capacity as the sole U.S. representative and investor in INTELSAT and Inmarsat.

Executive branch agencies have portfolios in trade policy (the USTR and Commerce Department), management of the federal government spectrum (the Commerce Department's NTIA), and telecommunications policy (the Department of State and NTIA). While these agencies typically do not regulate, they can impact regulatory policy primarily through interagency coordination and by long-range planning. This frees them to consider issues in a larger context and presents the opportunity, not always realized, to affect long-term policy making and strategy.

---

2. Administrative Procedure Act, as amended, 5 U.S.C. Secs. 701–706 (1990).

The Department of State

The State Department manages the foreign policy and international relations agenda of the United States. Substantive issues like telecommunications and aviation are subsumed within the department's broad interpretation of its mission. In application this means that career foreign service officers in the State Department will acquire a temporary portfolio in international telecommunications and work with some civil servants dedicated to a career in telecommunications policy. Such a blend of general foreign relations and specific telecommunications policy skills sometimes does not result in the best work product. A 1993 reorganization within the State Department further diluted in-house expertise in the subject when the separate Bureau of International Communications and Information Policy (CIP) merged with the much larger Bureau of Economic and Social Affairs.

The goals of diplomacy and international comity with foreign nations can conflict with policy initiatives of other executive branch agencies. Depending on the personalities of the leaders at the FCC and NTIA, turf battles and spirited interagency debates can develop. Some critics allege that the State Department tends to emphasize "international comity" and cooperation over the nation's sometimes parochial concerns. Wisely or not, foreign policy concerns can dampen the tone and stridency of U.S. advocacy.

CIP attempts to coordinate international telecommunications and trade policy with all other involved agencies. It assembles the delegations that represent the United States at international forums like the ITU, often conferring the status of ambassador to the delegation head. On matters involving INTELSAT and Inmarsat, it issues instructions to Comsat on what it should advocate and how it must vote on matters of national interest.

Despite perennial calls for better coordination between CIP, NTIA, and the FCC, the policy-making process can become disjointed. The potential for this is exacerbated by the fact that different congressional agencies have oversight and budgetary responsibility for each agency. So far, proposals to unify the policy-making process and make it more coherent have failed to gain significant congressional attention.[3]

The National Telecommunications and Information Administration

NTIA, an agency within the Department of Commerce, serves as the executive branch's principal voice in telecommunications policy.[4] The effectiveness of

---

3. See, for example, Congress of the United States, Office of Technology Assessment, "Jurisdictional Issues in the Formulation and Implementation of National Communication Policy," Chap. 13 in *Critical Connections—Communications for the Future*, January 1990.

4. "The [Commerce] Department's telecommunications policymaking functions are centered in

this agency depends in large part on the visibility and stature of its head, because it has no regulatory or licensing portfolio. Likewise, NTIA has only a small staff of 15 to 20 involved in aspects of international policy making.

Initially, NTIA constituted a department within the Office of the President. As the Office of Telecommunications Policy (OTP), it had a closer reporting line to the President, perhaps commensurate with the importance of telecommunications to the national economy. However, NTIA was spun off in 1978 to become a part of the Commerce Department. In 1993, NTIA was threatened with a further downgrading in visibility and proximity to the President by a proposal to make it a part of the National Institute of Science and Technology, formerly known as the National Bureau of Standards.

NTIA participates in the policy-making process principally through advocacy in FCC proceedings, and by studying issues like the National Information Infrastructure[5] and preparing thoughtful reports. Its statutorily conferred responsibilities address coordination and registration of the federal government's use of frequency spectrum,[6] including the identification of government-used spectrum that can be transferred to private use as ordered by Congress[7] and participation in the oversight of Comsat's activities in INTELSAT and Inmarsat.[8]

---

the National Telecommunications and Information Administration (NTIA), which was created pursuant to Reorganization Plan No. 1 of 1977, with the responsibilities of the Secretary of Commerce under Executive Order 12046 [Fed. Reg. 13349-13357 (29 March 1978)] delegated to it." ("International and Domestic Policymaking in the Year 2000," Chap. 9 in *NTIA Telecom 2000: Charting the Course for a New Century*, NTIA Spec. Pub. 88-21, Washington, D.C.: GPO, October 1988, p. 168 (hereafter cited as *Telecom 2000*).)

5. See Information Infrastructure Task Force, Ronald H. Brown, Chairman, *National Information Infrastructure: Agenda for Action*, Washington, D.C.: GPO, 15 September 1993; *National Information Infrastructure: Progress Report September 1993–1994*, Washington, D.C.: GPO, September 1994.

6. Under Sec. 305 of the Communications Act of 1934, as amended, the President retains the authority to assign frequencies to all radio stations belonging to the federal government. The President has delegated authority to the secretary of commerce, who has in turn delegated it to NTIA. (See Exec. Order No. 12046, as amended, 3 C.F.R. 158 (1978), reprinted in 47 U.S.C. Sec. 305 app. at 127 (1989); U.S. Dept. of Commerce, Dept. Organization Orders 10-10 and 25-7.) The applicable executive order and an Office of Management and Budget Circular No. A-qq "provide NTIA with the power to assign frequencies and approve the spectrum needs for new systems." (U.S. Dept. of Commerce, National Telecommunications and Information Administration, *U.S. Spectrum Management Policy: Agenda for the Future*, 20 NTIA Spec. Pub. 91-23, Washington, D.C.: GPO, February 1991.)

7. In 1993 Congress directed the secretary of commerce to identify at least 200 MHz of spectrum currently allocated on a primary basis for federal government use which is not required for present or identifiable future use by the federal government, and which is most likely to have the greatest potential for productive uses and public benefit if allocated for nonfederal use. (Omnibus Budget Reconciliation Act of 1993, Pub. L. No. 103-66, Title VI, § 6001(a)(3), 107 Stat. 312 (approved 10 August 1993); see also H.R. Rep. No. 103-213, 103rd Cong., 1st Sess. (1993).)On 10 February 1994, the secretary of commerce released *Preliminary Spectrum Reallocation Report (Preliminary Report)* NTIA Special Publication 94-27, identifying spectrum for reallocation

Most of NTIA's responsibilities are shared with other agencies. The FCC has greater technical expertise and a larger staff. The State Department leads on foreign policy issues, and other trade offices in the Department of Commerce and the USTR hold more direct authority to formulate trade policy. Nevertheless, NTIA's studies and advocacy documents help shape the international telecommunications and trade policy agenda. Particularly in the trade area, staff technical expertise augments the more generalist USTR on telecommunications facilities, services, and equipment matters.

If the FCC's constituency constitutes its licensees and the public that use licensee services, NTIA's natural allies are manufacturers and carriers concerned with market access and competition issues. However, NTIA has a less direct impact on any single enterprise in view of the fact that it confers no licenses and has no regulatory function. It accrues clout when pocketbook and industrial leadership issues dominate foreign policy, national security, and interagency turf concerns.[9]

United States Trade Representative

The USTR serves as the lead agency for coordinating negotiations with foreign nations on trade matters.[10] Telecommunications has become a significant component in the USTR's portfolio.[11] Because of the technical complexity of the issues and the generalist nature of its staff, the USTR regularly consults with other agencies and includes representatives of these agencies on its negotiating delegations. However, USTR staff have expressed concern when the FCC attempts unilaterally to carve out a trade portfolio. For example, in the 1980s the FCC proposed to monitor U.S. carrier procurement of foreign manufactured

---

from federal government use to private sector, including local government, use. (See also Report to Ronald H. Brown Secretary, U.S. Department of Commerce Regarding the Preliminary Spectrum Reallocation Report, 9 FCC Rcd. 6793 (1994).)

8. "Functions relating to international communications satellite systems, vested in the President by the Communications Satellite Act of 1962, and the Inmarsat Act of 1978 have also been delegated to NTIA." (*Telecom 2000*, p. 169.)

9. See, for example, S. Brotman, *The Council of Communications Advisers—The Right Place for Executive Branch Communications Policymaking, the Right Time for Change*, Washington, D.C.: Annenberg Washington Program, 1989.

10. See Trade and Tariff Act of 1984, Pub. L. No. 98-573, 98 Stat. 2948 (directing the President to negotiate reductions in foreign nontariff barriers), codified at 19 U.S.C. Sec. 2122 (1985); see also the Omnibus Trade and Competitiveness Act of 1988, Pub. L. No. 100-418, 1988 U.S. Code Cong. & Admin. News 1107 (1988).

11. See T. Howell et al., "International Competition in the Information Technologies: Foreign Government Intervention and the U.S. Response," *Stanford J. International Law*, Vol. 22, 1986, p. 215; and Note, "Opening Doors in Foreign Market Trade Through the Telecommunications Trade Bill," [1989] *Brigham Young University Law Review*, p. 639.

equipment and to scrutinize closely the operations of foreign-owned carriers in the U.S. The FCC articulated a model view of the international telecommunications and trade marketplace [2] and had strongly signaled its desire to use regulatory oversight as leverage to achieve its goals. To promote market access[12] opportunities for U.S. carriers and equipment manufacturers, the FCC proposed more extensive regulatory burdens on foreign carriers operating in the U.S. and thereby subject to FCC jurisdiction if U.S. enterprises were unable to access markets in the foreign carriers' home country.

The USTR strongly opposed the FCC's trade initiative as deviating from the commitment to national treatment. The FCC's final order backed off from a confrontation with the USTR on trade policy, opting instead to require reports only from the handful of foreign-owned and operated telecommunications carriers operating in the United States. In effect, the FCC agreed to refrain from using regulation to leverage market access opportunities for U.S. carriers.

In 1995, the FCC again proposed to use its regulatory authority to secure parity of market access between U.S. and foreign carriers operating within the U.S. [3]. While the agency appears to defer to the USTR's lead on trade policy, it appears that the FCC seeks to establish its own portfolio particularly for leveraging market access opportunities. In the context of establishing a national trade policy, the USTR must consult with other agencies. But in doing so, it must come to terms with the different constituencies represented. When the State Department participates, foreign policy concerns and appreciation for international comity are incorporated. When two agencies of the Department of Commerce participate (NTIA and the International Trade Administration), domestic industry concerns are voiced, sometimes not necessarily by the same person, or with the same perspective. When the FCC's technical expertise is tapped, the trade delegation or task force runs the risk of bolstering the FCC's interest in pursuing its own agenda.

The potential for turf battles, multiple constituencies and positions, and uncoordinated initiatives rises with the number of agencies and individuals involved. In the worst-case scenario, foreign governments with a better ability to reign in disparate agencies and players can "divide and conquer" the U.S. interagency policy-making process. At best, foreign governments have a number of agencies to visit and convince.

---

12. "[W]e believe that it is our duty to assess whether the regulations or practices of certain foreign administrations, such as restrictions on the availability of telecommunications equipment or the use of the telecommunications network, may have an adverse impact on our ability to meet the public interest goals set forth in Section 1 of the [Communications] Act." (Regulatory Policies and International Telecommunications, CC Docket No. 86-494, Notice of Inquiry and Proposed Rulemaking, 2 FCC Rcd. 1022, 1023 (1987).) The Commission sought to establish an "ideal" international telecommunications model based on four key elements: open entry, nondiscrimination, technological innovation, and international comity. (Ibid., p. 1022.)

17.1.4 Federal Communications Commission

The Communications Act of 1934 vests the FCC with authority to regulate inter-state and foreign communications by wire and radio. The FCC's public interest mandate propels a regulatory portfolio into every aspect of international tele-communications and at least peripherally into matters of international trade policy as well. A separate International Bureau was created in 1994 to consoli-date international functions that were distributed throughout the FCC. The In-ternational Bureau handles all satellite issues, spectrum coordination with other nations, preparations for FCC participation at ITU and other bilateral or multilateral conferences, and international policy matters.

The FCC operates as an independent regulatory agency, separate from the executive branch and the legislature. However, the President appoints the agency's five commissioners (three from the President's political party) subject to Senate consent. Congress votes operating funds and conducts regular over-sight hearings to ensure that the agency accommodates the legislature's interest even in the absence of new laws.

FCC rules, regulations, and policies substantially affect the terms, conditions, and profitability under which international telecommunications service providers operate. The Communications Act confers ample flexibility for the FCC to fashion new rules, regulations, and policies as the public interest so dictates. While subject to judicial review, the actions of this expert agency usually pass muster unless the FCC has violated procedural and fairness requirements [4].

The Communications Act of 1934 authorizes the FCC to license facilities and service providers and to ensure that they operate in the public interest. The power to license and regulate also means that the FCC has ongoing responsibili-ties, pursuant to Title II of the Communications Act, to ensure that service providers do not operate in an unreasonable or discriminatory manner. Like-wise, the FCC may require a licensee to operate as a common carrier,[13] thereby imposing a duty to hold itself indifferent to the public and to serve all who seek service, usually under a public contract known as a *tariff.*

17.2 U.S POLICIES PRIOR TO DEREGULATION

Until the 1970s, U.S. international telecommunications policy making closely paralleled the philosophy and approach of most nations. It placed a premium

---

13. The Communications Act of 1934, as amended, defines a common carrier as "any person en-gaged as a common carrier for hire, in interstate or foreign communication by wire or radio." (47 U.S.C. Sec. 153(h) (1990).) The Communications Act requires common carriers to provide service "upon reasonable request," at tariffed rates that are "just and reasonable" and "without unjust or unreasonable discrimination." (47 U.S.C. Sec. 201(a), (b), and 202(a).) See Note, "Re-defining 'Common Carrier': The FCC's Attempt at Deregulation by Redefinition." [1987] *Duke Law J.*, p. 501.

on industrial stability and universal, sometimes subsidized service. Prior to aggressive efforts to duplicate domestic marketplace deregulatory successes, the United States shared the view that natural monopolies would optimally serve the major international telecommunications market segments. While privately owned, USISCs were limited in the type of markets they could enter and the terms and conditions under which they could offer services.

U.S. international telecommunications policy established a number of service, market, technological, and geographical dichotomies that collectively restricted competition and market access. The FCC imposed restrictions on who could provide international services, and where service could originate and terminate. The FCC created a voice/data dichotomy whereby only AT&T could provide switched voice services and only a small group of IRCs could provide text services like telegrams and telexes.

The United States was divided into locations where international record traffic could originate (i.e., gateway cities), with the rest of the nation considered the hinterland, where customers would have to secure services of a domestic carrier to route traffic to a gateway. This end-on-end routing arrangement generated higher costs to consumers. For record services, users in the hinterland had to pay one carrier, the Western Union Telegraph Company, a separate domestic carriage charge and another IRC a separate international carriage charge. As part of an antitrust court case settlement, Western Union agreed to limit the scope of its services to domestic record traffic, including the carriage of traffic to and from the international gateways of other carriers.

The FCC also imposed policies that blunted the potential for competition by multiple service providers using the same transmission technology (intramodal competition) or different transmission media (intermodal competition). Until the middle 1980s, the FCC dutifully adhered to the international consensus of consortium and cooperative planning, deployment, circuit activation,[14] pricing, and maintenance of international facilities. The FCC accepted the premise that nations needed to coordinate on facilities planning and investment to

---

14. "The Commission has made decisions affecting the distribution of circuits among available international facilities nearly since the advent of communications satellites in 1965." (Policy for the Distribution of United States International Carrier Circuits Among Available Facilities During the Post-1988 Period, CC Docket No. 87-67, Notice of Proposed Rulemaking. 2 FCC Rcd. 2109 (1987), policy abandoned, 3 FCC Rcd. 2156 (1988).) For an example of how the balanced loading policy was applied, see ITT Cable and Radio, Inc.—Puerto Rico, 5 FCC 2d 823 (1966); AT&T, 7 FCC 2d 959 (1967) (FCC determination of the proper mix of submarine cable and satellite facilities for Puerto Rico; Comsat, 29 FCC 2d 252 (1971) ("proportional loading" ordered at the rate of five Atlantic Ocean region INTELSAT satellites circuits for every one TAT-5 submarine cable circuit activated). But see also Communications Satellite Corp., 5 FCC Rcd. 5952 (1990) (evidencing reduced scrutiny and greater sensitivity to amount of time for consideration of a Comsat investment obligation in INTELSAT prior to scheduled vote by the cooperative's signatories); accord, Communications Satellite Corp., DA 92-955 (rel. 23 July 1992) (authorizing Comsat investment in the INTELSAT K satellite); see also 5 FCC Rcd. 7358, 7393, 7344.

achieve scale economies, avoid wasteful duplication of investment, and to se-cure the benefits of averaging costs over a larger user base.

Only the Soviet Union's Intersputnik system provided some degree of in-tramodal satellite competition with INTELSAT for transoceanic routes. The spillover signals of domestic satellites did provide a limited, regional service alternative to INTELSAT, but only if a nation agreed to permit market access. The nearly universal decision to blend the cost of cable and satellite facilities into a single composite rate meant that users had no opportunity to select the cheaper of the two media, or perhaps to pay a premium for all cable service, if satellite signal echoing presented a problem.

U.S. avoidance of intramodal competition included legislation that created an international satellite monopoly. The Communications Act of 1962 created Comsat and authorized it to serve as the nation's sole investor and participant in INTELSAT. Comsat received a legislative franchise to serve as a carrier's carrier (i.e., middleman) for INTELSAT international satellite capacity. No other enter-prise can deal directly with INTELSAT and carriers and the federal government qualify as "authorized users" who can acquire wholesale capacity directly from Comsat. Users acquire retail capacity further down the distribution chain.

This structure supported the creation of multiple market tiers with a single facilities operator, INTELSAT, a single U.S. middleman, Comsat, and a select group of wholesalers, the USISCs. Channeling satellite capacity under such a system resulted in a number of pricing mark ups by intermediaries who pro-vided little enhancement to the basic transmission capacity and whose adminis-trative functions primarily involved order taking.

Until the middle and late 1970s, the FCC accepted the world view that resale of leased lines presented an unreasonable threat to carrier profitability and universal service. Resale is an exercise in arbitrage: securing bulk capacity intended for use by a single large-volume user and subdividing it for use by a number of customers with lower individual capacity requirements. The FCC grew to believe that such a function reduced the potential for facilities-based carriers to discriminate against small volume users and could stimulate the benefits of competition without the delay and expense of parallel facilities con-struction. Figure 17.1 graphically depicts the U.S. telecommunications policy-making process.

17.3   A HISTORY OF PERVASIVE GOVERNMENT REGULATION AND OVERSIGHT

Before embracing a deregulatory philosophy, starting in the late 1970s and early 1980s, the FCC established industrial policies that had the effect of segmenting the international telecommunications marketplace and subjecting all segments to pervasive regulation. At least initially such division may have had a techno-

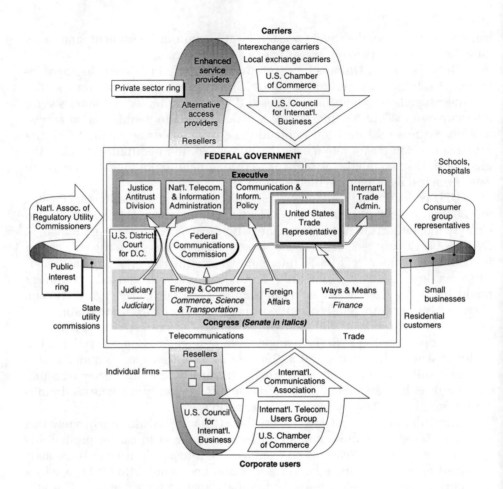

Figure 17.1   U.S. telecommunications policy structure. (*Source:* Office of Technology Assessment, 1993.)

logical justification: until the arrival of high-frequency radio, which could transmit voice and record signals, the submarine cable medium could only transmit textual messages via Morse code. However, any technological justification for segregating carriers into only voice and only record service markets ended well before 1980 when the FCC abandoned the policy.[15] The FCC finally permitted AT&T to provide record services [5] and the IRCs to provide voice services [6].

---

15. See American Telephone and Telegraph Co., 37 FCC 2d 1151 (1964) (TAT-4 voice/record dichotomy).

Actions like this resulted from an increasingly prevalent view that government need not intrude upon the interplay of market forces, because such involvement probably did more harm than good. Segmenting markets and foreclosing market access (e.g., denying AT&T the opportunity to provide record services) ostensibly "protected" users from monopolization by a carrier dominant in another market (i.e., AT&T's near monopoly in voice services). But instead of foreclosing additional monopolization, market segmentation policies prevented competition and allowed carriers to maintain high rates.

It took an act of Congress to reauthorize Western Union to provide international services,[16] a market it had relinquished to secure government consent to the acquisition of a key competitor.[17] Limiting a dominant carrier like Western Union also served an FCC policy seeking to maintain a dichotomy between hinterland and gateway localities[18] [7], but in application it financially disadvantaged users in localities not authorized to make direct international connections, regardless of demand.

Dichotomies based on service (voice/record), and location (gateway/hinterland) outlasted any technological justification. For example, new generations

---

16. See Record Carrier Competition Act of 1981, Pub. L. No. 97-130, 95 Stat. 1687 (1981), implemented in Interconnection Arrangements Between and Among the Domestic and Int'l Record Carriers, Interim Order, 89 FCC 2d 928 (1982), implementing tariffs rejected, 91 FCC 2d 483 (1982), modified on recon., 93 FCC 2d 845 (1983). For an analysis of the costs incurred and charges by Western Union to interconnect with international record carriers, see the Western Union Telegraph Co., 95 FCC 2d 881 (1983), aff'd sub nom., FTC Communications, Inc. v. FCC, 750 F.2d 226 (2d Cir. 1984); The Western Union Telegraph Co., FCC 86-190 (rel. 2 May 1986) (ordering the international record carriers to reimburse Western Union for service whose 15% discount was later deemed unjustified), on recon., 2 FCC Rcd. 2999 (1987), on further recon., 3 FCC Rcd. 2597 (1988); see also R. Frieden, "International Telecommunications and the Federal Communications Commission," *Columbia J. Transnational Law*, Vol. 21, No. 3, 1983, pp. 423, 466–485.

17. In 1943, Congress enacted Section 222 of the Communications Act to permit Western Union to acquire its major competitor, the Postal Telegraph Co. Western Union was required to divest itself of international record carrier operations in exchange for congressionally sanctioned antitrust immunity. For a history of Section 222, prior to enactment of the Record Carrier Competition Act, see Western Union Tel. Co. New Telex Serv. Arrangements via Mexico and Canada, 75 FCC 2d 461 (1979), vacated, ITT World Comms., Inc. v. FCC, 635 F.2d 32 (2d Cir. 1980) (FCC authorization of Western Union to provide international service via Mexico and Canada deemed a violation of Sec. 222).

18. Until enactment of the Record Carrier Competition Act of 1981, the FCC had established a geographical dichotomy between gateway cities, where international traffic could originate, and the hinterland, where foreign-destined traffic had to first access a gateway via a domestic carrier for subsequent retransmission via an international record carrier. (See *Domestic Public Message Serv.*, 71 FCC 2d 471 (1979), aff'd sub nom., Western Union Tel. Co. v. FCC, 665 F.2d 1126 (D.C. Cir. 1981) (expanding the domestic carrier set to include Graphnet, Inc., but maintaining the domestic/international service dichotomy).

of submarine cables[19] and radio facilities made it possible for international carriers to provide both voice and record services. The IRCs invested in these new cables, thereby providing them with the technological wherewithal to provide voice services in competition with AT&T. Yet the FCC perpetuated the voice/record dichotomy, presumably to ensure that AT&T did not dominate record markets as it had voice markets. While designed to guard against AT&T leveraging of market power into record services, the policy substantially reduced intramodal competition among carriers using the same transmission medium they collectively owned. Even without such a policy, the action by all carriers to form cable consortia and satellite cooperatives discouraged price competition, primarily because each carrier had roughly the same transmission capacity costs. AT&T's disqualification from record service markets made it that much easier for the handful of IRCs to form an oligopoly and agree not to engage in price competition. Likewise, the absence of IRC voice services in effect guaranteed that AT&T could maintain its monopoly.

17.4 THE DEREGULATORY CAMPAIGN (1980 TO PRESENT)

In the late 1970s and early 1980s, the United States began to abandon support for market segmentation and regulated monopolies. In 1980 the FCC resolved to foster "an improved international communications system with more choices for consumers, more diverse service offerings, and lower rates" [8]. The FCC embraced the view that reduced government oversight and marketplace intervention would enhance consumer welfare, reduce rates, promote competition, spur innovation, foster service diversity, and achieve a more efficient telecommunications marketplace.

Many U.S. officials viewed international telecommunications policy as unnecessarily lagging and diverging from already completed domestic deregulatory initiatives. They hoped that unilateral action could set a deregulatory foundation that foreign counterparts would embrace. The industry and consumer welfare enhancing dividends achieved in the United States over the last 25 years[20] presumably could apply internationally.

---

19. "AT&T, the dominant domestic telephone carrier, initiated transatlantic telephone message service in 1927 by high frequency radio. In 1956, AT&T developed a reliable underwater repeater, and together with the British government, laid the first transatlantic cable (TAT-1)." (Overseas Communications Services, CC Docket No. 80-632, 84 FCC 2d 622, 623 (1980).)

20. See, for example, Competitive Carrier, 77 F.C.C.2d 308 (1979), First Report and Order, 85 F.C.C.2d 1 (1980), Further Notice, 84 F.C.C.2d 445 (1981), Second Report and Order, 91 F.C.C.2d 59, Third Report and Order, 48 Fed. Reg. 46791 (1983), Fourth Report and Order, 95 F.C.C.2d 554, Fifth Report and Order, 98 F.C.C.2d 1191, Sixth Report and Order, 99 F.C.C.2d 1020 (1985), rev'd and remanded sub nom. MCI Telecom. Corp. v. FCC, 765 F.2d 1186 (D.C. Cir. 1985); Hush-A-Phone v. United States, 239 F.2d 266 (D.C. Cir. 1956), on remand sub nom.

However, international policy making driven from domestic experience fails to consider different and unavoidable forces and philosophies. Prior to its philosophical transformation, the FCC ignored evidence that marketplace forces could work in international telecommunications. Its newfound deregulatory zeal may have compounded problems, because efforts to make up for lost time typically backfired, since the regulatory counterparts and legislators in other nations did not match the FCC's initiatives. Representatives from other nations refused to be rushed or threatened by initiatives that they viewed as requiring study to confirm the absence of harm to incumbent carriers and their essential service mission.

Even now, virtually all governments have chosen not to relinquish oversight responsibilities in international telecommunications. While contemplating the free interplay of marketplace and technological forces, even the FCC has recognized the need to consider countervailing factors that work against facilities-based carrier competition and new regulatory classifications, like private, noncommon carriers[21] that promote market entry by enterprises substantially free of government oversight. Many U.S. deregulatory initiatives failed to achieve the support from other nations necessary to achieve new investment and market access opportunities. While theoretically supportable and often bolstered by empirical proof of enhanced consumer welfare in the U.S., many FCC initiatives have been deemed too bold, risky, and threatening by other governments.

Officials at the FCC have recognized belatedly the futility in applying a domestic market–oriented deregulatory model to the international telecommunications environment. In the 1980s, the FCC initiated a number of rulemakings

---

Hush-A-Phone v. AT&T, 22 FCC 112 (1957) (establishing right of customers to attach acoustical devices to their telephones); Use of the Carterfone Device in Message Toll Service, 13 FCC 2d 420 (1968) (establishing public's right to attach electronic devices to the telephone); Allocation of Frequencies in the Bands Above 890 Mc. 27 FCC 359 (1959), on recon., 29 FCC 825 (allocation of frequency spectrum for competing microwave radio services); Microwave Communication, Inc., 18 FCC 2d 953 (1969) on recon., 21 FCC 2d 190 (authorizing common carrier competition); Regulatory Policies Concerning Shared Use and Resale of Common Carrier Services Facilities, 47 FCC 2d 644 (1974), 48 FCC 2d 1077 (1974), 60 FCC 2d 261 (1976), modified, 61 FCC 2d 70 (1976), further modification, 62 FCC 2d 588, aff'd sub nom. AT&T v. FCC, 572 F.2d 17 (2nd Cir. 1977), cert. den., 439 U.S. 875 (1978) (authorization for shared use and resale of long-distance lines which created the opportunity for arbitrage); MCI Telecom. Corp. v. FCC, 561 F.2d 365 (D.C. Cir. 1977), cert. den., 434 U.S. 1040 (1978) mandate enforced, 580 F.2d 590 (D.C. Cir. 1978), cert. den., 439 U.S. 980 (1978) (creating right of alternative carriers to access the public switched telephone network and thereby provide alternative long-distance telephone service).

21. See, for example, Tel-Optik, Limited, 100 FCC 2d 1033 (1985) (authorization of a private, noncommon carrier international fiber-optic submarine cable operator); and Pan American Satellite Corp., 101 FCC 2d 1318 (1985) (authorization of a private, non-common carrier international satellite operator). The Record Carrier Competition Act of 1981, Pub. L. No. 97-130, 95 Stat. 1687 (1981) authorized the Western Union Telegraph Company's return to international services after years of prohibited access.

to undo the complex web of protectionist policies it had designed to create market segments, service dichotomies, and limits to competition. While the FCC could undo domestic policies, which retarded competition, it generally failed to convince other nations of the need for speedy and dramatic change. Until nations perceived the benefit of privatization, liberalization, and deregulation, FCC edicts could do little to foster international telecommunications policies based on theoretical premises or even domestic experiences.

For example, the FCC could provide concrete evidence of public dividends accruing from authorizing enterprises to resell bulk private line services.[22] The FCC's proresale domestic policy included large-volume calling services such as WATS lines, initially made available only to individual companies generating high volumes of long-distance calls [9]. Private line resale involves the aggregation of small-volume users whose collective demand qualifies for a bulk volume discount otherwise unavailable to any single small-volume user. The FCC viewed this option as putting pressure on "underlying [facilities-based] carrier[s]...to realign the relationship between unit and bulk prices to make that relationship wholly cost-based" [10].

When the FCC first attempted to apply the resale concept internationally [11], PTTs universally rejected it. They viewed the FCC as authorizing "pirates" who would jeopardize PTT revenue streams and threaten sovereignty.[23] U.S. could enter the international market without PTT authorization simply by leasing the facilities of other carriers that had foreign operating agreements. Foreign carriers threatened to eliminate international private line tariff options used by resellers, and forcefully objected to the concept. Over the last 15 years, a small but growing number of nations have recognized the public benefits in having market entrants who lease lines and either add value and enhancements to them (e.g., IVANs) or provide a brokerage (arbitrage) function for basic services. Even nations with less than cutting-edge regulatory policies (e.g., Japan, Korea, Chile, and Sweden) support market entry by resellers that enhance basic capacity. Canada, the United Kingdom, Australia, and New Zealand have joined the U.S. in supporting basic service resale.

---

22. See Regulatory Policies Concerning Resale and Shared Use of Common Carrier Services and Facilities, 60 FCC 2d 261, modified, 61 FCC 2d 70, further modifications, 62 FCC 2d 588 (1977), aff'd sub nom., AT&T v. FCC, 572 F.2d 17 (2nd Cir. 1978), cert. denied, 439 U.S. 875 (1978).

23. The ITU's Consultative Committee on International Telegraph and Telephone has issued quite limiting Recommendations on shared use and resale of leased lines. (See CCITT, Vol. II, Fascicle II.I, *General Tariff Principles, Charging and Accounting in International Telecommunications Services, Recommendations of the D Series* (Blue Book), Geneva, 1989.) Collectively, these Recommendations suggest significant restrictions on the availability of lines and the type of user group qualifying to share. The Recommendations propose that carriers have the right to refuse service to customers who violate such restrictions.

17.5 THEMES IN U.S. INTERNATIONAL TELECOMMUNICATIONS AND TRADE POLICY

Perhaps even at the risk of being overbearing, the United States has pursued international telecommunications policies that foster competition through market entry and decreased regulatory oversight. While most foreign counterparts now recognize the benefits accruing from recalibrating the scope of regulation to stimulate innovation and efficiency, they are reluctant to promote broader consumer choices and downward rate pressure if it threatens the PTT's ability to generate revenues needed for subsidizing deliberately underpriced services. Unlike the FCC's campaign to foster cost-based pricing, other nations establish clear or implicit obligations on the PTT to underwrite certain offerings such as POTS and postal operations or to contribute to the general treasury. Low-cost and ubiquitous POTS supports national cohesiveness and security. Likewise it helps promote positive network externalities, the economic view that the value of service and the network providing it increase with the number of subscribers and the ability to link urban and rural users. Many U.S. initiatives have achieved less than universal support from foreign counterparts because of the real or perceived challenge to the ability of PTTs to carry out their service mission.

17.5.1 Facilities-Based Competition

The U.S. has encouraged market entry by facilities-based carriers as a way to expedite the introduction of technological innovations [12]. Many nations want to reserve a basic services monopoly for the PTT or its privatized replacement. These nations do not purposefully block innovation, nor do they want to stifle competition throughout the industry. However, they do believe basic services constitute a natural monopoly; that is, economic analysis supports the view that a single enterprise can most efficiently operate because it can reduce costs by serving a large user base. Such nations believe that limits on the scope of permissible competition are necessary to ensure that the incumbent carrier can achieve social goals such as universal service and the below-cost provision of POTS.

17.5.2 Cost Deaveraging

The U.S. has sought to promote transmission facility pricing as a function of the particular medium used, without averaging or forced use of more costly or less efficient facilities. The FCC no longer requires "balanced loading" of cable and satellite circuits by USISCs. Likewise, the FCC has abandoned a policy that required carriers to average costs of different facilities into a composite rate [13]. Some nations support cost averaging between high- and low-density routes and between cable and satellite facilities, because they generate comparatively less traffic, or operate in high-cost areas. Other nations, including ones generating high volumes, believe that averaging supports global connectivity and stands as one of they key reasons for creating worldwide cooperatives like INTELSAT.

17.5.3   Private Networking and Access to the PSTN

The U.S. strongly supports the opportunity for users to engineer their own private network solutions to particular needs, including access to foreign PSTNs. Nations have grave concerns about the extent to which end users and entrepreneurs can engineer networks that access the PSTN. The PSTN constitutes a bottleneck, that is, an essential facility over which most traffic traverses with no widely available alternative. Carriers controlling the PSTN have great leverage in setting charges for access. If a PTT has a duty to provide POTS at less than fully compensatory rates, then it must generate higher revenues from other services. The ability to charge rates well in excess of costs in large part depends on the PTT's control of PSTN access. PTTs fear that new service providers, like resellers, and users will find ways to access the PSTN without detection, or at lower rates than designated for the type of service provided. PTTs also worry that market entrants and users will migrate switched service traffic off high-cost public networks onto cheaper private ones. If providers of highly profitable international message telephone traffic (i.e., resellers and private line operators) can avoid paying high rates for access to the PSTN, then the PTT may have lower revenues available to subsidize POTS. Accordingly, most nations refrain from supporting any market entry or liberalization initiative that jeopardizes captive traffic streams allegedly necessary to maintain a system that makes it possible to provide some services at subsidized levels.

17.5.4   Alternatives to Satellite Cooperatives and Cable Consortia

The U.S. also has aggressively sought to promote market entry by:

- The transborder services of domestic satellites;[24]

---

24. See, for example, Transborder Video Services, 88 FCC 2d 258 (1891). But in Communications Satellite Corp. v. FCC, 836 F.2d 623 (D.C. Cir. 1988), the D.C. Circuit vacated an FCC order granting the use of a domestic satellite for transborder services to and from Jamaica on grounds that the FCC failed to articulate how it was uneconomical or impractical to use the INTELSAT cooperative. The Earth station in Jamaica was subsequently sold and reconfigured to provide international service via INTELSAT. In 1995, the FCC proposed to merge its transborder domestic satellite and international separate system policies. See Amendment to the Commission's Regulatory Policies Covering Domestic Fixed Satellites and Separate International Satellite Systems, IB Docket No. 95–41, FCC 85–146, 1995 West Law 240658 (FCC) (rel. April 23, 1995).

- Facilities-based satellite alternatives to INTELSAT and Inmarsat [14];
- Private cable competitors to common carrier consortia facilities [15].

The FCC authorized the limited international, transborder use of domestic satellites whose footprints spill over into adjacent nations.[25] In 1985, the FCC authorized the first dedicated international satellite system separate from INTELSAT [16]. To promote competition in the provision of international services delivered via satellite, the FCC also unbundled regulatory consideration of the space segment from the Earth station services and determined that the latter could be provided independently by entities instead of an FCC-organized consortium of international carriers [17].[26] Recently, the FCC allowed limited two-way international Earth station operations by private carriers [18]. However, after preliminarily considering the option for end users and international carriers to achieve direct cost-based access to INTELSAT capacity, the FCC concluded that Comsat still deserved an exclusive middleman role.[27]

---

25. See Transborder Satellite Video Services, 88 FCC 2d 258 (1981).

26. Previously, the FCC had ordered the bundling of satellite capacity with Earth station services, with the latter available only from a Modification of Policy on Ownership and Operation of U.S. Earth Stations That Operate With the INTELSAT Global Communications Satellite System, Comsat-managed facility owned by a consortium of international carriers. (See Ownership and Operation of Earth Stations, Second Report and Order, 5 FCC 2d. 812 (1966); and also R. Frieden, "Getting Closer to the Source: New Policies for International Satellite Access," *Federal Communications Law J.*, Vol. 37, 1985, p. 293.) Previously, the FCC designated Comsat as a 50% owner and manager of all continental U.S. Earth stations. (Amendment of Pt. 25 of the Commission's Rules and Regulations With Respect to Ownership and Operation of International Earth Stations in the United States for Use in Connection With the Proposed Global Commercial Communications Satellite System, Docket No. 15735, Second Report and Order, 5 FCC 2d 812 (1966).) The other 50% was allocated to all other USISCs on the basis of their relative use of the Earth stations. AT&T subsequently acquired complete ownership of the ESOC facilities in the continental U.S. (Am. Tel. & Tel. Co., File No. I-T-C-87-109, 4 FCC Rcd. 2327 (1989).)

27. The Communications Satellite Act of 1962 established Comsat as the sole U.S. signatory to IN-TELSAT and specified that only certain special authorized users (e.g., the military and international telecommunications common carriers), shall have the right to deal directly with Comsat to procure capacity on a wholesale basis. (Authorized User Policy, 4 FCC 2d 421 (1966) recon. granted in part, 6 FCC 2d 593 (1967)(Authorized User-I).) Nonauthorized users (e.g., end users) must acquire INTELSAT capacity on a retail basis from authorized users. The Authorized User-I policy specified that Comsat shall serve as the "carriers' carrier," on a common carrier basis. In 1982, the FCC modified Authorized User-I to permit Comsat to provide international satellite capacity directly to a broader set of carriers and users who could arrange with Comsat to have service made available at a carrier's Earth station, and to offer retail, end-to-end service through a separate subsidiary. (*Authorized User Policy*, 90 FCC 2d 1394 (1982) (Authorized User II).) That decision was vacated and remanded to the FCC, because the FCC failed to consider contemporaneously two related issues: (1) whether any carrier should have the right of direct access to INTELSAT capacity, thereby eliminating the Comsat middleman markup; and (2) whether carriers could operate their own Earth stations independent of the Comsat managed facilities owned by a few major carriers. (Consideration of Modifications to Its Authorized User

While refusing to allow a variety of carriers to access INTELSAT directly [19] without the signatory services and markup of Comsat, the FCC did allow a larger group of carriers deal directly as authorized users with Comsat rather than face a double markup of INTELSAT capacity by both Comsat and the previously limited consortium of international carriers with Comsat as Earth station facility manager and majority owner. It also took a hard look at the corporate structure and performance of the COMSAT Corporation, the sole U.S. participant in the INTELSAT and Inmarsat international satellite cooperatives.[28] In 1984, the FCC also authorized the first private international cable system, Tel-Optik, Ltd., now owned by Sprint International and Cable & Wireless of the United Kingdom, whose fiber-optic cable PTAT-1 links the United States, Bermuda, Ireland, and the United Kingdom and serves as an alternative to the conventional multiple-carrier consortia system [20].

Facilities-based competition can achieve a fundamental change in the industrial structure of international telecommunications. While the U.S. was instrumental in forming international cooperatives and supporting their exclusivity, it now believes that with adequate safeguards, the marketplace can support competition even for some switched services that access the PSTN. Most nations strongly objected to this scenario and only with the passage of time and the appearance of limited separate system competition have they begun to accept that conditional competition will not prevent incumbents from achieving their service mission.

Facilities-based competition and access to the PSTN constitute key U.S. policy predicates that few nations endorse because of the perception that they would handicap the ability of national carriers to achieve social and pricing goals. Most nations do not endorse and have not implemented policy initiatives that promote facilities-based competition, cost deaveraging, private network access to the PSTN, and private carriers separate from the conventional satellite

---

Policy, ITT World Communications, Inc. v. FCC, 725 F.2d 732 (D.C. Cir. 1984).) After further proceedings, the FCC in effect readopted its Authorized User-II policy. (*Authorized User Policy*, 97 FCC 2d 296 (1984), reaff'd, 99 FCC 2d 177 (1985) (Authorized User-III).) Contemporaneously, the FCC rejected direct access for carriers (97 FCC 2d 296 (1984)), but authorized independent Earth station ownership (100 FCC 2d 250 (1984).) This time, the D.C. Circuit Court of Appeals affirmed the FCC's action. (Western Union International, Inc. v. FCC, 804 F.2d 1280 (D.C. Cir. 1986).

28. See Comsat Study-Implementation of Sec. 505 of the International Maritime Telecommunications Act, 77 FCC 2d 564 (1980); Changes in the Corporate Structure and Operations of the Communications Satellite Corp., Notice of Proposed Rulemaking, 81 FCC 2d 287 (1980), First Report and Order, 90 FCC 2d 1159 (1982), recon. denied, 93 FCC 2d 701 (1983), Second Report and Order, 97 FCC 2d 145 (1984), on recon., 99 FCC 2d 1040 (1984). The FCC concluded that Comsat should be permitted to engage in nonjurisdictional (i.e., competitive) non-INTELSAT/Inmarsat signatory activities if such ventures were "not inconsistent with its statutory mission" and provided it modified its corporate structure to establish a separate retail services subsidiary and revised cost allocation formulas.

cooperative and cable consortia structure. However, an increasing number of nations have privatized the PTT, or at least created a corporate structure. The EU has established a 1998 deadline for elimination of a facilities-based telecommunications monopoly. Enhanced and even basic resale of leased lines has become a more acceptable vehicle for promoting some degree of competition. Depending on one's viewpoint, the FCC deserves credit for persevering in its deregulatory campaign which has achieved success, or credit should be shared with economists, advisors, World Bank officials, commercial lenders, and larger cast that have collectively embraced competition, deregulation, and liberalization as essential predicates for an effective telecommunication infrastructure.

17.5.5 Fair Trade in Telecommunications

As part of its campaign to foster competition, the U.S. has engaged other nations in a dialogue with an eye toward encouraging market entry by multiple carriers and equipment manufacturers. Only through market entry can end-user rates drop significantly, in part because monopoly carriers may have no incentive to negotiate lower accounting rates, the level of competition set by carrier correspondents to reflect the cost of providing service. Likewise, market entry is essential to reduce a growing trade deficit in telecommunications equipment. The U.S. pursues this goal through greater advocacy to eliminate closed procurements, discriminatory testing and certification requirements, nondisclosure of technical interface details, discriminatory duties on imports, and so on.[29]

To achieve a more receptive hearing for any of the above policies, the United States has attempted to appeal to the growing business orientation of many PTTs and the evolving responsiveness to consumer requirements by other nations' regulatory and policy-making bodies. While it remains true that no policy initiative may achieve acceptance without thorough ventilation of the issues and PTT cooperation, financial matters increasingly predominate over loftier notions of sovereignty and international comity. PTTs and government ministries increasingly realize that a robust and fairly priced telecommunications network is a sine qua non to commercial development. Both existing and prospective employers, which heavily rely on telecommunications facilities and services, can "vote with their feet" to other, more accommodating nations.

---

29. See, for example, *U.S. Telecommunications in a Global Economy: Competitiveness at a Crossroads*, Report from the Secretary of Commerce to the Congress and the President of the United States, August 1990. Similarly, the EEC and other nations appear inclined as well to reduce barriers to efficiently operating markets. (See Towards a Dynamic European Economy—Green Paper on the Development of the Common Market for Telecommunications Services and Equipment, COM(87) 290 final, 30 June 1987; see also, R. Frieden, "Open Telecommunications Network Policies for Europe: Promises and Pitfalls," *Jurimetrics*, Vol. 31, November 1990, pp. 319–328.)

Nations increasingly question the rationale for favoring national heroes and for insulating incumbent carriers from competition. Protectionist policies may make it possible for them to remain profitable in the short term even as their long-term viability may become doubtful. Without pressure to innovate and streamline, these enterprises may become inefficient and noncompetitive outside their closed markets. Such policies leave the nation vulnerable to trade disputes, customer migration, and comparatively inadequate service.

On the other hand, nations must find ways to underwrite substantial investments to upgrade, rather than maintain, existing networks. The allure of ISDN[30] and new intelligent systems[31] results in part from the PTTs' view that a wider services wingspan will spread costs over a broader base and enable the PTT to exploit economies of scale and scope. However, a too-visible PTT campaign to incorporate, if not monopolize, information services may backfire as large-volume users of private networks authorized by government regulators vigorously object to any migration to one-size-fits-all public networks.

FCC Strategies

The FCC has pursued a number of different strategies to promote open and competitive international telecommunications markets. Its greatest success occurs when it can dismantle policies, rules, or regulations that insulate U.S. carriers and manufacturers from competing. It has achieved far less success in securing parallel deregulatory initiatives in other nations, particularly when it threatens retaliatory sanctions on foreign ventures operating in the U.S.

For example, the FCC heightened its attention to the net outflow of over $3 billion per year in international toll revenue settlements with foreign carriers.[32] Mindful of the contribution telecommunications services and equipment sales can have on the national economy, the FCC believed that it could leverage its power to monitor and comparatively regulate foreign companies operating in the U.S. more heavily to secure lower accounting rates. The FCC proposed to

---

30. See Integrated Services Digital Networks (ISDN), Notice of Inquiry, 94 FCC 2d 1289 (1983), First Report, 98 FCC 2d 249 (1984).

31. See Intelligent Networks, CC Docket No. 91-346, Notice of Inquiry, 6 FCC 7256 (1991).

32. See International Accounting Rates and the Balance of Payments Deficit in Telecommunications Services, Report of the Common Carrier Bureau to the Federal Communications Commission (12 December 1988); see also Uniform Settlement Rates, 84 FCC 2d 121 (1980); Implementation and Scope of the International Settlements Policy for Parallel International Routes, Notice of Proposed Rulemaking, 50 Fed. Reg. 28,418 (1985), Report and Order, 59 Rad. Reg.2d (P&F) 982 (1986), on partial recon., 2 FCC Rcd. 1118 (1987), on further recon., 3 FCC Rcd. 1614 (1988); Regulation of International Accounting Rates, CC Docket No. 90-337, Notice of Proposed Rulemaking, 5 FCC Rcd. 4948 (1990), Report and Order, Phase I, 6 FCC Rcd. 3552 (1991), Further Notice of Proposed Rulemaking, Phase II, 6 FCC Rcd. 3434 (1991), First Report and Order, Phase II, 7 FCC Rcd. 559 (1992).

scrutinize foreign carriers' and manufacturers' U.S. operations, but substantially curbed the scope of its oversight and the reporting requirements it imposed[33] in response to concerns voiced by the executive branch that such selective scrutiny would violate the U.S. commitment to national treatment.

The FCC also reduced regulatory oversight of international carriers and service providers who operate in a competitive marketplace [21].[34] The FCC narrowed the scope of conventional common carrier regulation by creating a new regulatory category for nondominant carriers (i.e., carriers lacking the power to affect the price or supply of a service). Such carriers qualify for streamlined regulation, including a shorter time period between filing for a new service or price change and FCC authorization. The FCC now imposes conventional common carrier regulation only to carriers with a foreign ownership majority, AT&T on a service and nation-specific basis, carriers dominant outside the continental United States, and Comsat in its INTELSAT/Inmarsat signatory functions.

The FCC may have overestimated the potential for other nations to adopt a similar strategy for reducing or eliminating regulation. Without parallel deregulatory action abroad, FCC initiatives have the effect of promoting domestic market access without securing commensurate opportunities in other nations. The FCC recognized this outcome by conditioning the opportunity for foreign-owned carriers operating in the U.S. to qualify for nondominant carrier status. Carriers with majority foreign ownership can qualify for nondominant status on a route-by-route basis upon demonstrating that their parent carrier does not have bottleneck control in the foreign home country.

17.5.6 Resistance to U.S. Procompetitive Initiatives

Many of the international telecommunications and trade policy initiatives in the United States result from interest in broader global market access and the

---

33. See Regulatory Policies and International Telecommunications, CC Docket No. 86-494, Report and Order and Supplemental Notice of Inquiry, FCC 88-71 (25 March 1988), 4 FCC Rcd. 323 (1989) (substantially trimming back the scope of reporting requirements), 7 FCC Rcd. 1715 (1992) (proposing further reductions in reporting requirements of foreign-owned or controlled carriers).

34. In 1991, the FCC issued a Notice of Proposed Rulemaking to consider a proposal to deem foreign-owned or controlled carriers to be "dominant" only when providing services to locations where they control access to bottleneck telecommunications networks. (Petition for Rulemaking to Modify the Commission's Regulation of International Common Carrier Services Provided by Foreign-Owned U.S. Common Carriers, 6 FCC Rcd. 290 (1991) CC Docket No. 91-260, Notice of Proposed Rulemaking, 7 FCC Rcd. 577 (1992); see also Regulatory Policies and International Telecommunications, CC Docket No. 86-494, 7 FCC Rcd. 1715 (1992) (proposal to reduce further reporting requirements of foreign-owned or controlled carriers); Market Entry and Regulation of Affiliated Enterprises, IB Docket No. 95–22, Notice of Proposed Rulemaking, 10 FCC Rcd. 4844 (1995) (proposing to use Secs. 214 and 310 of the Communications Act as leverage to secure equivalent market access).

belief that enhanced domestic consumer welfare from deregulation can apply internationally.[35] However, international telecommunications and trade matters do not have symmetry with any one nation's domestic policies. Despite empirical success in the U.S., deregulation does not necessarily discredit the traditional PTT model, nor does it trigger liberalization and dismantling PTTs. Furthermore, the multilateral and cooperative nature of international telecommunications and trade policy militate against simply extending one nation's policies abroad. Accordingly, the pace of international policy liberalization may lack the speed U.S. deregulators and entrepreneurs would prefer.

The FCC's mandate to serve the public interest increasingly points that agency toward deregulation. Other nations' public interest assessments have supported the view that the PTT constitutes a natural monopoly best situated to pursue social goals like universal service. Notwithstanding a long-term predisposition toward the regulated monopoly model, many nations now recognize that stimulating a more businesslike approach in the telecommunications sector will better serve the national interest than simply maintaining the status quo. These nations will consider abandoning government ownership of the PTT and centralized planning. But few have yet to join with the United States, United Kingdom, and New Zealand in the view that private enterprise, in large part, can provide *all* telecommunication facilities, services, and equipment at reasonable rates while also serving government-articulated social objectives. Even the U.K. government reserves the right to cast its "golden share" and veto any corporate decision of British Telecom deemed inconsistent with the national interest. The United Kingdom also established the Office of Telecommunications with substantial regulatory authority.[36]

17.6 REFORMING U.S. TELECOMMUNICATIONS POLICY MAKING

Infrequently, federal agencies involved in international telecommunications undertake a comprehensive examination of both the process and substance of U.S. policy making.[37] The findings over the years can be summarized as having identified three "fundamental defects" [22,23]:

---

35. The FCC believes its procompetitive, open-entry policies will foster "an improved international communications system with more choices for consumers, more diverse service offerings and lower rates." (Am. Tel. & Tel. Co., 75 FCC 2d 682, 288 (1980) (removal of restriction on the use of international telephone lines for data applications), aff'd sub nom., Western Union International, Inc. v. FCC, 673 F.2d 539 (D.C. Cir. 1982).)

36. See Gist, P., "The Role of Oftel," *Telecommunications Policy*, Vol. 14, February 1990, pp. 26–51).

37. See National Telecommunications and Information Administration, Comprehensive Examination of U.S. Regulation of International Telecommunications Services, Docket No. 921251-2351, Notice of Inquiry, 58 Fed. Reg. No. 10, 4846-4861 (15 January 1993). For an analysis of

First, its outlook often tends to be reactive and skewed toward achieving short-term objectives rather than proactively establishing and working to reach long-range goals...Second, the allocation of financial and staff resources among a variety of particular agencies, most with only limited interests or responsibilities in communications and information policies, does not foster the development of effective communications policy or policymakers...Finally, the current structure makes leadership—and effective Congressional oversight—difficult...[leading to] a fragmented...hastily-coordinated, lowest common denominator.[38]

In application, the U.S. federal agencies with jurisdiction over international telecommunication and trade matters often fail to speak effectively with one voice or to coordinate:

- Rigorous examination of the international telecommunications marketplace using both data routinely collected and additional specially requested data essential for reasoned decision making;
- Long-term policy advocacy with foreign counterparts that avoids overbearingness on one hand and passivity on the other;
- The use of domestic deregulation as a tool to secure reciprocal liberalization abroad, rather than merely to pursue unilateral philosophical goals at home;
- Regulatory and policy reform that will forestall the need for users to resort to self-help made possible by technological innovations and the inability of carriers to identify the origin of calls.

The policy-making products generated by the U.S. multiagency process can miss the mark by being too aggressive or in other instances by being too cautious. Examples of the former category occur when the U.S. fails to consider the consequences of initiatives, such as market entry by resellers and facilities-based carriers, on the ability of PTTs to achieve their service mission. Examples of the latter category occur when the U.S. successfully completes a deregulatory program at home and expects other nations to adopt similar initiatives without dialogue and advocacy in international forums like the ITU.

---

the numerous previous examinations, see S. Brotman, *The Council of Communications Advisers*, Washington, D.C.: Annenberg Washington Program in Communications Policy Studies, May 1989, pp. 6–14; and *Telecom 2000*, pp. 163–184.

38. *Accord Final Report, President's Task Force on Communications Policy* ("Rostow Report"), 7 December 1968, Washington, D.C.: GPO, pp. 21–22: Neither the President nor individual executive branch agencies have access to "a source of coordinated and comprehensive policy advice. As a result, the Executive Branch has difficulty presenting a coherent and consistent position on policy problems."

Over the course of 25 years, the FCC has unbundled telecommunications equipment from service,[39] permitted tariff flexibility, and reduced carriers' regulatory duties [24],[40] and shown increasing reluctance to impose structural requirements for guarding against anticompetitive practices [25]. It has abandoned a variety of regulatory policies that segmented the market into mutually exclusive segments based on dichotomies, including domestic/international,[41] voice/record [26], satellite/cable, dominant/nondominant, carrier/end user, and gateway/nongateway categories.[42]

39. The FCC mandated interconnection of equipment to the PSTN subject to a registration program to ensure no technical harm. (See Interstate and Foreign Message Toll Telephone, First Report and Order, 56 FCC 2d 593, modified on recon., 58 FCC 2d 716 (1976); Second Report and Order, 58 FCC 2d 736 (1976), aff'd sub nom., North Carolina Utility Commission v. FCC, 522 F.2d 1036 (4th Cir.), cert. den., 434 U.S. 874 (1977).

40. However, foreign-owned or controlled carriers providing international services from the United States are considered "dominant" because of their potential to leverage bottleneck control at home to secure concessions in the United States. The FCC requires such carriers to bear greater administrative and substantive filing requirements. In December 1991, the FCC proposed to refine its policy by suggesting that dominant carrier status apply on a route-by-route basis as a function of whether the carrier has the ability to discriminate against other nonaffiliated U.S. carriers through control of bottleneck facilities and services at foreign locations. (Petition for Rulemaking to Modify the Commission's Regulation of International Common Carrier Services Provided by Foreign Owned U.S. Common Carriers, 6 FCC Rcd. 290 (1991) CC Docket No. 91-260, Notice of Proposed Rulemaking, 7 FCC Rcd. 577 (1992).) See also Regulatory Policies and International Telecommunications, CC Docket No. 86-494, 7 FCC Rcd. 1715 (1992) (proposal to reduce further reporting requirements of foreign-owned or controlled carriers). The FCC continues to require all U.S. international carriers to seek authorization to operate under Section 214 of the Communications Act. Under the new rules, all Section 214 applicants would have to certify whether they control or are under control of a provider of telecommunication services in a foreign market they seek to serve.

41. See, for example, Int'l Record Carriers' Scope of Operations (Gateways), 76 FCC 2d 115 (1980), aff'd sub nom., Western Union Tel. Co. v. FCC, 665 F.2d 1126 (D.C. Cir. 1981) (expanding domestic locations where international record carriers could accept traffic).

42. See, for example, Modification of Policy on Ownership and Operation of U.S. Earth Stations That Operate With the INTELSAT Global Communications System, 100 FCC 2d. 250 (1984) (authorizing Earth station ownership independent of a Comsat-led consortium); Licensing Under Title III of the Communications Act of 1934, as amended of Private Transmit/Receive Earth Stations Operating With the INTELSAT Global Communications Satellite System, 3 FCC Rcd. 1585 (1988), aff'd sub nom., TRT Telecommunications Corp. v. FCC, 876 F.2d 135 (D.C. Cir. 1989) (authorizing end users to operate international transmit/receive Earth stations independently of the carrier providing the space segment); International Record Carriers' Scope of Operations, 38 FCC 2d 543, 545 (1972) (describing the gateway concept), relaxed to include new 21 gateway cities, 76 FCC 2d 115 (1980), on recon., 80 FCC 2d 303 (1980), aff'd sub nom, Western Union Tel. Co. v. FCC, 665 F.2d 1112 (D.C. Cir. 1981). Until enactment of the Record Carrier Competition Act of 1981, the FCC had established a geographical dichotomy between *gateway*, cities where international traffic could originate and the *hinterland*, where foreign-destined traffic had to first access a gateway via a domestic carrier for subsequent retransmission via an international record carrier. (See Domestic Public Message Serv., 71 FCC 2d 471 (1979), aff'd sub nom., Western Union Tel. Co. v. FCC, 665 F.2d 1126 (D.C. Cir. 1981) (expand-

Many of the U.S. policy initiatives if embraced internationally could strain technological, physical, and marketplace resources available in single nations and shared by all nations. For example, the U.S. domestic open skies policy [27], which encourages market entry by qualified satellite applicants, could result in severe overcrowding of the geosynchronous orbital arc. The proliferation of satellites could threaten the ability of the ITU to resolve interference problems, which in turn would bolster the claims of lesser developed nations that developed countries have locked up too much of a limited and shared global resource.

The profusion of satellites does not necessarily guarantee that the market will absorb all additional transponder capacity without significant price cutting. Should capacity gluts and discounting arise, one could allege that an open skies policy promotes waste and inefficiency. The planning and deployment of shared international transmission facilities through cooperatives like INTELSAT could reflect the need to rationalize inventory with demand while spreading costs and risks over a large user population. Rather than constituting collusion or market preemption, the coordinated deployment and use of facilities may serve interests in efficient loading and shared access to a properly sized transmission medium.

Despite concerns for efficiency and the potential for wasteful duplication of resources, nations have begun to join the United States in supporting competition, including private carrier alternatives to submarine cable consortia and satellite cooperatives.[43] But the general view remains that competition can go but so far and that a laissez faire regulatory approach invites destructive competition[44] (i.e., temporary price reductions, but higher rates in the longer term as carriers exit the market and survivors recoup losses in the absence of rigorous competition). Most nations continue to support monopolies for basic services in domestic and international markets. Likewise, most nations monitor or actively participate in the planning for new capacity and the loading of cables and satellites with traffic.

The PTAT-1 cable [28] in the North Atlantic and North Pacific Cable [29] demonstrate that private carrier transmission facility alternatives may not result in wasteful duplication of facilities. Price wars have not occurred, nor are in-

---

ing the domestic carrier set to include Graphnet, Inc., but maintaining the domestic/international service dichotomy).)

43. See, for example, Toward Europe-Wide Systems and Services—Green Paper on a Common Approach in the Field of Satellite Communications in the European Community, COM(90) 490 final, November 20, 1990 (proposal for substantial liberalization in Earth station ownership and access to satellite capacity).

44. Destructive competition generates short-term price reductions followed by vastly higher rates as survivors of a market shakeout raise rates to recoup prior losses. Survivors may operate more efficiently or simply have deeper pockets, enabling them to tolerate longer the mounting financial losses that occurred when competition reduced rates to noncompensatory levels.

cumbent carriers unable to provide essential, noncompetitive services. Indeed, some of the carriers that have invested in traditional submarine cable and satellite ventures have agreed to acquire capacity or at least interconnect with alternative carriers, because they provide faster and cheaper circuit restoration options and some minor facility cost savings as well.

Notwithstanding cost savings, circuit redundancy, and enhanced circuit restoration options, many nations refrain from supporting alternative carrier and transmission facility options. U.S. carriers, which lack the common carrier status, present a new type of regulatory status unfamiliar or unattractive to foreign carriers and governments. Such private carriers may be perceived as lacking the proper bona fides, the commitment to long-term service availability, circuit restoration options, and reliability. Notwithstanding attractive prices, private carriers like Private Trans-Atlantic Telecommunications System, Inc., initially had difficulty in attracting carriers to end users. A more established carrier, Sprint, acquired the company [30], thereby generating greater credibility and acceptability for the venture.

17.7 REFORMS IN THE COORDINATION OF POLICY: THE MEETING OF PRINCIPALS

The international telecommunications policy-making process requires coordination and collaboration among federal agencies in addition to FCC administrative rulemakings. The latter obligates the government to operate "in the sunshine," with appropriate public notice and opportunities for the public to participate in the process. But the policy coordination process requires agencies with different constituencies and expertise to meet in private and to reach consensus on matters affecting the national interest and foreign relations. Should the agencies fail to reach consensus as a function of different perspectives and expertise or because of turf concerns, it may make sense to consolidate the executive branch functions into a council that can serve as a neutral and high-level broker to undertake long-range planning, manage policy coordination, and have a direct reporting channel to the President so that telecommunications policy may ascend to the level necessary for effective advocacy.

Without agency collaboration or consolidation of functions and power, foreign counterparts can continue to forum-shop, opting to pursue bilateral discussions with individual agencies perhaps on the basis of substantive weakness, disinterest, or shared viewpoints. Foreign governments have grown adept at dividing and conquering the fragmented U.S. policy-making process, primarily with an eye toward delaying initiatives that would make their markets more competitive and open.[45]

---

45. However, the ability of foreign governments to forum-shop does not justify any hasty or superficial reorganization. The growing importance of international telecommunications necessitates

The United States has achieved success in interagency consensus building through an underused and irregularly scheduled Meetings of Principals. Face-to-face meetings of the FCC chairman, NTIA associate administrator, and State Department director of the Bureau of International Communication and Information Policy build on the particular strengths each agency brings to the table. This process is particularly important when the United States must speak with one voice (e.g., at conferences of the ITU, assemblies of parties of INTELSAT and Inmarsat, and other international forums).

The Meeting of Principals has a successful, if limited, track record. When the United States sought to promote limited competition by satellite networks like PanAmSat, separate from INTELSAT, success was achieved through direct involvement by the principals rather than relying on staff issuance of instructions to Comsat, the exclusive representative of the United States to INTELSAT. The instructional process involves the government issuance of orders to Comsat on how to serve the national interest in INTELSAT and Inmarsat matters going beyond Comsat's business interest.

Without the direct involvement of principals and candid face-to-face discussions with senior officials of Comsat, the United States would have lacked an effective advocate of positions not in Comsat's corporate self-interest, but truly in the national interest. Comsat was directed to use its considerable influence in INTELSAT to persuade the cooperative to avoid using procedures designed to guard against economic and technical harm from separate systems as the vehicle to veto any degree of competition.

The agencies have used this process for extraordinary jawboning of Comsat and uncustomary consensus building. It could become a more frequent and regular process by which the principals and senior staff can share concerns, coordinate positions, and resolve petty turf concerns.

17.8 SETTING A FOUNDATION FOR PARITY THROUGH RECIPROCITY

For the most part, parity in market access is a trade negotiation objective. Nations may use trade concepts like national treatment to demonstrate that they do not discriminate against foreign enterprises, but such references can overstate

---

heightened attention and bolstered visibility in government, outcomes not likely to occur should telecommunications remain a relatively insignificant subject with the Departments of Commerce and State. The federal agencies should consider how to elevate international telecommunications issues in the policy-making process. They should also address how to engage in long-range planning as a substitute for ad hoc rulemaking that responds to chronic problems that have grown acute. To start this kind of endeavor, the federal agencies should consider H. Geller, *The Federal Structure for Telecommunications Policy*, Washington, D.C.: The Benton Foundation, 1989.

the actual extent of access. A nation may have closed procurements and monopolies making its markets inaccessible to both domestic and foreign companies.

The concept of parity of market access in telecommunications is not new. As far back as 1921, with enactment of the Submarine Cable Landing License Act, 47 U.S.C. Sections 34–39, the U.S. Congress has considered the importance of reciprocity in aspects of telecommunications. If a foreign nation permits a U.S. enterprise to land a submarine cable, then the United States will authorize enterprises from that nation to do so on U.S. soil. While some might want to see an "open seas" policy (i.e., any carrier can install and operate a submarine cable) to parallel its open skies satellite policy, the trade, industry, and national defense concerns of other nations prevent this. Any unilateral overture by the United States to further open markets could exacerbate the dichotomy in treatment. Yet any consideration of using regulatory safeguards to close the gap typically generates allegations of discrimination and denial of national treatment.

No nation can afford to open a market completely where other nations have policies that blunt comparative advantages. The matter of market access requires the patience and pragmatism displayed in international aviation, where market access opportunity results after painstaking negotiations with foreign governments who recognize the monetary value of market access. In aviation, only the Netherlands and Canada have accepted a U.S. invitation for a joint open skies policy, where any carrier of either nation can serve the other nation. Even with U.S. market access well exceeding the value of access to the foreign nation, such that reciprocal access does not have equivalent monetary value, all other nations tenaciously cling to negotiated arrangements that specify routes and carriers.

The international telecommunications marketplace does not significantly deviate from the aviation model. With limited exceptions, U.S. carriers cannot establish an operating presence in foreign locales to receive and manage traffic incountry. Artificially high accounting rates confer a financial windfall to foreign carriers, developing and developed alike. Foreign carriers appear to enjoy superior bargaining leverage and may have some opportunities to whipsaw U.S. carriers (i.e., play one carrier against another to secure monetary concessions). It appears that U.S. officials first ignored the accounting rate settlement problem and subsequently took an aggressive and activist posture. In international telecommunications, compromise and accommodation require ongoing multiyear dialogue to achieve even incremental progress.

The proper balance may lie in linking deregulation and further market access opportunities with reciprocal opportunities in specific countries. In 1992, the FCC adopted such a market- and country-specific deregulatory initiative. It now will selectively eliminate a dominant carrier designation of foreign carriers operating in the U.S. who do not have or abuse bottleneck control in their home locales.[46]

---

46. See Regulation of International Common Carrier Services, CC Docket No. 91-360, 7 FCC Rcd. 7331 (1992).

17.9 RECALIBRATING MARKET ACCESS OPPORTUNITIES IN THE U.S.

In view of mounting trade deficits in telecommunications and no apparent progress in fostering parity of access, national treatment may have become a red herring. The FCC and other federal agencies have focused their attention on parity of treatment between and among national and foreign enterprises, without any linkage to how this matter is addressed in foreign countries. In turn, the FCC and other agencies are reminded of their moral duty not to discriminate, something many foreign governments do on purpose.

It appears that the FCC, in conjunction with other agencies with a trade portfolio, will now consider using regulation as market access leverage and selective deregulation as a reward to progressive nations in the following areas:

- Establishing an operating presence in foreign locales, particularly for ventures that do not use frequency spectrum (e.g., installation of switches, resale of leased lines, construction of fiber-optic networks);
- Acquiring whole circuits and the ability to operate a network independent of the host PTT;
- Providing cost-based and nondiscriminatory interconnection rights, particularly for international value-added networks and market entrants;
- Reducing accounting rates to align them closer to actual costs;
- Relaxing alien ownership restrictions.

17.10 USERS' ABILITY TO RESORT TO SELF-HELP NECESSITATES ACTION

Accounting rate evasion, bypassing incumbent carrier facilities, creative traffic routing, and a host of technology-aided user schemes mean that national governments have limited time remaining to manage timely and incremental reforms. Without expedited consideration by the federal agencies and their counterparts on a bilateral and multilateral process, user-directed self-help schemes will override the ability of the ITU and other forums to manage change.

The international accounting rate issue provides an instructive case study. For years, the United States seemingly could afford to ignore a process increasingly out of synch with changing circumstances. But the pace at which the United States financially suffers from neglect has substantially increased as the net settlement deficit grew almost $2 billion to over $3 billion in just 4 years.[47]

---

47. See Regulation of International Accounting Rates, CC Docket No. 90-337, Phase I, Order on Reconsideration, 7 FCC Rcd. 8049, 8051 (1992).

The population of the United States, its multicultural composition, and its multinational corporate enterprises contribute to approximately half of the accounting rate settlement deficit incurred by U.S. carriers. However, the other half reflects the fact that foreign countries retard demand for switched telephone services into the United States. They do this by imposing excessively high collection charges, promoting outbound private lines in lieu of switched services, refusing to authorize facilities-based competition and inbound reseller access, and insisting on retaining high accounting rates despite technological innovations that have substantially reduced the per-unit cost of service.[48] Without conscientious efforts by carriers and governments to reduce accounting rates, users will take matters into their own hands by using a variety of legal, illegal, and gray-area tactics to avoid accounting rate liability while still securing access to the PSTN.

17.11 CONCLUSION

The U.S. track record in international telecommunications policy shows a long period of pervasive regulation that segmented markets, tolerated or promoted monopolies, and used public utility ratemaking. In the late 1970s and early 1980s, deregulatory initiatives gathered momentum, but for some issues the U.S. government seemed to have allowed anachronistic policies to remain (e.g., service dichotomies that artificially segmented the market and limited the number of carriers authorized to provide service).

Without coordinated efforts by governments to foster cost-based services, diversity, and competition, high-volume users will have incentives to find ways for avoiding excessive rates, service limitations, and carrier unresponsiveness. Already both legitimate and questionable enterprises provide new and creative solutions to service or price barriers erected by incumbents. This market will grow as more users and entrepreneurs direct their attention to bypassing the current regime.

Similar kinds of domestic bypass threats forced the FCC to revamp the pricing of local facilities interconnection to avoid encouraging users to pursue less efficient but cheaper routing options. The international telecommunications marketplace has the same kind of tariff and pricing anomalies. Yet in this instance the FCC cannot unilaterally act to remedy an international matter involving sovereign nations. Accordingly, national regulatory authorities need to revamp the terms and conditions for market access. The failure to do so threatens the very foundation by which sovereign carrier correspondents match half

---

48. See R. Frieden, "International Toll Revenue Division: Tackling the Inequities and Inefficiencies," *Telecommunications Policy*, Vol. 17, April 1993, pp. 221–233.

circuits. Without immediate and conscientious action to foster competition, or at least to eliminate unreasonable service limitations, such as the inability to use a leased line to route both voice and data traffic, the sovereign customer will resort to self-help that in time could divest carriers of network control and the ability to fully control market access and network interconnection. Unequal market access has enabled some carriers and governments to extract a larger share of the financial benefits accruing from deregulation. Heightened efforts to foster parity of market access will balance the flow of benefits and will also reduce incentives for users to resort to self-help through black or gray market options.

The impact of the different and inconsistent national regulatory policies is enormous. No nation can extend their domestic policies and regulations. Looks are deceiving—what appears to be a market access initiative may do just the opposite:

- Open network architecture (U.S.) versus open network provision (EC);
- GATT and the concept of national treatment;
- Standardization and equipment testing/certification (e.g., ETSI).

Even the U.S. has protectionist policies:

- Alien ownership restrictions and greater regulatory burdens;
- Monopoly creation (e.g., Comsat, AMSC);
- Limited, conditional, and incremental approach to change (e.g., separate system competition with INTELSAT);
- Policies have often favored maintaining the Rule of Multiple Cs: balanced loading, USG participation as observers in the consultative process, sponsoring the AT&T-Comsat circuit activation agreement, no mandatory IRU/MAUO conveyancing, uniform settlements policy.

U.S. procompetitive initiatives often fail to achieve foreign support:

- Separate system policies;
- Shared use and resale;
- Private carriage as opposed to common carriage;
- Transborder satellite operations;
- CPE unbundling;
- Cost-based charges for access to the PSTN;
- *Computer Inquiries*;
- ISDN interfaces.

References

[1] Market Entry and Regulation of Foreign-Affiliated Entities, IB Docket No. 95-22, Notice of Proposed Rulemaking, 10 FCC Rcd. 4844 (1995).

[2] Regulatory Policies and International Telecommunications, CC Docket No. No. 86-4994, Notice of Inquiry and Proposed Rulemaking, 2 FCC Rcd. 1022 (1987); Report and Order and Supplemental Notice of Inquiry, FCC 88-71 (rel. 25 March 1988); on recon., 4 FCC Rcd. 323 (1989).

[3] Market Entry and Regulation of Foreign-Affiliated Entities, IB Docket No. 95-22, Notice of Proposed Rulemaking, FCC 95-53, 1995 FCC Lexis 1124 (rel. 17 February 1995).

[4] Ibid., Sec. 706(2)(a).

[5] American Telephone and Telegraph Co., 75 FCC 2d 682 (1980); Overseas Communications Services, 84 FCC 2d 622 (1980), modified, 92 FCC 2d 641 (1982) aff'd sub nom., Western Union Int'l v. FCC, 673 F.2d 539 (D.C. Cir. 1982).

[6] Western Union Int'l, Inc., 76 FCC 2d 166 (1980), aff'd sub nom., Western Union Int'l v. FCC, 673 F. 2d 539 (D.C. Cir. 1982).

[7] International Record Carriers' Scope of Operations, 38 FCC 2d 543, 545 (1972) (describing the gateway concept), relaxed to include new 21 gateway cities, 76 FCC 2d 115 (1980), on recon., 80 FCC 2d 303 (1980), aff'd sub nom., Western Union Tel. Co. v. FCC, 665 F.2d 1112 (D.C. Cir. 1981).

[8] American Telephone and Telegraph Co., 74 FCC 2d 682, 688 (1980) aff'd sub nom., Western Union International, Inc. v FCC, 673 F.2d 539 (D.C. Cir. 1982) (removal of restrictions on the use of international telephone lines for data applications).

[9] Regulatory Policies Concerning Resale and Shared Use of Common Carrier Domestic Public Switched Network Services, 83 FCC 2d 167 (1980), on recon., 86 FCC 2d 820 (1981).

[10] Regulatory Policies Concerning Resale and Shared Use of Common Carrier Services and Facilities, 60 FCC 2d, pp. 298–99.

[11] International Resale Policy, 77 FCC 2d 831 (1980), proceeding terminated in Regulation of International Accounting Rates, CC Docket No. 90-337, Phase II, First Report and Order, 7 FCC Rcd. 559 (1992).

[12] Graphnet Sys., Inc., 63 FCC 2d 402 (1977), aff'd sub nom., ITT World Communications, Inc. v. FCC, 595 F.2d 597 (2d Cir.), on remand, 71 FCC 2d 1066 (1979); International Relay, Inc., 77 FCC 2d 819 (1980), on recon., 82 FCC 2d 41 (1980).

[13] Policy for the Distribution of United States International Carrier Circuits Among Available Facilities During the Post-1988 Period, CC Docket No. 87-67, Notice of Proposed Rulemaking, 2 FCC Rcd. 2109 (1987), Report and Order, 3 FCC Rcd. 2156 (1988) (abandoning circuit distribution guidelines).

[14] Pan American Satellite Corp., 101 FCC 2d 1318 (1985) (conditional grant), 2 FCC Rcd. 7011 (1987) (final C-band authority), 3 FCC Rcd. 677 (1988) (final Ku-band authority).

[15] Tel-Optik, Ltd., 100 FCC 2d 1033 (1985).

[16] Application of Pan American Satellite Corporation for Authority to Construct, Launch and Operate a North Atlantic Region Satellite System, 101 FCC 2d 1201 (1985); see also, Establishment of Satellite Systems Providing International Communications, 101 FCC 2d 1046 (1985).

[17] Modification of Policy on Ownership and Operation of U.S. Earth Stations That Operate With the INTELSAT Global Communications Satellite System, 100 FCC 2d 250 (1984), aff'd sub nom., Western Union Int'l v. FCC, 804 F.2d 1280 (D.C. Cir. 1986).

[18] Licensing Under Title II of the Communications Act of 1934, as Amended, of Private Transmit/Receive Earth Stations, 3 FCC Rcd. 1585 (1988) (declaratory ruling authorizing licensure of non-common carrier Earth stations under Title III of the Communications Act notwithstanding Section 201(c)(7) of the Communications Satellite Act which only expressly recog-

nized Comsat and other authorized user/carriers), aff'd sub nom., TRT Telecommunications, Inc. v. FCC, 876 F.2d 135 (D.C. Cir. 1989); policy applied in Reuters Information Services, Inc., 4 FCC Rcd. 5982 (1989).

[19] Authorized User Policy, 97 FCC 2d 296 (1984), reaff'd, 99 FCC 2d 177 (1985), aff'd sub nom., Western Union Int'l v. FCC, 804 F.2d 1280 (D.C. Cir. 1986).

[20] Tel-Optik, Ltd. (Private Submarine Cable Decision), 100 FCC 2d 1033 (1985).

[21] International Competitive Policies, 102 FCC 2d 812 (1985), on recon., 60 Rad. Reg. 2d (P&F) 1435 (1986).

[22] Telecom 2000, p. 165.

[23] Ibid., pp. 165–166.

[24] International Competitive Carrier, 102 FCC 2d 812 (1985), recon. den., 60 Rad. Reg. 2d (P&F) 1435 (1986).

[25] Third Computer Inquiry, CC Docket No. 85-229, Phase I, Report and Order, 104 FCC 2d 958 (1986), mod. on recon., 2 FCC Rcd, 3035 (1987), further recon., 3 FCC Rcd. 1135 (1988), second further recon., 4 FCC Rcd. 5927 (1989), Phase II, CC Docket No. 85-229, 2 FCC Rcd. 3072 (1987), on recon., 3 FCC Rcd. 1150 (1988), partially reversed and remanded sub nom., California v. FCC, 905 F.2d 1217 (9th Cir. 1990); on remand, CC Docket No. 90-368, 5 FCC Rcd. 7719 (1990), recon., 7 FCC Rcd. 909 (1992) petitions for review pending, California v. FCC, No. 90-70336 (9th Cir. filed 5 July 1990); see also Bell Operating Company Safeguards and Tier 1 Local Exchange Company Safeguards, 6 FCC Rcd. 7571 (1991); Filing and Review of Open Network Architecture Plans, CC Docket No. 88-2, Phase I, 4 FCC Rcd. 1 (1988), recon., 5 FCC Rcd. 3084 (1990), 5 FCC Rcd. 3103 (1990), erratum, 5 FCC Rcd. 4045 (1990), recon., 8 FCC Rcd. 97 (1993). Compare Comsat Study—Implementation of Sec. 505 of the International Maritime Satellite Telecommunications Act, 77 FCC 2d 564 (1980); and Changes in the Corporate Structure and Operations, 90 FCC 2d 1159 (1982), Second Mem. Op. and Order, 97 FCC 2d 145 (1984) (analysis of Comsat's corporate structure with an eye toward segregating competitive ventures into separate subsidiaries).

[26] Inquiry Into Policy To Be Followed in Future Authorization of Overseas Service, Notice of Inquiry, 36 FCC 2d 605 (1972), Report and Order, 57 FCC 2d 705 (1976), reviewed sub nom., ITT World Communications, Inc. v. FCC, 555 F.2d 1125 (2d Cir. 1977), policy applied, 75 FCC 2d 682 (1980), aff'd sub nom., Western Union Int'l v. FCC, 673 F.2d 539 (D.C. Cir. 1982) (allowing AT&T to provide data services); Western Union Int'l, Inc., 76 FCC 2d 166 (1980) (lifting restrictions on voice services provided by record carriers); Overseas Communications Services, 92 FCC 2d 641 (1982) (authorizing AT&T's entry into the international record services market).

[27] Domestic Communications Satellite Facilities, 35 FCC 2d 844 (1970).

[28] Tel-Optik, Ltd., 100 FCC 2d 1033 (1985).

[29] Pacific Telecom Cable, Inc., 2 FCC Rcd. 2686 (1987) (conditional authorization), 4 FCC Rcd. 8061 (1989) (final authorization and Cable Landing License grant upon showing that domestic entities had reasonable opportunity to participate in the planning, manufacture, installation, operation, and maintenance of the North Pacific Cable).

[30] U.S. Sprint Communications Co., Application to Operate the PTAT-1 Cable, 4 FCC Rcd. 6279 (1989) (approving application to operate the PTAT-1 cable); Private Trans-Atlantic Telecommunications System, Inc., 4 FCC Rcd. 5077 (approving transfer of Cable Landing License).

Prevailing Trends in International Telecommunications

18

Thanks in large part to initiatives of the United States and the United Kingdom, the international telecommunications enterprises have become more business-minded and policy makers have embraced limited alternatives to an absolute PTT monopoly. No longer can it be concluded that an absolute dichotomy exists between a "U.S. industry of entirely private companies facing a non-U.S. industry of government entities" [1]. Instead, PTTs are shedding their public utility, solely national orientation and scanning the world for investment and joint-venturing opportunities.

In growing numbers, PTTs have transformed into private or quasiprivate ventures aiming to build new telecommunications infrastructures for cable television, fiber-optic communications, mobile telecommunications, and digital switching at home and abroad. Both incumbent carriers and their government regulators recognize that strategic alliances in telecommunications, even if targeting foreign markets, stimulate the former PTT to become more efficient, competitive, and innovative at a time when market entry and deregulatory initiatives at home may cut captive revenue streams.

Significant changes in governmental policies and regulatory philosophies have occurred, spurred in large part by the impact of rapid technological innovation and the growing number of options available to sophisticated corporate users. A new environment has evolved that forces old-standing telephone monopolies previously "trapped under regulatory ice for a century...[to come] back to life as revolutionary new technologies merge the telephone with the computer and mandate rapid change" [2].

The incentive to experiment, innovate, and liberalize has increasingly less to do with economic theories and more to do with the dynamics of the computer-driven information age. Even if a PTT telecommunications manager would prefer to dispute the advice of a financial magazine author "to free up... [the] telecommunications sector or watch [the national] economy stagnate" [2], it would be harder to dispute the similar advice of an official publication of the European Organization for Economic Cooperation and Development (OECD) [3]:

> The rate of innovation in telecommunications transmission and switching equipment and in computerized information processing techniques has accelerated over the last several decades...Reflecting underlying technological change, traditional industry boundaries are changing. The marketplace is in a state of flux and small, medium and large-sized suppliers of equipment and services are proliferating, all seeking to capture a share of existing and new markets...Consensus with respect to policies, regulations, and standards applicable to telecommunications networks and services is becoming more difficult to achieve.
>
> The implications of a failure to develop a national and international policy environment that encourages the development of new services would be far-reaching.

International telecommunications and information processing has changed more in the last 10 years than in the preceding century. With increasing volatility and velocity, technological innovations and policy initiatives challenge the PTT model and the applicability of descriptive words like *consultation*, *cooperatives*, and *clubbiness*.

At some point, change will have become so institutionalized that the new world order can rightfully claim dominance over the old school. For the time being, however, one should consider recent changes in the context of forward-looking trends. Enough time has passed to support the view that changes are inevitable, regardless of a nation's political philosophy, telephone line penetration, and level of economic development. But more time must pass before these trends become institutions.

18.1 INCREASING CHALLENGES TO THE PTT MODEL

The PTT monopoly model will face increasing challenge as liberalization, privatization, and other experiments accrue measurable public benefits. Similarly, arbitrary service restrictions will lose both political support and economic justification as users find technological vehicles to evade PTT monitoring and enforcement of such restrictions. For example, PTT prohibitions on private line access to the PSTN are unenforceable when legislation orders the deregulation of CPE and its separation from telephone service. Users can install private branch exchanges that have the capability to "leak" into the PSTN by linking private lines with an inventory of local lines.

As PTTs and foreign carriers diversify and invest outside their home territories, government-mandated business orientation becomes perhaps a part of the PTTs' internal culture. When British Telecom acquired ownership interests in U.S. companies involved in electronic messaging, long-distance telephone, paging, cellular radio, and enhanced services [4],[1] it developed a global orientation, more reliant on open markets and strategic foreign alliances. Such alliances, whether through joint ventures or acquisitions, require open markets and the free flow of venture capital across borders. For every British Telecom foreign initiative, there exists a comparable alliance involving a U.S. company, some of which have targeted U.K. companies and markets.

Telecommunications equipment manufacturing initiatives by newly industrialized nations in the Pacific basin provide ample evidence that no nation's carrier or telecommunications industry can rest on its laurels. Lower labor costs combined with industrial policies targeting telecommunications for global marketplace initiatives means that national heroes will have to become more efficient because cheaper alternatives may become irresistible to users and governments may find it politically foolish to continue insulating the national hero from competition.

Perhaps the future threatens the very existence of the closed PTT model and policies aiming to maintain a closed market, or at least one characterized by substantial entry barriers. At the very least, the formerly cozy and often overprotected world of telecommunications will experience heightened pressure for carriers to bring all service prices closer to cost. Supracompetitive rates (i.e., prices above what a competitive market would support) may become unsustainable. The PTT network monopoly cannot operate with a business-as-usual attitude in view of broadened CPE choices, technological innovation, and the merger of telecommunications and information services. PTTs may not have unlimited opportunities to price services as a function of internal social policies if consumers can bypass overpriced services and erect their own networks.

Rather than exploit demand inelasticities of the few users who can afford to use an overpriced network (e.g., the foreign visitor who needs to call the home office), PTTs may find that new digital technologies and the information

---

1. Sec. 310 of the Communications Act, 47 U.S.C. Sec. 310 (1990), imposes restrictions on the extent to which foreign governments and private entities can invest in FCC licensees. Sec. 310(a) applies to any station license and prohibits ownership by foreign governments and their representatives. Sec. 310(b) only applies to a limited class of radio licenses and does not apply to private carrier operations. See Separate Satellite Systems, 101 FCC 2d 1046 (1985); Orion Satellite Corp., 6 FCC Rcd. 4201 (1991) (granting final approval of a financial structure involving a limited partnership where foreign investors, including foreign government participation, could own up to 87.5% in equity, provided the U.S. licensee retains control of the satellite facilities). For an extensive analysis of the Communications Act restrictions on foreign ownership and ways to evade such limitations, see R. Gavillet et al., "Structuring Foreign Investments in FCC Licensees Under Section 310(B) of the Communications Act," *California Western Law Review*, Vol. 27, 1990, pp. 7–50.

age promote policies supporting both competition and stimulated demand for incumbent carrier services (e.g., access to the PSTN). Even at the risk of losing a captive user population for some markets, PTTs may gain more than enough replacement revenues through new information services that still rely on underlying PTT facilities and services (e.g., the PSTN).

Some nations will cling to the status quo for as long as possible. The debate on liberalization will take place in the policy-making forums of the ITU and elsewhere even as the marketplace conducts its own referendum. Nations opening up markets will have an impact on indigenous manufacturers by eliminating captive markets. Loss of captive revenue streams (e.g., from the overpriced international telephone calls and closed procurements) means that incumbents must find replacement revenue sources and become more efficient. Even the need to conduct such a search can run counter to many entrenched, old-school managers. Nevertheless, the combination of official U.S. advocacy, sometimes shared by the United Kingdom and other nations, and corporate joint ventures and acquisitions points to gathering deregulatory momentum.

Global markets demand online, 24-hour-a-day telecommunications networks. When bankers, currency traders, and stockholders count on easy access to foreign markets, wherever they may be, national policies must strive to accommodate or risk alienating powerful constituencies. Governments must respond to the twin imperatives of technology and consumer sovereignty. Technology can distribute network control to end users through rooftop satellite dishes, modem-equipped computer terminals, and functionality previously resident only at the telephone company switching office. As a result, parochial policies over time must give way to a broader reliance on the marketplace and how individual carriers and national policies fit into the composite picture.

18.2 TECHNOLOGY CONVERGENCE, DIGITIZATION, AND COMPRESSION

Technological innovations serve as the catalyst for many of the trends shaping international telecommunications. While one cannot say that technology defines industrial structure or policy, more widespread access to innovations makes it possible for substantial changes to the status quo.

The direction and manner by which telecommunication technologies are evolving challenge conventional categories. Policy makers cannot hope to successfully compartmentalize innovations into fixed classifications serving as the basis for applying different levels of government regulation and approved market access. Nevertheless, many regulators strive to separate basic services from enhanced services by reserving a monopoly in the former for the incumbent PTT and allowing competition in the latter. All enhanced service operators use "plain vanilla" transport of traffic as building blocks over which technological enhancements ride. But the technologies that can enhance a basic service can also improve transport facilities and inject intelligence throughout the basic services network rather than in a particular location or piece of equipment.

Accordingly, it becomes difficult, if not impossible, for regulators to assign a particular regulatory classification for each technological innovation. Any such assignment would have the potential of preventing certain operators from applying a technological innovation to improve or diversify both basic and enhanced services.

Technological innovations also make it less feasible to categorize a technology by the industry sector or market it mainly serves. They make market segments less separate. For example, the conversion of signal transmission functions from an analog to digital format means that a single bit stream can carry voice, data, facsimile, news, information, and entertainment. A single transmission conduit makes it less possible to differentiate between:

- Broadcasting and downloading information;
- Entertaining and informing;
- Passive carriage and active programming;
- Advertising and catalog shopping;
- Cable television and telephone service.

When a technological innovation can serve several previously discrete markets, one can expect that operators in any one segment will consider entering additional, adjacent markets. Digital signal processing, which makes it possible to transmit hundreds of television channels simultaneously, also improves telecommunication services. This creates the potential for the cable television operator to serve new telecommunications markets like local exchange telephone service, personal communication networks connecting various cellular radio transmitting sites to central switching facilities, and alternative local access services that link telephone callers with long-distance telephone companies. It also creates the potential, in the absence of regulatory restrictions, for telephone companies to enter cable television, entertainment, and information delivery markets.

Innovations in data compression technology, coupled with digital transmission and deployment of broadband fiber-optic cable facilities, make massive increases in bandwidth and throughput feasible. Real-time access to video programming and remote databases stimulate the creation of additional bit streams and new ways to make the information available.

Compression technologies expand the volume of databases accessible, promote the ability for television broadcasting to include advanced, higher definition options, and make it more likely that telephone companies will serve video markets. The holy grail for incumbents as well as newcomers has become the ability to provide voice, video, online and offline data services, and other utility functions on an integrated and interactive basis using a single transmission and signal processing platform. To make it possible for one or more players to realize this strategic plan, regulations and policy must change. But perhaps even more important, a major change in marketplace and technological orientation must also occur: governments must abandon policies that perpetuate monopolies and reserve services so that full and fair competition can evolve.

18.3 HEIGHTENED RECOGNITION OF CONSUMER SOVEREIGNTY

Speaking at a 1992 conference on international telecommunications, Clay Whitehead, former director of the White House predecessor to the NTIA, proclaimed that "Karl Marx is alive and well, and living in the hearts and minds of PTT managers." The theoretical underpinning of a natural monopoly in telecommunications has lost popular and governmental support because of the attitude reflected in Mr. Whitehead's comment. Technological innovations prevent a single enterprise from claiming unimpeachable economies of scale, and a new businesslike passion spreads throughout the world.

Consumers have increasing latitude to vote with their feet and currency: to locate new installations in nations willing to make new services available and to deal flexibly with charges. Likewise, consumers can leverage their potential to bypass expensive PTT services and to migrate traffic and facilities from nations with less responsive carriers. As never before, international carriers have to compete. While they have yet to provoke a price war, carriers compete on service and vie particularly for multinational enterprises that require systems integration and network management functions. Carriers strive mightily to devise new offerings that convince the multinational enterprise to establish an incountry hub that will manage facilities and control routing for an entire region.

Consumers have greater sovereignty now, because they can choose from a larger number of carriers due to market entry and the growing distance insensitivity in traffic routing. The latter trend means that carriers previously considered too far away for consideration can now compete for traffic and network management projects (e.g., using an Australian carrier to route traffic between the United States and Japan). Loose or nonexistent prohibitions on discrimination among similarly situated users mean that some carriers can "sharpen their pencils" when responding to Request for Proposals from price-elastic, large-volume users. The absence of tariff filing requirements in some countries permits the private carrier to tailor a customized service arrangement that will not become known to others.

The balance of power has shifted from the carrier to the consumer in many transactions, particularly where users have options and traffic volumes desired by a number of carriers. All carriers, even ones insulated from direct competition, must accept the trend of heightened consumer sovereignty. High-volume, corporate users have begun to look at telecommunications as a cost center requiring close management and attention to cost-cutting maneuvers. Technological innovations have enabled users to consider and employ new options, including ones not offered by the incumbent. Some of these innovations enable sophisticated users to devise routing schemes that avoid overpriced services while still finding ways to leak into the PSTN, often as an unmetered, virtually free local call.

18.4 USERS HAVE GROWING AND DIVERSIFIED REQUIREMENTS

Users have expressed a growing need for telecommunications that is visual, personalized, mobile, intelligent, integrated, broadband, and digital. The visual component means that people desire to communicate through several senses, not just conventional auditory or textual means. New services accommodating these requirements include videoconferencing, virtual reality, high-definition imaging, telemedicine, and computer-assisted design and manufacturing.

The personalized components mean that users have particular requirements not well served by "plain vanilla" offerings. The array of services will have to proliferate further, and carriers may have to offer services on an a la carte, building-block basis thereby permitting users to customize services using an open architecture of telecommunications resources. Personalized services require carriers to relinquish a degree of centralized control and also to provide more user-friendly features that facilitate network usage (e.g., smart cards that facilitate metered network usage via pay telephones without currency problems).

The mobile component has grown in importance as users want ubiquitous access while on the move. Concerns for efficiency, responsiveness, and work productivity spur developments that enable mobile users to tap into mainframes and use computing and communicating devices while on the road.

The intelligent aspect of future networks signifies the development of advanced features that provide greater functionality, diverse features, and user control of the network. The vision of an ISDN provides the basis for satisfying vastly higher bandwidth requirements on a nearly ubiquitous basis.

One should not dismiss the overall trend of growing bandwidth requirements because of the premature or unsuccessful rollout of initial ISDN services or the current view that few residences will have requirements to load the two bearer channels and one data channel (approximately 144 Kbps) in ISDN's basic rate interface. In the near term, telephone companies will broaden their marketing wingspan to include cable television-type services.

We can expect to see growing demand in six existing applications: voice, bulk data transfer, electronic mail, facsimile, compressed viodeoconferencing, and internetworking of information processing networks, also known as *local area networks*. We also can anticipate new bandwidth guzzling applications like video imaging, file transfer, multimedia, and scientific computing.

18.5 GLOBALIZATION

Carriers have begun to expand their geographical horizons. In developed countries, domestic markets may have reached near saturation with limited prospects for growth, and the government may have authorized facilities-based competition or resale with the likely result of lost market share. International markets present growth opportunities well in excess of what most carriers have available at home. Incumbent carriers have a growing interest in serving new profit centers.

Integrated economies propel a global orientation. Many of the largest multinational enterprises seek assistance from carriers in network management so that company personnel can concentrate on *their* industry. These companies would prefer to limit the number of carriers with which they must do business, if anything to simplify the line of responsibility for network management. For a carrier to provide such one-stop shopping, it must establish a global presence or strategic alliances with carriers in other regions of the world. Since market access and foreign ownership restrictions may limit the opportunity to establish a direct operating presence, geographically diverse carriers are more likely to collaborate in joint ventures. A forward-looking trend may be increased advocacy in trade forums for the right to establish a foreign country operating presence, thereby obviating the need to share traffic and revenues with the local carrier.

18.6 LIBERALIZATION/DEREGULATION/PRIVATIZATION

An increasing number of governments have decided to revamp the telecommunications industrial structure and regulatory process. Some nations have privatized the incumbent carrier. A larger number have retained government ownership, but have sought to bring a more businesslike approach to the provision of telecommunication equipment and services. This is done primarily by authorizing competition in some markets, reducing regulatory oversight, and freeing the incumbent carrier to respond to competition.

Deregulatory, procompetitive, and efficiency-enhancing initiatives result when nations recognize that revamping the telecommunications sector can stimulate the national economy. A strong correlation exists between having a state-of-the-art telecommunications infrastructure and the ability to participate in information age markets like financial services and data processing.

Incumbent PTTs have lost the economic argument that they constitute a natural monopoly and that a single telecommunications franchisee can achieve economies of scale while serving social objectives like providing universal service at subsidized rates. Competition can be messy and will result in winners and losers. But more nations now believe that market entrants can generate consumer dividends rather than duplicate investment in plant, resulting in waste.

18.7 NOMADIC AND FUNCTIONAL PRIVATE NETWORKING

Customers typically care more about whether carriers can satisfy requirements than how they do it and the location of their network. With sufficient pricing incentives and assurances of service reliability and ease in reconfiguration, users will migrate to "virtual" private networks (i.e., capacity partitioned from public switched facilities by software applications). Carriers that cannot deliver on all three requirements risk losing users to the facilities of carriers than can provide better network management and price. At best, they will have to accept the consequences that most high-volume customers will want to maintain *closed user arrangements* separate from public switched services.

Changing customer requirements means that carriers must provide a speedy way to reconfigure networks. Intolerance of network outages means that carriers must also offer alternative and redundant routing that can be activated quickly. Carriers providing satellite routings typically secure restoration capacity on cables and other satellites. Carriers providing cable traditionally have relied on satellites to provide restoration as alternative cable routings have not existed. A new trend involves "self-healing" cable networks that have built-in more than one complete fiber-optic cable routing to the final destination.

Carriers initially devised private lines as an incentive for large-volume users to commit more traffic to telecommunications instead of postal and other media. Because new telecommunication facilities came online in large chunks of capacity, they typically began operation with plenty of dormant capacity. The marginal cost of activating additional circuitry in these facilities for usage-insensitive, unmetered service approached zero, provided the users of such private line service did not simply migrate traffic from other metered services.

Carriers and their high-volume customers have achieved a "win-win" arrangement with private line availability. The carrier could activate circuits without crowding out metered service users. Additionally, it could provide a cost savings to a relatively small set of users able to generate enough demand for multiple circuits of point-to-point service. Users have embraced the private line option, because it cost less, provided for customized, intracorporate networks, and supported a higher level of security.

Carriers willingly offer the private line option, provided they can limit availability to the most price-elastic users generating the most traffic volumes. Resale, shared use, and arbitrage options prevent the carrier from limiting private line availability and also preclude the carrier from offering preferential terms and conditions to a select group of users. With the introduction of government-mandated resale and technological innovations that have enabled users to devise shared-use arrangements irrespective of their legality, carriers are less keen on providing private lines.

To limit their revenue loss exposure from expanded access to private lines, carriers have devised new services that integrate the latest technological breakthroughs with public switched networks. Some of these innovations create a virtual private line out of switched public services. With software, a carrier can partition capacity from the PSTN, endow it with special features, and market it possibly on a metered basis to high-volume users who have previously used unmetered private lines.

Customers care more about functionality than the terms and conditions set forth in a tariff. Carriers can achieve migration from usage-insensitive services to metered services if the new service provides for greater network reliability and ease in reconfiguration at rates approximating the average private line rate. Software-defined networks promote speedy redeployment in cases of outages and changes in consumer requirements. On the other hand, conventional private lines are hardware-oriented and deployed with the expectation that they will remain in place, unchanged, for a significantly long time period. Given a heightened emphasis on quick deployment and reconfiguration, many customers have opted for new virtual private line networks.

18.8 TELECOMMUNICATIONS AS A COMMODITY SUBJECT TO NARROWING MARGINS

The old world order in international telecommunications blended elements of diplomacy and international relations with joint business ventures. Satellite cooperatives like INTELSAT and Inmarsat were created by intergovernmental agreement. Even submarine cable consortia integrated a diplomatic style with regularly scheduled consultations to project demand and schedule new facility deployment.

This high-minded and leisurely approach is juxtaposed with the prevailing view that telecommunications simply transports bit streams. While carriers have developed reputations based on efficiency and responsiveness, the services they render increasingly become a building block upon which customization takes place either by the carrier or the customer. This view characterizes telecommunications as a commodity subject to fierce competition with narrowing profit margins.

International telecommunications certainly has become more competitive, primarily resulting from technological innovations available for application by carrier and customer alike. Consumers have ascended the learning curve and have become adept at negotiating service and equipment bargains.

Large-volume users and the system integration companies they have retained use experts in tariff analysis and least-cost routing to devise strategies for minimizing cost and accounting rate liability. These specialists determine whether users can migrate to new, cheaper services that still meet all particular telecommunication requirements. While carriers must cater to these requirements, they run the risk of creating services with lower margins and the inability to prevent migration from more profitable services.

Government deregulatory initiatives may expand the potential for such cannibalization of high-margin services. More nations are considering policies that would expand opportunities for users to share facilities and for businesses to serve as capacity brokers. Underlying facilities-based carriers fear that they will lose the ability to price discriminate between users as a function of their traffic requirements. If a broker can lease discounted, bulk capacity at rates previously available only to single entities and resell subsets of that capacity to a number of users who individually lack traffic volumes qualifying for discounts, the facilities-based carriers will not be able to limit service bargains to an exclusive group of users. A broader set of users can now tap technological innovations to achieve least-cost routing and to exploit opportunities to bypass overpriced services.

18.9 STRATEGIC PARTNERS SHARE GOALS, BUT OFTEN COMPETE IN OTHER MARKETS

The international telecommunications market is fast losing the mutual exclusivity fashioned by foreign correspondents matching half circuits. Carriers want to acquire and manage whole circuits. They want to install switching facilities on foreign soil, which would enable them to provide end-to-end routing without the assistance of a strategic partner who would expect a share of the proceeds.

Carriers facing competition at home and the likely prospect of declining market share look abroad for new profit centers. Often, foreign ownership restrictions and the comparative advantage enjoyed by a national hero or regional carrier favor alliances among carriers. Many of the newly formed ventures contain carriers chosen primarily for their geographical market share and only secondarily for their expertise in provisioning networks.

Global alliances emphasize the ability of participants to serve the diverse requirements of multinational enterprises. The growing complexity of intracorporate networks favors one-stop shopping, turnkey networking and systems design by experts. Users will pay more for complete end-to-end network management by carriers on the assumption that they will be able to concentrate on their line of business, confident that the network manager can solve outage conditions on a timely basis.

Just because enterprises find it mutually advantageous to strike global or regional alliances does not attenuate their interest in exclusively capturing new market share where possible. The new world order in telecommunications rejects the previous view that carriers could consider incountry users as captive consumers. The old world concept of sovereign foreign correspondents linking half circuits has given way to a variety of new routing arrangements where the line of demarcation is less clear in terms of who is providing the service and what services foreign carriers can provide. Foreign carriers have devised new strategies to capture larger market share, achieved not by stimulating demand, but by migrating traffic streams to their control.

18.10 HUBBING

Hubbing represents the attempt to set up an operational presence where traffic from throughout a region routes into and out of facilities of a single carrier. This strategy follows the airline hubbing concept: build an efficient and high-speed facility through which traffic transits rather than route directly from point to point. In aviation, a number of international routes are served via hubs like New York, London, Frankfurt, Tokyo, and Singapore rather than on a direct basis. Hubbing achieves scale efficiencies by aggregating traffic from a number of origination points. In telecommunications, the traffic aggregation function at a hub will likely result in cost savings as traffic from a number of origination points is collected and routed to a single foreign destination over a higher capacity line with a lower per-unit cost.

18.11 POACHING

Carriers also seek to establish an operating presence outside their home country. Currently, restrictions on foreign ownership in most countries limit the manner in which carriers can establish hubs and other facilities. Few countries even permit foreign carriers to establish anything beyond a sales office to facilitate order taking in the foreign country.

Global carriers like AT&T, British Telecom, and Cable & Wirless have undertaken an aggressive campaign to reduce foreign operating restrictions so that they can install switches, originate or transit traffic, and use leased lines to erect global networks under their direct supervision.[2] Foreign carriers using these options can engineer services that appear to the user as if the carrier has direct and

2. See, for example, "U.S. International Carriers Oppose Grant of BT-NA Resale Applications, Say U.K. Interconnection Not Equivalent," *Telecommunications Reports*, Vol. 59, No. 18, 26 April 1993, pp. 22–24; and "AT&T Files 'Me Too' FCC Application for U.S./U.K. Resale," ibid., p. 21.

unimpeded control over facilities and lines throughout the world. Even if the foreign carrier interconnects leased private lines with facilities it owns, the end result is a seamless network capable of providing value-adding, software-defined enhanced services that large-volume corporate users seek. Carriers offering global network services can capture and manage a large multinational corporation's entire telecommunications requirements rather than only provide the lines needed to originate and terminate traffic in one country.

International carriers strive mightily to demonstrate their networking sophistication to highly prized multinational enterprises. Carriers can no longer assume that domestic companies will choose them to manage traffic functions abroad or even incountry. Some customers will opt to construct dedicated facilities or negotiate and lease lines from a number of carriers. Others will separate domestic traffic from international traffic and contract with different carriers. Still others will cut better deals with foreign carriers and assign to them the network management functions for both domestic and international routes. When the national carrier loses a key customer, it may allege "poaching" and predatory pricing. The winning carrier would view the outcome as the product of superior competitive skills.

18.12 INCREASED NETWORK COMPLEXITY REQUIRES MULTIPLE ROUTES AND CARRIERS

The old world order in international telecommunications sought to ensure network reliability by mandating carrier use of both satellite and submarine cable media. Regulatory agencies required balanced loading of facilities to ensure that carriers would activate circuits in the currently more expensive medium, with an eye toward supporting redundancy, security, and service reliability. Policy makers opted to blunt intermodal competition and market-based resource allocation, presumably because circuit activation based on the comparative merits of various transmission media might discourage investment of promising, but currently more expensive media. Such intervention in circuit activation stemmed from the belief that regulators had to prop up a new, more costly technology even if it lacked marketplace support.

The new world order shows that sovereign consumers demand facility redundancy and alternative routing. Governments do not have to mandate self-healing networks that can speedily route around an outage. Customers demand it. Historically, satellite users have willingly paid more for backup "restoration" transponders, preferably on a different satellite. Submarine cable users have expected satellite restoration in the event of a cable cut.

Increased corporate networking requirements, including more widespread and higher volume international traffic, have encouraged carriers to bolster routing options. Submarine cables in all ocean regions will soon have other cables as a real-time option for routing traffic during an outage.

18.13 EXPANDED OPTIONS FOR MOBILE USERS

Technological innovations in mobile communications present the near-term prospect for service any time, any place via handheld terminals. The unprecedented marketplace success of cellular radio and other mobile technologies confirm our desire to stay in touch while on the move. But for the time being, only terrestrial islands of local, cellular, and special mobile radio services exist. Even when nationwide cellular roaming becomes possible, a variety of different operating standards will limit the prospect for using the same transceiver when traveling abroad.

Low- and middle-Earth-orbiting satellite projects will make personal communications global in scope. These ventures include a constellation of nongeostationary orbiting satellites providing an interoperating array of beams that illuminate the entire globe. Individually and collectively, these systems aim to provide ubiquitous, wireless, digital coverage to pocket-sized telephones.

The concept of global PCS provides users with flexible communication options free of cords, using terrestrial radio options, where available, augmented by more expensive satellite-delivered services. Numerous logistical and regulatory problems must be resolved to make this vision a reality, but by most accounts the demand exists. A number of firms have plans to expand the availability of terrestrial mobile radio options by reducing the size of transmission cells from several miles, as is the case in cellular radio, to several hundred yards, as will be the case with micro- and picocell-delivered personal communications networks.

18.14 ENTERPRISE NETWORKING

Multinational corporations have rapidly expanded the scope and complexity of their international networking requirements. Carriers have responded with a broader range of technologies and networking products, particularly ones that operate over fiber-optic cables and satisfy the high-speed data networking requirements and burstiness of most data networks. Carriers will also need to provide enterprise networks that enable customers to link different types of local area networks operating under different standards and protocols.

The complexity, sophistication, security, and bandwidth requirements of enterprise networks strain the functional characteristics of publicly available services like international direct distance dialing and switched 56-/64-Kbps data lines. Carriers will have to deploy new standardized technologies like frame relay, synchronous optical networking (SONET), ATM, and fiber distributed data interface (FDDI) to upgrade the set of available private line, virtual private line, and public services.

Many enterprise network users have opted to rely on outsourcers and systems integrators for management services, but few have confidence that any organization, even the carrier providing part of the network, can provide problem-free management. Such functions are inherently difficult, because user requirements are often individualized and subject to change. Enterprise networking must provide the flexibility for quickly implementing changes as network requirements evolve, including the shift upward or downward in the amount of throughout (bandwidth and transmission rate) at any single location. It must also provide reliability, meaning that alternative and redundant routing must be incorporated into the network topology.

Multinational enterprise networks also require the participation of many national carriers with significantly different levels of expertise, network quality, and service diversity. The absence of an international standard or a carrier's decision not to deploy a state of the art and possibly unrefined technology can mean that an enterprise network will have interconnection and internetworking problems.

The variables affecting where to establish a regional hub for an enterprise network include price, feature availability, and qualifications of the prospective hub manager. Hubbing requires traffic engineering studies and the ability of the manager to work with other carriers in the region to accommodate the user's requirements. Users of enterprise networks may have the ability to extract rate concessions, bargain for a higher level of service quality and maintenance, and secure favorable terms for active facilities management.

18.15 A WORD OF CAUTION: THE NEW WORLD ORDER MAY HAVE BEEN OVERSOLD

A forward-looking trend analysis tends to look for change, thereby running the risk of wishful thinking and self-fulfilling prophesies. Currently, one would overstate the scope of change by concluding that the PTT model has no relevance and that nations throughout the world have decided to privatize the PTT, deregulate, liberalize operational rules for incumbent carriers, and encourage a global orientation.

The pace of change has accelerated and the international telecommunications marketplace looks more like a market than a government-operated venture like the postal system. But the fundamental nature of international telecommunications has not changed, particularly the costly social goals nations typically want carriers to achieve. Carriers must strive for universal service, especially in nations where the infrastructure is unreliable or nonexistent in rural areas. Even private carriers must accept some degree of social engineering that for telecommunications means the deliberate underpricing of some services and the overpricing of others.

Many nations still have fewer than 10 telephones per 100 inhabitants. In many developing countries, residents still joke about half of the population waiting for a telephone and the other half waiting for dial tone. When considering telecommunications trends, one may concentrate on what developed nations are doing to satisfy the complex requirements of multinational enterprises. However, a large portion of the world, while keen on attracting and serving foreign corporations, has to pursue an ambitious schedule of projects to improve the basic telecommunications infrastructure. Policy makers in these countries have to balance efforts to attract foreign investment in overlay networks to serve the expatriate and business community, with investments in basic telephony for the general populace. A nation like Mexico, which has opted for privatization, has the burden of ensuring that the newly corporatized telephone company does not abandon infrastructure and service improvements to squeeze out higher short-term profits and shareholder dividends.

The new world order promises greater volatility, change, and innovation. But old world imperatives have not gone away, because most nations have not come close to achieving social goals, and the citizenry will not let the government forget it.

Finally, the top 10 trends in international telecommunications are:

1. *Heightened recognition of consumer sovereignty.* New marketplace imperatives challenge the notion that Karl Marx is alive and well—in the hearts and minds of PTT managers.
2. *Globalization.* Integrated economies and the search for new profit centers propel an international orientation;
3. *Privatization.* Increasingly, nations seek ways to stimulate efficiency in the provision of telecommunication equipment and service. Note, however, that governments may simply opt to create a private as opposed to public monopoly.
4. *Liberalization/deregulation.* Many nations that have privatized or corporatized the PTT have also reduced regulatory burdens on the incumbent and barriers to market entry.
5. *Nomadic and functional private networking.* Customers care more about functionality than geography- or regulator-mandated divisions of the market. They require network reliability, ease in reconfiguration, and volume discounts. Carriers who cannot deliver on all three requirements will see users more inclined and better able to "vote with their feet" as companies relocate or rely on new hubbing facilities provided by carriers operating in more competitive and flexible business environments.

6. *Sophisticated users will sniff out telecom bargains everywhere.* Large-volume users, or their outsourcers, know about least-cost routing, bypass opportunities, accounting rate evasion, shared use and resale, and a variety of other white, gray, or black market tactics. Carriers may have to introduce new services, like virtual or software-defined networks, even though they may cannibalize existing services by migrating existing users to new services that generate lower revenues.

7. *Even strategic partners will poach customers.* Notwithstanding all the talk and press about strategic alliances, individual carriers will take steps to capture a larger percentage of international, regional, and transiting traffic.

8. *Strategic alliances and one-stop shopping will proliferate.* Despite false starts and hype, a number of strategic alliances will serve multinational, high-volume corporate users.

9. *Wireless networking will grow in importance.* Wireless telecommunications will serve applications with mobile users or ones willing to pay for service otherwise unavailable or unable to achieve the same level of productivity enhancement.

10. *Increased network complexity requires multiple routes and carriers.* Technological necessity, in addition to economic theories, support network redundancy, alternative routes, and a multicarrier environment.

References

[1] Goldberg, H., "One-Hundred and Twenty Years of International Communications," *Federal Communications Law J.*, Vol. 37, 1985, pp. 131, 132.

[2] Grisby, "Global Report: Telecommunications," *Financial World*, 18 April 1989, p. 32.

[3] Organization for Economic Cooperation and Development, *Telecommunications Network-Based Services: Policy Implications*, Paris, 1989, p. 13.

[4] McCaw Cellular Communications, Inc., Concerning Compliance With Sec. 310(4) of the Communications Act (alien ownership) With Regard to an Investment in McCaw by British Telecom, 4 FCC Rcd. 3784 (1989).

Appendix A
Transmission Capacity Forecast

Table A.1
Estimate of Voice Grade Capacity in the Atlantic Ocean Region (1993 Baseline)

Name of Cable (FCC Site)	Total Bearer Circuits	Equivalent Voice Paths
CANTAT-2	2,000	10,000
TAT6-7	2,969	11,876
TAT-8	7,560	30,240 (4:1 DCME)
PTAT-1 100 FCC2d 1033 (1985)	17,010	68,040
TAT-9 4 FCC Rcd. 1129 (1988)	15,255	61,020
TAT-10 7 FCC Rcd. 445 (1992)	30,240	151,200
TAT-11 7 FCC Rcd. 136 (1992)	22,680	113,400

Table A.2
Estimate of Near-Term Additions of Voice Grade Capacity in the Atlantic Ocean Region
(1994–1996)

Name of Cable (FCC Site)	Total Bearer Circuits	Equivalent Voice Paths
CANTAT-2	2,000	10,000
TAT 6-7	2,969	11,876
TAT-8	7,560	30,240
PTAT-1	17,010	68,040
TAT-9	15,255	61,020
TAT-10	22,680	113,400
TAT-11	22,680	113,400
CANTAT-3	60,480	302,400
COLUMBUS-2/TCS-1 8 FCC Rcd. 5263 (1993)	Min. 7,560/max. 12,096	Min. 37,800/max. 60,480
TAT-12/TAT-13 8 FCC Rcd. 4810 (1993)	120,000	600,000
PTAT-2	60,480	302,400

Table A.3
Estimate of Voice Grade Capacity in the Pacific Ocean Region (1993 Baseline)

Name of Cable (FCC Site)	Total Bearer Circuits	Equivalent Voice Paths
HAW-3/TPC-2	169	845
HAW-4/TPC-3	7,560	30,240
NPC-1	17,000	68,000
HAW-5 5 FCC Rcd. 7344 (1990); connects with PacRim E/W, 5 FCC Rcd. 7331, 7362 (1990); GPT, HJK, Tasman-2, 6 FCC Rcd. 2958 (1991), etc.	15,120	75,600
TPC-4 4 FCC Rcd. 8042 (1992)	15,255	61,020 (assumes 4:1 DCME)

Table A.4

Estimate of Voice Grade Capacity in the Pacific Ocean Region (1994–1996)

Name of Cable (FCC Site)	Total Bearer Circuits	Equivalent Voice Paths
HAW-3/TPC-2	169	845
HAW-4/TPC-3	7,560	37,800
NPC-1	17,000	68,000
HAW-5 5 FCC Rcd. 7344 (1990); connects with PacRim E/W, 5 FCC Rcd. 7331, 7362 (1990); GPT, HJK, Tasman-2, 6 FCC Rcd. 2958 (1991), etc.	15,120	75,600
TPC-4 4 FCC Rcd. 8042 (1992)	15,255	61,020 (assumes 4:1 DCME)
TPC-5/6 7 FCC Rcd. 7758	120,960	604,800
APCN	Est. 120,960	Est. 604,800
SE-ME-WE-3 or FLAG	Est. 120,960	Est. 604,800
CANPAC-1	Est. 120.960	Est. 604,800

Estimate of Near-Term Demand for Voice Grade Capacity in the Atlantic Ocean Region
(1994–1996)

Documents filed with the FCC estimate an aggregate demand of 29,696 bearer circuits (148,480 virtual voice paths) in 1996. The FCC calculates that this constitutes 74% of the estimated total capacity available. The 1996 estimate of total AOR cable capacity in 1996 is 200,649 virtual voice paths (40,130 bearer circuits). FCC estimates for the year 2000 have ranged from 498,578 to 583,870 virtual voice paths.

Source: 8 FCC Rcd. 4810 (1993) and carrier filings.

Table A.5
Estimate of Near-Term Demand for Voice Grade Capacity in Pacific Ocean Region (1994–1996)

Year	Demand in MAUOs	Equivalent Voice Paths
1996	16,508	82,540
1997	20,998	104,990
1998	23,998	119,990
1999	27,798	138,990
2000	31,695	158,475

Source: 7 FCC Rcd. 7758 (1992) and carrier filings.

Table A.6
Carrier Estimates of Demand Ramp-Up for New AOR Submarine Cables

Year	Cable	Est. of Demand Half-MAUOs
1992	TAT-9	7,090
	TAT-10	2,203
1993	TAT-9	8,324 (81% of capacity)
	Total AOR Cables	14,648
1996	TAT-10	8,119
	TAT-11	1,220
1997	All AOR cables except TAT-12/13	29,696
	Total AOR Cables	44,539
2000	Total AOR Cables	56,774

Table A.7
Carrier Estimates of Demand Ramp-Up for New AOR Submarine Cables

Year	Cable	Est. of Demand Half-MAUOs
1998	TPC-5	23,998
2000	TPC-5	31,695 (52.4% of total capacity)

Table A.8
Estimate of Submarine Cable Costs Atlantic Ocean Region

Cable	Cost per Half MAUO
TAT-8	$22,200 per half MAUO
TAT-9	$10,000–14,000 per half MAUO
TAT-10	$10,000 per half MAUO
TAT-11	$9,600 per half MAUO
TAT-12/13	$6,300 per half MAUO original capital cost $3,100 per half MAUO at design capacity $1,240 per voice grade channel with DCME

Table A.9
Estimate of Submarine Cable Costs Pacific Ocean Region

Cable	Cost per Half MAUO
HAW-4/TPC-3	$40,000 per half MAUO
TPC-4	$11,545 per half MAUO at design capacity

TPC-5	Original capital cost:
	  Guam-Japan—$3,000
	  U.S. mainland-Hawaii-Guam—$11,900
	  U.S.-Japan—$9,000
	At design capacity:
	  U.S. mainland-Hawaii-Guam—$8,000
	  U.S.-Japan—$4,900
	Voice grade channel with DCME:
	  U.S.-Japan—$1,960

Appendix B
Telephone Line Penetration in Selected Countries

Table B.1
Highest Teledensities in Selected Countries (1993)

Country	Lines per 100 in Population	Total Main Lines (millions)
Sweden	68.3	5.9
Switzerland	62.8	4.3
Canada	59.7	16.3*
Denmark	58.9	3.1
United States	57.8	148.1
Luxembourg	56.7	0.2
Finland	54.9	2.8
Iceland	54.7	0.1
France	53.9	30.8
Norway	52.9	2.3
Netherlands	50.3	7.6
Hong Kong	50.2	3.0
Australia	47.8	8.5
Greece	46.9	4.7
Japan	46.2	57.7*
Germany	46.2	36.9
United Kingdom	46.1	26.6
New Zealand	45.8	1.6
Belgium	44.0	4.4
Cyprus	43.2	0.3
Italy	41.8	24.2

* Based on data from 1992.
Source: ITU.

Table B.2
Lowest Teledensities in Selected Countries (1993)

Country	Lines Per 100 in Population	Total Main Lines (millions)
Bangladesh	0.2	0.3
India	0.9	8.0
Sri Lanka	0.9	0.2
Indonesia	1.0	1.8
Pakistan	1.3	1.6
Philippines	1.3	0.9
China	1.5	17.3
Morocco	3.0	0.8
Peru	3.0	0.7
Cuba	3.2	0.3*
Thailand	3.7	2.2
Algeria	3.9	1.1
Egypt	4.3	2.4
Tunisia	4.9	0.4
Ecuador	5.4	0.6
Iran	5.8	3.6
Brazil	7.4	11.7
Columbia	8.2	2.8*
Mexico	8.4	7.6
Saudi Arabia	9.5	1.6
South Africa	9.5	3.7
Venezuela	9.9	2.1

* Based on data from 1992.
Source: ITU.

Table B.3
International Message Telephone Traffic Growth for Selected Countries (1990–1992)

Country	Growth Rate
Argentina	39.0%*
Columbia	16.6%
Mexico	27.4%
Venezuela	16.4%
Bulgaria	26.6%*
Czechoslovakia	39.0%*
Greece	18.4%
Italy	18.8%
Portugal	16.4%
United Kingdom	6.1%
Israel	13.9%
Saudi Arabia	13.3%*
South Africa	9.4%*
Turkey	19.6%
China	44.3%*
Hong Kong	24.9%
India	33.0%
Indonesia	18.1%*
Japan	23.0%
Malaysia	22.1%
Singapore	19.8%*
South Korea	27.5%
Taiwan	31.9%
Thailand	19.8%*
Unweighted average	*22.72%*

*   Rate reflects 1991–1992 only.

Glossary

Accounting rate A unit of currency negotiated by international carriers as the amount of compensation to cover the cost of completing an international call. This amount includes all satellite or submarine cable transmission costs and the use of domestic tail circuits used to link an international gateway with the call originator and recipient. Carriers typically divide the accounting rate in half when settling accounts.

Ad valorem tariff A tariff calculated as a percentage of the value of foreign goods clearing customs.

Additional Plenipotentiary Conference (APP) An extraordinary Plenipotentiary Conference of the International Telecommunication Union held in 1992 to consider recommendations on how to streamline operations. The AAP significantly revised the Constitution and Convention of the ITU.

Administrative Council An elected body of the ITU that performs executive board functions, including the scheduling of conferences.

Administrative Procedure Act (APA) An act of the U.S. Congress setting forth required procedures that administrative and regulatory agencies must use to provide for public participation and due process in the setting of rules and policies.

Advance Publication The process by which nations inform other nations of future satellite orbital arc requirements through submission of information about the satellite and the desired orbital location. The Radio Regulations Board, formerly known as the International Frequency Registration Board of the ITU provides the forum for dissemination of orbital arc requirements and avoidance of resolution of actual or potential interference between satellites.

Algorithm The application of a computing principal that sets the foundation for processing or transmitting data.

Alternative dispute resolution (ADR)   The use of conflict resolution alternatives to litigation and conventional administrative channels, including negotiations managed by a private facilitator or mediator.

American National Standards Institute (ANSI)   A nonprofit organization addressing standard setting, primarily by certifying other expert bodies to formulate standards in a narrow area of expertise.

Analog   Information transmitted by varying modulation over a radio carrier wave. Voice and music originate in analog form and are converted for transmission over digital facilities like fiber-optic cables.

Arbitrage   A brokering function where a business acquires bulk capacity and resells it to individual users who singularly could not qualify for the discounts accruing to large-volume customers.

Arianespace   The world's first commercial space transport company, established in 1980 by 36 European aerospace and electronics companies, 13 major banks, and the French space agency. The company launches satellites from a facility in Kourou, French Guyana.

Asia-Pacific   The region encompassing the Pacific Rim nations and countries bordering the Indian Ocean. This region has experienced the greatest economic growth, which in turn has stimulated explosive development of telecommunications infrastructures.

Assembly of Parties   The forum where representatives of nations that have associated with the INTELSAT Agreement and Inmarsat Convention meet to address major policy issues involving the cooperatives.

Association of Southeast Asian Nations (ASEAN)   A forum of Southeast Asian nations for addressing issues of mutual economic, technical, and trade issues. Many of the ASEAN nations have leased satellite capacity on the Palapa satellite system operated by the Indonesian government.

Asynchronous transfer mode (ATM)   An advanced form of data transport whereby users can dynamically select the amount of bandwidth and transmission speed required and change the assignment as conditions warrant.

Atlantic Ocean Region (AOR)   One of three major geographical regions, identified by a prominent body of water, that outline the largest possible satellite footprint. A satellite hovering above the equator midway between North America and Europe will have a footprint illuminating all of the Atlantic Ocean and most of North, Central, and South America; Europe; and Africa.

Automatic number identification (ANI)   A feature in advanced telecommunication routing that provides the calling party's telephone number as the call is set up. Telephone companies have packaged this feature as a way to screen calls in residences and a marketing and billing tool in commercial applications.

Autonomy The assertion of independence and sovereignty, which if recognized by other nations confers the right to participate in the ITU, INTELSAT, and other forums where policy and investments decisions are made.

Balanced loading A regulatory policy requiring international carriers to activate satellite and submarine cables on a prescribed ratio or an even basis with an eye toward bolstering use of the more expensive or developing transmission technology.

Band A term that refers to a specific range of frequency spectrum. For example, the C-band refers to frequencies used to uplink and download traffic to satellites) around 6 GHz uplink and 4 GHz downlink.

Bandwidth The total range of frequencies necessary to accommodate a spectrum using signal without distortion. Typically, bandwidth requirements grow as the amount of information to be transmitted increases. For example, an AM radio signal contains 10 kHz, while an FM channel contains 200 kHz and a broadcast television signal contains 6 MHz.

Bandwidth on demand The ability of users to adjust up or down the amount of capacity leased from a carrier, primarily through new technologies that make it easier to aggregate bit streams.

Barriers to market entry Structural, regulatory, and trade limitations to competition by domestic or foreign enterprises. Such barriers include prohibitions or limitations on the amount of foreign investment in a telecommunication service provider, reserving service monopolies for the incumbent carrier, imposing duties and other financial impediments on the equipment or services, and subjecting foreign carriers and manufacturers to more burdensome certification, testing, and licensing burdens.

Basic rate interface (BRI) The least common denominator of capacity in ISDNs, composed of two 64-Kbps bearer circuits and one 16-Kbps data channel.

Basic services The switching, routing, and transmission of voice or data provided by facilities-based carriers traditionally subject to common carrier regulation. Enhanced-service providers use basic services such as "plain vanilla" building blocks over which addition services and features are added.

Basic serving arrangement (BSA) A term used by the FCC in its *Third Computer Inquiry* to identify the components in generic connections between enhanced-service providers and facilities-based carriers providing basic services. BSAs consist of the access links between facilities as well as the transport, routing, and functions.

Basic serving elements (BSE)   A term used by the FCC in its *Third Computer Inquiry* to identify optional network features, available on an unbundled, a la carte basis such as ANI.

Bearer circuit   A transmission pathway that can be subdivided into channels of less bandwidth or data throughput through multiplexing. For example, a bearer circuit with a total capacity of 64 Kbps can be subdivided into four channels with the capacity to transmit one voice grade channel.

Beggar-thy-neighbor policy   Attempts by one country to reduce unemployment and to increase domestic output by raising tariffs and erecting nontariff barriers to reduce imports. This strategy has proven risky because it can provoke retaliation.

Bell Communications Research, Inc. (Bellcore)   The research arm primarily owned by the Bell Operating Companies.

Bell operating company (BOC)   The telephone service operating companies of the regional holding companies divested from AT&T. For example, the regional holding company BellSouth operates two BOCs: Southern Bell Telephone and South Central Bell Telephone.

Bilateral trade agreement   A formal or informal agreement between two nations.

Bit   A binary digit, the smallest unit of measurement in the transmission of data having only two values (0 or 1).

Bit error rate   The extent to which a digital transmission network generates an error. The rate can be reduced through error detection and correction.

Bit rate (also known as *throughput*)   The speed at which bits are transmitted, usually expressed in terms of bits per second. For example a reasonably fast modem can handle data at the rate of 14,400 bps.

Boomerang box   A device that provides dial tone to callers in another country, thereby providing them a virtual presence in a location with lower outbound international long-distance telephone rates.

Boresight   The center point on Earth where a satellite's radiated transmissions are strongest.

Bottleneck   A facility or portion of a route where traffic aggregates, often according the operator or service provider the opportunity to charge monopoly rates and to engage in anticompetitive practices.

Boycott   A refusal to deal commercially or otherwise with a person, firm, or country.

British Telecom (BT) The dominant local and international carrier in the United Kingdom that in 1994 acquired a 20% ownership share in MCI.

Build, operate, and transfer (BOT) A method for infrastructure development in which a foreign enterprise agrees to build a facility, operate it for a specified time period, and then transfer title to the national government or carrier who will then take over operations after having had time to develop operational and management expertise.

Build, own, and operate (BOO) A method for infrastructure development in which a foreign enterprise agrees to build and operate a facility. This arrangement creates a service franchise and the incentive for the operator to upgrade and maintain facilities without fear of nationalization.

Build, transfer, and operate (BTO) A method for infrastructure development in which a foreign enterprise agrees to build a facility, transfer title to the national government or carrier, and operate the facility at a profit for a specified time period after which a domestic carrier may take over operations. This arrangement enables the government to maintain a greater degree of facilities control than if a private foreign entity held title.

Bundled The process by which two or more possibly segregated features are offered jointly.

Bypass The use of alternative facilities or services to what incumbent carriers offer, such as the use of cable television systems or digital termination systems to access interexchange carrier facilities rather than the incumbent telephone company's wireline network.

Byte An intermediate unit of data transmission capacity composed of 8 bits.

C-band The portion of the radio spectrum (Earth to space at 6 GHz; space to Earth at 4 GHz) allocated for satellites providing service between fixed points or broadcast services.

Cablehead The oceanfront location where a submarine cable makes a landfall and where power, amplification, and multiplexing functions take place.

Call-back services Providing outbound international long-distance calling capabilities to customers in another country through the use of a device that processes a request for dial tone. Call-back service can be initiated by a conventional inbound international long-distance call that is terminated before triggering a toll charge.

Cellular radio A terrestrial radio service designed primarily for mobile applications using microwave transmitters whose low-powered operations enable frequency reuse and integrated service throughout a region.

Central office A telephone company facility that provides centralized management of switching, routing, and line transport functions that may traverse other facilities closer to the end user.

Circuit switching Physically linking two or more points via dedicated circuits.

Code-division multiple access (CDMA) A process for deriving more throughput and accommodating increasing demand for service by subdividing spectrum into various code sequences.

Collection rate The end-user charge for service, typically set out in a tariff.

Colocation The physical interconnection of lines and equipment owned and operated by different carriers typically on the premises of the major incumbent carrier.

Command, control, communications, and intelligence (C3I) Tactical requirements of the military.

Committee on the Peaceful Uses of Outer Space (COPUOS) A committee of the United Nations that addresses issues pertaining to space exploration, settlement, and orbiting objects, including satellites.

Commercialization of space Eliminating government subsidies and management so that private ventures can compete fully and fairly in space-related ventures.

Common carrier A legal and regulatory classification that requires a telecommunications facility or service provider to serve any user within a certificated geographical region and to provide service in a nondiscriminatory manner, typically through public tariffs.

Commonwealth of Independent States (CIS) An affiliation of sovereign states that formerly comprised the Soviet Union.

Communications Act of 1934 The primary U.S. law establishing the FCC and the general scope of broadcast and common carrier regulations.

Communications Satellite Act of 1962 The U.S. law that created Comsat Corporation as the sole signatory to INTELSAT and established general satellite policy.

Comparably efficient interconnection (CEI) The requirement of the FCC that the BOCs provide a plan demonstrating that they will not discriminate against unaffiliated enhanced-services providers when the BOC decides to provide similar services. While the *Second Computer Inquiry* required the BOCs to provide enhanced services through a separate subsidiary, the *Third Computer Inquiry* eliminated structural separate provided the BOCs filed a CEI plan for each enhanced service.

Comparative advantage A fundamental international trade concept that views nations or regions having the ability to produce certain goods or services more efficiently than other nations. By emphasizing production where a comparative advantage exists, nations can efficiently use available natural and human resources and trade for goods and services in which the nation has a comparative disadvantage.

Compatibility The ability of users and carriers to interconnect equipment, lines, and facilities while maintaining services with a reasonable degree of reliability and quality.

Complementary products Products that add to the value and utility of a product, such as sugar with coffee and modems with personal computers.

Compression The application of techniques for reducing the amount of frequency spectrum or channel capacity needed to derive a circuit. For example, 4 to 1 compression provides the means for deriving four channels where previously only one channel was available.

Comsat Corporation The sole U.S. signatory (investor) in INTELSAT and Inmarsat, which operates as a "carrier's carrier" by leasing wholesale satellite capacity to other carriers for subsequent retailing to end users.

Computer Inquiries A set of proceedings of the FCC beginning in the 1970s designed to erect a regulatory system that permits facilities-based carriers to enter enhanced-services markets without cross-subsidization by users of the carrier's basic services.

Conference of European Post and Telecommunications Administration (CEPT) An assembly of European postal and telecommunications regulatory authorities aiming to harmonize standards, policies, and initiatives for the ITU and other global forums. Diversifying constituencies and the establishment of a separate standard-setting body, the ETSI, are challenging the ability of CEPT to establish significant and uniform policies.

Conscious parallelism The deliberate matching of prices and services by carriers to avoid more aggressive competition and price wars.

Consent decree A remedy in an antitrust case in which the proceeding concludes without a verdict in exchange for an agreement by the defendant to refrain from continuing to engage in certain activities or practices. AT&T agreed to consent decrees in 1956 and 1982 to settle antitrust suits.

Consultation process Scheduled meetings of international carriers and government agencies to evaluate facilities demand and determine where and when to deploy additional capacity.

Consultative process　The requirement in Article XIV of the INTELSAT Agreement and Article VII of the Inmarsat Convention that parties to these agreements consult with the cooperative to ensure that another satellite system will not cause technical or economic harm.

Continental United States (CONUS)　Continental United States; the lower 48 states.

Convergence　The merger of technologies that previously served discrete markets leading to integrated, additional offerings; for example, computer terminals will serve entertainment applications (in addition to information processing) and television sets will serve information-processing applications (in addition to entertainment).

Coordination　The process by which nations, carriers, and other service providers meet to resolve potential conflicts, including the potential for radio interference. The INTELSAT Agreement requires nations that have agreed to become parties to the organization to demonstrate the absence of technical and economic harm to INTELSAT when authorizing separate satellite systems. The Radio Regulations of the ITU require nations to coordinate the use of frequencies and the satellite orbital arc.

Corporatization　The conversion of a government-owned carrier into a more businesslike enterprise, with or without a change in ownership.

Cross-subsidization　Using revenues accrued from one service to underwrite the provision of another service at less than fully compensatory rates.

Customer premises equipment (CPE)　Telecommunication equipment, including telephones and private branch exchanges, located on user premises.

Customer proprietary network information (CPNI)　Information about a customer's basic services usage that can provide marketing leads to enhanced-service providers. The *Computer Inquiries* established rules requiring the BOCs to withhold such information from enhanced-service affiliates or competitors unless authorized to do so by the user.

Data Network Identification Codes (DNIC)　ITU-recognized coding system for identifying and routing traffic to specific network operators.

Data over voice/data under voice (DOV/DUV)　A technique for transmitting voice and data over a single channel through multiplexing or other signal splitting technology.

De facto standard setting　The creation of a dominant or single standard by the interplay of market forces rather than promulgation of a standard through the standard-setting process.

De jure standard setting The formation of a standard by law or through the rulemaking process of the appropriate regulatory agency.

Dedicated Reserved for the use of one or more specified users. For example, a private line or satellite transponder may be dedicated for use by a particular lessee. Nondedicated capacity is provided on a virtual or on-demand basis.

Demand assigned multiple access (DAMA) A queuing method by which a larger number of users may share transmission capacity by allocating on an as-needed basis.

Destructive competition Short-term price competition at noncompensatory levels resulting in market exit by some enterprises and the potential for survivors to raise rates above competitive levels to recoup prior losses.

Deutsche Telekom (DT) The exclusive telecommunications service provider in Germany likely to privatize by 1998.

Digital The use of a binary form, composed of on and off pulses to represent the continuously varying signals of images and sounds. Information, entertainment, voice traffic, and other forms of communication can be encoded, stored, processed, and transmitted in a numeric as opposed to analog form.

Digital audio broadcasting (DAB) The transmission of audio signals in a higher quality digital format rather than the conventional analog method.

Digital European Cordless Telephone (DECT) A European standard for a new generation of cordless telephones and wireless private branch exchanges.

Digital termination systems (DTS) Digital microwave transmission systems that provide a bypass alternative to conventional local exchange carrier wireline facilities.

Digitization The use of computer-readable bit streams for transmitting, switching, and routing information.

Direct access The opportunity for end users and service providers to acquire satellite capacity directly from the operator, such as INTELSAT and EUTELSAT, instead of having to deal with an intermediary.

Direct broadcast satellite The use of a satellite to transmit programming, including audio and video, directly to end users equipped with receiving dishes and the necessary electronic components (also known as direct-to-home broadcasting).

Divestiture Severing part of a corporation and creating a separate business entity on a voluntary basis or as part of an antitrust settlement or verdict.

Dominant carrier   An FCC classification of common carriers who because of their market power require closer regulatory scrutiny. Such carriers have less flexibility to change tariff terms and conditions. The classification applies to AT&T, foreign-owned carriers on a route-specific basis, Comsat in its INTELSAT and Inmarsat signatory functions, and carriers operating the only Earth station in an off-shore point (e.g., Guam).

Downlink   The transmission of traffic from a satellite to receiving locations.

Duopoly   A monopoly shared by two enterprises.

E-1   A European standard for transmission capacity handling 2.048 Mbps.

Earth station   Terrestrial equipment used to transmit and receive satellite telecommunications. Some Earth stations provide receive-only functions.

Economies of scale   A measurement of economic efficiency that identifies who produces a good or service at the lowest per-unit cost and the optimal amount of production.

Economies of scope   A measurement for assessing how efficiently an enterprise will operate when serving adjacent markets (e.g., whether a telephone company can efficiently use its network to provide cable television and information services).

Elasticity of demand   The intensity of preference for a good or service based on user reaction to an increase or decrease in price.

Elasticity of supply   How consumers react to changes in the availability of a good or service.

Electronic data interchange   The use of telecommunications and information processing to conduct business transactions, often in an integrated network combining different media, such as voice, text, and data processing.

Electronic Data Systems (EDS)   A Texas-based company that provides systems integration and one-stop-shopping services in telecommunications and information processing.

Electronic funds transfer   The use of telecommunications and information processing to achieve a transfer of money, typically from one bank to another.

E-mail   Computer-mediated text communications.

End office   The telephone company switching facility closest to a particular user.

End-on-end routing   The segmentation of a route into two or more segments typically priced individually.

End-to-end routing   A complete traffic routing arrangement typically priced at one composite rate.

Enhanced services   Enhancements to basic common carrier transmission services involving computer processing that acts on the code, content, protocol, or format of the information in such a way as to change the output and possibly also to store it for subsequent retrieval and manipulation.

Enterprise networks   Diversified, customized, and complex international telecommunications and information networks primarily used by multinational enterprises to serve particular requirements. These networks often require the assistance of system integrators and outsourcers who plan, procure, and manage the network.

Erlang   A unit for measuring telecommunications traffic and capacity needed to support an acceptable level of service during the busiest usage times.

European Committee for Electrotechnical Standardization (CENELEC)   A regional standard-setting body in electrical and electronic matters.

European Committee for Standardization (CEN)   A regional standard-setting body in all fields except electrical and electronic matters.

European Economic Community (EEC)   A trade organization and governance structure for European nations.

European Free Trade Association (EFTA)   A trading association of European nations that are not part of the EEC, including the Nordic countries, Austria, and Switzerland.

European Telecommunications Standards Institute (ETSI)   A regional standard-setting body organized by the EC.

European Telecommunications Satellite Organization (EUTELSAT)   A regional satellite cooperative providing satellites for delivery of telecommunications and video programming.

European Union (EU)   The union of European nations formerly affiliated in a less integrated European Community.

Federal Communications Commission (FCC)   The expert regulatory agency created by federal law to allocate, allot, and assign spectrum and to oversee broadcasting and common carrier telecommunications.

Feeder link   A radio link to and from satellites for conveying information, including the tracking, telemetry, and network control needed to maintain a satellite in proper orbit.

Fixed satellite service The use of satellites to provide service between users at fixed locations.

Footprint The range of geographical coverage of a satellite transmission.

Forbearance The FCC-articulated concept of refraining from regulating certain common carriers lacking market power, thereby affording greater flexibility in pricing and providing service.

Foreign correspondent A carrier that has entered into a foreign operating agreement with another carrier, entitling it to originate and terminate international traffic.

Frame relay A new data transmission technology that quickly switches and routes digital packets with a low bit error rate.

Fully distributed costs (FDC) Costing that includes a contribution to fixed costs by all consumers.

Future Public Land Mobile Telecommunication Service (FPLMTS) A concept for accommodating the growing telecommunications requirements of mobile users with additional spectrum allocations and new services. Terrestrial systems will provide most of FPLMTS, augmented by some satellite operations.

Gateway A satellite Earth station or submarine cablehead where domestic facilities access international transmission facilities.

General Agreement on Tariffs and Trade (GATT) A multilateral agreement by most nations to reduce barriers to trade.

General Agreement on Trade in Services (GATS) The process for extending GATT trading principals to services in addition to goods.

Geostationary The location above the Earth where launched objects appear stationary relative to Earth. Satellites become geostationary at 22,300 miles in altitude. Placing satellites in geostationary orbit above the equator maximizes the scope of geographical coverage.

Gigahertz (GHz) A measure of radio frequency equivalent to 1 billion hertz (cycles per second).

Global information infrastructure (GII) A concept that globally extends national initiatives to promote widespread deployment and access to broadband digital technologies and services requiring high-speed digital transmission (e.g., video programming, telemedicine, large data file transport, and Internet access).

Global Systems for Mobile Communications (GSM) A proposed standard for digital cellular radio systems that originated in Europe.

Globalization Expanding the trading, marketing, and operational scope of an enterprise; for example, PTTs facing lost market share domestically as a result of market entry may seek market share in foreign markets.

Green paper A vehicle for publishing future policy objectives. The Commission of the European Community prepared a green paper in 1987 articulating telecommunication harmonization goals as part of the Single Europe Act.

Gross national product (GNP) An aggregate measure of income generated from all goods and services.

Half circuit The smallest unit of international transmission capacity matched by foreign correspondents to route traffic.

Harmful interference Spectrum usage that endangers, degrades, or deteriorates the proper functioning of another registered spectrum use.

Heads of Agreement A preliminary agreement by governments and business enterprises articulating general terms followed by a more comprehensive document.

High-definition television (HDTV) The development of a video transmission and production standard calling for higher resolution and better sound quality.

High-Level Committee (HLC) An ad hoc committee established by the Administrative Council of the ITU in response to a Plenipotentiary Conference held in 1989 seeking recommendations on ways to streamline the ITU, make it more responsive to consumer requirements, and improve operations. The HLC reported its findings to a special meeting of the ITU, which adopted most recommendations.

Hubbing The aggregation of traffic from throughout a region, thereby achieving circuit-loading efficiency and economies of scale. Hubbing in telecommunications is analogous to airport hubbing, where long-haul routes are loaded from a number of short-haul traffic originating in nearby localities.

Inband signaling The customary process of including switching and routing information as headers preceding the call, increasingly replaced by out-of-band signaling.

Incentive regulation Alternatives to the conventional rate base regulation designed to reward innovation and efficiency by allowing carriers to capture financial gains rather than automatically flow them to ratepayers through refunds or lower rates.

Inclined orbit Satellite orbits that conserve station-keeping fuel to extend the operational life of a satellite by allowing it to deviate slightly from a geosynchronous orbit, thereby requiring Earth station tracking.

Incumbent carriers Carriers that have installed an extensive network and heretofore have faced limited, if any, competition.

Indefeasible right of user (IRU) A method for conveying the rights, but not the title, to use international telecommunications capacity.

Infrastructure An essential system to the public health and welfare, such as water, sewage, electricity, currency, telecommunications, and roadways.

Institute of Electrical and Electronics Engineers (IEEE) An international organization that participates in the telecommunication standard-setting process.

Integrated services digital network (ISDN) A telecommunications standard envisioning interconnected digital networks capable of simultaneous delivery of voice and data services.

Intelligent network The architecture and plans for a future telecommunications infrastructure with advanced features made possible by expanded network switching, signal processing, and intelligence.

Inter-American Telecommunications Conference (CITEL) An assembly of nations that participate in the Organization of American States, with a mission to address radio and telecommunications issues, including ones to be addressed by upcoming conferences of the ITU.

Interactive The ability of users to request, manipulate, process, and change data through online commands. Interactivity converts one-way information sources into two-way media.

Interconnect The physical connection of equipment and lines to secure a complete route and service arrangement. Users physically interconnect CPE to a telecommunications network through a plug physically attached to a jack.

Interdepartment Radio Advisory Committee (IRAC) A committee, organized under the auspices of the Commerce Department and composed of 20 to 25 representatives from various U.S. federal agencies, with a mission to coordinate spectrum management issues affecting the U.S. government.

Interexchange carrier (IXC) A provider of long-haul telecommunication services.

Interexchange telecommunications Long-haul services that link local exchanges. In the United States, the Modification of Final Judgment precludes BOCs from providing those interexchange services that cross local access and transport boundaries.

Interface A shared boundary, for example, between CPE and the telecommunication plant owned and operated by a telephone company.

Intermediate circular orbit (ICO) The installation of satellites into nongeostationary orbits closer to Earth to support applications using small, low-powered handheld terminals.

International Civil Aviation Organization (ICAO) A specialized agency of the United Nations that establishes rules, standards, and policies affecting international aviation, including navigation and telecommunications issues.

International comity Recognition that each nation has sovereignty and that mutually beneficial results accrue when nations relinquish a degree of independent decision making to reach a single consensus policy.

International Frequency Registration Board (IFRB) The part of the ITU's radiocommunications sector, now known as the Radio Regulation Board, that registers spectrum and orbital arc uses and resolves potential interference problems.

International Maritime Satellite Organization (Inmarsat) A cooperative of over 70 nations organized to provide ubiquitous maritime services, including safety and distress functions. The parties of Inmarsat recently endorsed an expanded function to include aeronautical and land mobile services on an ancillary basis.

International message telephone service (IMTS) Conventional dialup international long-distance telephone service.

International Radio Consultative Committee (CCIR) Committee of the ITU reformulated in 1993 with portions of its portfolio assigned to the newly formed radiocommunications and telecommunications standards sectors.

International record carrier (IRC) International carriers that concentrated on providing textual services such as telegraphs and telexes.

International Settlements Policy (ISP) An FCC policy requiring all U.S. carriers to apply identical accounting and settlement rates for IMTS. The FCC considers the ISP essential to prevent foreign carriers from whipsawing U.S. carriers.

International Standards Organization (ISO) A specialized agency of the United Nations with a broad standard-setting portfolio that includes data processing.

International Telecommunication Satellite Organization (INTELSAT) A cooperative of over 120 nations organized to provide ubiquitous international satellite capacity.

International Telecommunications Union (ITU) A specialized agency of the United Nations that formulates policy, regulations, and recommendations in telecommunications, including spectrum allocation, satellite orbital arc registration, telecommunication development, standard setting, and conflict avoidance and resolution.

International Telecommunications Users Group (INTUG)   A trade association of international telecommunication users.

International Telegraph and Telephone Consultative Committee (CCITT) Committee of the ITU reformulated in 1993 with portions of its portfolio assigned to the newly formed radiocommunications and telecommunications standards sectors.

International value-added network (IVAN)   An enterprise that enhances leased private lines with customized features (e.g., credit card verification and airline reservation systems).

Internet   A multifaceted international network of databases and e-mail users organized initially for military, scientific, and academic users.

Intersatellite link (ISL)   Extremely high-frequency communication links between satellites that substitute for a round trip from a satellite down to an Earth station and back up to another satellite.

Intersputnik   An international satellite cooperative organized by the former Soviet Union.

Japan Approvals Institute for Telecommunication Equipment   The centralized organization for certifying that equipment satisfies Japanese telecommunication equipment standards.

Ka-band   That relatively unused portion of the radio spectrum (30-GHz uplink/20-GHz downlink) allocated for new satellite requirements.

Kilohertz (KHz)   A measurement of radio waves equal to 1,000 cycles per second.

Kokusai Denshin Denwa (KDD)   The dominant Japanese international carrier.

Ku-band   An increasingly used portion of the radio spectrum (14-GHz uplink/11- to 12-GHz downlink) allocated for satellite requirements.

L-band   That portion of the frequency band (1–3 GHz) where a variety of mobile services are provided by geostationary and LEO satellites.

Leaky PBX   The ability of users to interconnect on-premises equipment to the public switched telecommunications network, thereby linking private lines designed to link two points only with often unmetered switched local services.

Lesser developed country (LDC) A classification of nations with economies and telecommunications infrastructure below that of industrialized nations; also known as *newly industrialized nations.*

Level competitive playing field The goal of market entrants who claim that regulators and other decision makers must take affirmative steps to reduce unfair market access opportunities accruing to incumbent carriers as a function of their earlier market entry, bottleneck control, customer base, and regulatory status.

Liberalization The relaxation of rules and service obligations, such as subsidized and underpriced local services, imposed on incumbent carriers, often occurring contemporaneously with privatization, deregulation, and market entry initiatives.

Line-side interconnection Interconnection that takes place at a low level in the switching hierarchy typically for end users.

Link budget The determination of whether an adequate signal can be received by calculating the transmission power and factors that attenuate the signal.

Link margin The estimated degree of signal strength above noise at a particular location under various circumstances.

Local access and transport area (LATA) A geographical region sharing common cultural, social, and economic interests within which the divested BOCs may provide local and interexchange services.

Local area network A network of work stations and personal computers linked via wiring installed by the operator or provided by the telephone company or via wireless applications.

Local exchange carrier (LEC) The provider of local services, often considered a natural monopoly and bottleneck through which most interexchange traffic traverses.

Local loop Local exchange facilities comprising the first and last legs of a telecommunications route.

Long-run incremental cost (LRIC) A measure of the additional cost incurred over the long term to provide a particular amount of a good or service. This measure accounts only for the additional costs incurred without regard to embedded, fixed costs that could be shared by additional users.

Low Earth orbit Nongeostationary orbit for satellites primarily providing mobile telecommunications to lightweight transceivers. LEO satellites operate only a few hundred miles above the Earth surface, thereby reducing launch costs and the power needed to transmit to and from the satellites. On the other hand, the closer proximity to Earth requires a larger number of satellites to achieve desired regional or global coverage.

Low-Earth-orbiting satellites (LEOs)   Satellites operating in orbits below the geostationary orbiting arc, where service can be provided to low-powered handheld terminals. Big LEOs provide a variety of voice and data services while small LEOs provide data and emergency position location services.

Market power   The ability to affect the price or supply of a good or service.

Memorandum of understanding (MOU)   A preliminary agreement setting forth basic terms and conditions typically followed up with a more comprehensive agreement.

Message telephone service (MTS)   Conventional dial-up voice telephone service, also known by the acronym POTS.

Middle Earth orbit (MEO)   Nongeostationary satellite orbit between low Earth orbits below and geostationary orbits above, where operators attempt to balance the benefits accruing from operating closer to Earth (e.g., access by lower-powered terminals)and operating high enough to provide broad geographical coverage.

Midocean   The concept that international half circuits are matched by foreign correspondents at the centerpoint of a submarine cable.

Minimum assignable unit of ownership (MAUO)   The smallest unit of submarine cable capacity, typically a 64-Kbps circuit with associated capacity for multiplexing, allocatable to an investor or available to others through an Indefeasible Right of Use conveyance.

Ministry of posts and telecommunications (MPT)   The government agency in many nations with telecommunications operational and regulatory responsibilities.

Ministry of Trade and Industry (MITI)   The   powerful   Japanese   ministry, which has trade and development portfolios.

Minitel   A diverse array of information services provided via France Telecom using a terminal device initially distributed to telephone subscribers in lieu of up-to-date printed directories.

Mobile-satellite service (MSS)   An ITU-recognized service designation for the provision of voice and data services to mobile users via geostationary or nongeostationary orbiting satellites.

Modem   A device that modulates and demodulates data signals enabling computers and other digital devices to operate over an analog network.

Modification of Final Judgment (MFJ)   Modifications to the 1956 consent decree agreed to by AT&T and the Justice Department, thereby settling an antitrust suit. The MFJ required AT&T to divest its BOCs in exchange for the opportunity to serve data processing markets. The spun-off BOCs were limited to local exchange services.

Most favored nation (MFN) A basic trade principle requiring nations associating with the GATT to extend to all nations any market access benefit accorded other nations.

Multimedia The integration of various media previously considered available via a separate pipeline or marketing channel (e.g., using the television for entertainment and new interactive data processing, consumer order entry, and utility monitoring).

Multinational enterprise A venture doing business in a number of nations, making it a candidate for turnkey services from systems integrators and outsourcers.

Multiplexing Subdividing a circuit into more than one channel to derive more throughout and capacity.

Multipoint Using telecommunication services to serve more than one physical location (e.g., originating a video program for delivery to numerous cable systems using the broad geographical coverage of a satellite).

National hero A domestic carrier or manufacturer benefiting from policies designed to protect it from foreign competition, including closed procurements and other policies that handicap or prohibit market entry.

National Information Initiative (NII) A plan for using government to stimulate primarily private ventures that will upgrade and expand the telecommunications and information-processing infrastructure in the U.S. The government initiative includes a vision for ubiquitous access to a feature-rich information highway.

National Institute of Science and Technology (NIST) An agency of the U.S. Department of Commerce that coordinates standard setting for the federal government and whose standards affect the parallel private standard-setting process.

National Telecommunications and Information Administration (NTIA) The U.S. executive agency, located in the Department of Commerce, that serves as the President's researcher and advisor on telecommunications and information-processing policy issues.

National Television Standards Committee (NTSC) The developer of a broadcast color television standard in the U.S. that is implemented in North America and Japan.

National treatment   Subjecting foreign manufacturers and service providers to the same regulatory treatment as are afforded domestic enterprises.

Natural monopoly   A single manufacturer or service provider who singularly can operate most efficiently and who can maximize scale economies.

Newly industrialized country (NIC)   A nation with an expanding economy and typically with growing telecommunications requirements.

New World Information Order (NWIO)   A vision for establishing closer parity of access to information between developed and developing nations.

Nippon Telephone and Telegraph (NTT)   The dominant local-exchange and long-distance carrier in Japan.

Node   A point in a telecommunication or information-processing network configuration where two or more lines, routes, or pathways come together (e.g., at a switch, Earth station, or private branch exchange).

Noncommon carrier   A classification that reduces or eliminates regulation in recognition of the carrier's lack of market dominance and the nonessentialness of the services offered.

Nontariff barrier (NTB)   Barriers to trade that are not documented in a customs duty or tariff.

North American Free Trade Agreement (NAFTA)   A trade agreement between the U.S., Canada, and Mexico resolving to reduce or eliminate barriers to trade between these nations, with an eye toward creating a regional trading bloc.

Number portability   The ability to designate a single telephone number for accessing individuals with several telecommunication devices operating in different locales.

Off-shore branching unit   A device located on the ocean floor that is used to interconnect submarine cables, thereby expanding coverage and providing for redundant or alternative routing of traffic.

Office of Telecommunications (Oftel)   The U.K. independent regulatory agency.

One+ dialing   A calling arrangement for direct connections with a prearranged billing commitment.

One-stop shopping   The provision of a number of functions, such as telecommunications network, design, procurement, and management, by a single enterprise as an alternative to the end user performing these functions or securing the services from a number of enterprises.

Open network architecture (ONA) The FCC-articulated blueprint by which the BOCs will revamp their facilities to provide for equal access to network facilities by affiliated and unaffiliated enhanced-service providers.

Open network provision (ONP) The EC-articulated blueprint for harmonizing local exchange facilities access within the EC.

Open systems interconnection (OSI) A seven-level model for organizing data processing and telecommunication functions.

Orbital arc Locations above Earth where satellites are located after launch. Most communication satellites occupy the geostationary orbital arc to maximize stability and geographical coverage.

Organization for Economic Cooperation and Development (OECD) An organization that supports shared views on economic development, including compiling statistics and polling member nations on telecommunications and information policies.

Organization of Petroleum Exporting Countries (OPEC) An organization seeking to manage the price and supply of petroleum.

Out-of-band signaling The use of a channel separate from the one carrying traffic to set up, supervise, and bill the call.

Packet switching A transmission technology that reduces messages and data into individually routed packets and reassembles them before reaching the final destination.

PanAmerican Satellite (PanAmSat) The first licensed U.S. facilities-based international satellite carrier to compete with INTELSAT.

Party A nation that has acceded to an international agreement, thereby accepting its terms.

Personal communication services (PCS) A variety of new terrestrial and satellite-delivered services for handheld terminals operating at low power.

Phased alternation by line (PAL) A broadcast color television standard implemented in Europe and most of Asia.

Plain old telephone service (POTS) Conventional telephone services provided to residential and small-business users.

Point-to-multipoint Traffic routing from a single location to many recipients (e.g., video programming distribution to numerous cable operators).

Post, telegraph, and telephone administration (PTT) The incumbent and often exclusive carrier in most nations that provides services on a ubiquitous basis, some of which may be subsidized.

Power flux density A measure of power radiated from a transmitter.

Preemption Assertion of jurisdiction by one regulatory authority that would replace and dislodge the assertion of jurisdiction by another agency.

Price cap A form of incentive regulation in which carriers and regulators agree to replace conventional rate base regulation for a cap on rates and the requirement that they drop by a certain percentage each year to reflect productivity improvements as offset by a measure of overall increased costs to producers. Carriers are permitted to capture all or part of increased profits.

Primary rate interface The large standard unit of capacity in an ISDN composed of twenty-three 64-Kbps bearer channels and one 16-Kbps data channel totaling 1.54 Mbps.

Primary status The designation for a particular radio service that qualifies users for maximum permissible protection from interference by users of services holding a lower priority.

Private branch exchange (PBX) A telephone line switching device located on the premises of users with a number of telephones and lines. The PBX switches and routes all inbound and outbound calls.

Private line Dedicated capacity designed to link a single user with requirements in two international locations. Carriers now provide "virtual" private lines by partitioning capacity from public networks through the application of software.

Privatization Wholly or partially converting the incumbent PTT to private ownership. The privatized enterprise may maintain a monopoly or face varying degrees of competition from market entrants.

Protocol A standard operating procedure or format.

Public switched telecommunications network (PSTN) The publicly available local and long-haul facilities of the incumbent carrier.

Public telecommunication organization (PTO) A new designation for PTT administrations.

Ramsey pricing Charging users on the basis of their demand elasticity (i.e., users with plenty of options, including dedicated facilities and leased lines from a number of carriers, qualify for rates below that charged users with fewer options).

Recognized private operating agency (RPOA) An ITU-approved designation of nongovernmental telecommunication service providers that confers official recognition and enhances the ability of such organizations to participate in ITU forums.

Redundancy The availability of backup capacity to restore service in the event of a facility outage or peak demand condition.

Regional African Satellite Communication System (RASCOM) An affiliation of African telecommunications administrations with plans to operate a dedicated satellite system after having leased capacity from INTELSAT.

Regional Bell operating company (RBOC) Upon divestiture from AT&T, the Bell operating companies were reformulated into seven new corporations representing a particular geographical region of the U.S.

Request for proposal (RFP) A tender offer soliciting bids by parties interested in securing a contract to perform some form of work.

Resale The acquisition of bulk transmission capacity and other services for subsequent resale to individual users who singularly do not generate the demand for large-capacity offerings. Resellers perform an arbitrage function and profit by acquiring discounted bulk capacity from underlying facilities-based carriers and repackaging it at rates less than what an individual user could secure directly from the carrier.

Research and Development in Advanced Communications in Europe (RACE) An EC initiative to expedite and emphasize telecommunications research and development.

Roamer An individual desiring to use a mobile telecommunication device (e.g., a cellular radiotelephone) while away from the location where service is usually provided.

Roll-off The manner in which signal strength drops off at locations increasingly distant from the targeted service location.

Rulemaking The process by which a U.S. administrative or regulatory agency establishes a binding rule, regulation, or policy. Procedural due process requirements obligate the agency to notify the public of proposed actions and to provide opportunities for participation through filed comments or testimony.

Satellite beam types Global-transmission beams have the maximum geographical illumination, approximately one-third of the entire globe; hemispheric-transmission beams are shaped to cover an entire hemisphere (e.g., North, Central, and South America or Europe); zone-transmission beams are shaped to cover one or more portions of a hemisphere with higher signal strength; spot-transmission beams are shaped to concentrate signal strength over relatively small but highly populated locations.

S-band The portion of the frequency spectrum around 2 GHz used for terrestrial microwave and some mobile-satellite applications.

Sender keep all   A toll revenue division arrangement whereby the originating carrier keeps all charges.

Sender pay all   A charging mechanism whereby the calling party pays all charges.

Separate system   A satellite system that is separate from the INTELSAT cooperative and provides international services.

Sequential color with memory (SECAM)   A French-developed broadcast television standard adopted in France, French-speaking African nations, and the former Soviet Union.

Settlement rate   The unit of currency and percentage division negotiated by carriers for settling accounts regarding carriage of inbound international traffic. Typically, international carriers on direct routings equally divide the negotiated rate.

Signaling system 7 (SS7)   An advanced method for providing software-defined signaling necessary to set up telecommunication links.

Signatory   A company or government that has executed a binding legal commitment (e.g., Comsat serves as the sole U.S. signatory to INTELSAT and Inmarsat).

Single Europe Act   Legislation adopted by the constituent members of the European Community agreeing to closer collaboration and harmonization of laws, policies, and regulations.

Smart card   A credit card–sized instrument containing microchips and associated electronics for providing credit and other data for facilitating transactions.

Software-defined network   The use of software to partition transmission capacity from public networks for private use, making it appear as the functional equivalent to capacity dedicated for a single user.

Sovereignty   Recognition of independence and legitimacy in a nation's determination of policies, laws, and regulations affecting its citizens and internal affairs.

Space station   A satellite or other device that operates in space.

Special drawing rights (SDR)   An aggregation of national currencies administered by the International Monetary Fund and often used by international carriers in settling toll revenue accounts.

Spectrum   A range of frequencies used in telecommunications.

Spectrum allocation   The designation of a frequency band to one or more specified services by the ITU and by individual nations.

Spectrum allotment The designation of specific channels of operation to particular localities.

Spectrum assignment The grant of operational authority that includes a license to operate on a specific frequency.

Spread spectrum A transmission technique whereby signals are transmitted over a range of spectrum at low power with messages handled by assigning a discrete sequence of code or by hopping over different frequencies.

Station keeping The use of jets aboard a satellite to keep it "on station" (i.e., in the prescribed orbital location with antennas properly aimed). A station-kept satellite eliminates costly tracking requirements for each Earth station pointing at the satellite. Satellites typically reach end of life when station-keeping fuel has been used up, even though the electronic and power generation components remain operational. Operators may conserve station-keeping fuel to extend a satellite's usable life through the use of maneuvers that allow the satellite to drift slightly.

Store and forward The use of computerization to store traffic temporarily for subsequent transmission in batches or efficiently loaded packets. A satellite may store messages until it reaches a point in its orbit where the traffic can be downlinked.

Strategic alliances Corporate business relationships designed to tap the comparative advantages and expertise of the participants as well as to achieve a broader geographical marketing presence.

Submarine Cable Landing License Act An Act of Congress requiring the FCC and State Department to grant licenses to land a submarine cable on U.S. soil. Foreign carriers will receive such grants only if U.S.-owned and operated submarine cables can make landfall in the foreign country.

Substitutable products Products that can replace and compete with each other (e.g., tea instead of coffee and Centrex instead of private branch exchanges).

Switch The portion of the telecommunications infrastructure where traffic is received by a device that identifies the intended destination and selects the routing for delivering traffic onward to that point.

Synchronous Optical Network (SONET) Protocols for operating fiber-optic transmission facilities at very high multimegabit speeds.

T-1 A unit of transmission capacity equal to 1.544 Mbps in throughout.

Tail circuits Domestic facilities used to transport traffic from an international caller to a gateway and from another gateway to the call recipient.

Tandem switch A telecommunication switching facility with management control over one or more end offices.

Tariff A contract for service setting out the terms and conditions under which the general public will secure service.

Telecommunications Development Bureau (BDT) A permanent organ of the International Telecommunication Union formed in 1989 to promote telecommunication development, including technical and managerial training.

Teledensity The penetration of telecommunication access lines typically measured per 100 inhabitants.

Telepoint A new generation of pay telephones and one-way wireless communications based on microcellular radio technology.

Teleport A satellite Earth station with extensive access to terrestrial facilities and a broad base of users situated on real estate developments adjacent to the Earth stations or at other locations.

Teletext Slow-speed transmission of textual information over the vertical blanking intervals of broadcast television signals.

Television receive-only (TVRO) Satellite dishes equipped for one-way reception of television signals.

Throughput The amount of data carried over a particular amount of capacity in a specified time span (see also bit rate).

Time-division multiple access (TDMA) A technique for deriving more throughput and accommodating increased demand by assigning channels to users in discrete blocks of time.

Traffic Messages, programming, files, and other intelligence requiring transport from one location to another.

Transborder data flow The transmission of data across national borders, alternatively viewed as a vehicle to share the wealth of information or a source for cultural imperialism.

Transceiver A device that contains components that transmit and receive radio signals.

Transiting The carriage of international traffic via the facilities of an intermediary third country between the sender and recipient.

Transponder The components in a satellite able to transmit and receive traffic. Satellite capacity is typically stated in the number of available 36-MHz equivalent transponders.

Treaty of Rome Treaty enacted in 1957 that created the European Community.

Trunking A computer-assisted process for queuing channel demands, thereby more efficiently using spectrum and accommodating episodic demands for service. For example, a trunked radio system will provide communication channels for users on an as-demanded basis from the available inventory.

Trunk side interconnection The interconnection of facilities and lines between carriers at a level higher up the hierarchy than typically accorded end users.

Unbundled Separation of services and equipment into discrete elements available on an individual, a la carte basis.

United States international service carrier (USISC) A U.S. carrier licensed by the FCC to provide international services.

United States Trade Representative (USTR) The U.S. executive branch agency responsible for engaging foreign nations in bilateral and multilateral trade negotiations.

Uplink The process of transmitting traffic from Earth to a satellite.

Value-added network (VAN) An enhanced-services provider that typically leases transmission capacity and adds value by performing customized services (e.g., credit card verification, electronic funds transfer, etc.).

Very-small-aperture terminal (VSAT) Very small satellite Earth stations, typically situated on user premises, which promote diversification in services, easier access, and lower costs.

Video dial tone The provision of transparent common carrier transmission capacity for delivery of video programming;

Videotext Slow-speed transmission of textual information over closed circuits.

Virtual networks Software-derived networks typically involving lines partitioned from public switched facilities.

Virtual private line Using software to partition capacity from switched public facilities to make it function the same as capacity dedicated to a single user.

Whipsawing The practice of a monopoly carrier to extract concessions from foreign correspondent carriers operating in a competitive market who must vie for return, inbound traffic from the monopoly carrier.

Wide area network (WAN) An interconnected network of personal computers and work stations situated in different locations.

Wide area telephone service (WATS) Toll-free calling for the call originator provided by many retailing and service organizations.

Windowing Segmenting the distribution of a product such as a movie into sequences based on consumer demand elasticities and time from initial availability. Movie producers attempt to extract the greatest amount of rent by calibrating availability and charges (e.g., $6–$8 at a theater, $2–$3 on rented video tape, and no charge to viewers on broadcast television).

Wireless local loop The use of unwired transmission facilities to provide connections to the wireline public switched telecommunications network. Many of these technologies also provide mobile services.

Wireless networks The use of radio technology to serve increasingly diverse applications, including mobile services and some applications heretofore provided via wireline facilities.

Wireline The conventional wire-based telecommunications infrastructure installed and managed by incumbent carriers.

World Administrative Radio Conference/World Radio Conference (WARC/WRC) Periodically convened conferences of ITU member nations to address regional or global spectrum planning and policy issues.

World Administrative Telegraph and Telephone Conference (WATTC) A periodically convened conference of ITU-member nations to address issues pertaining to the rules of the road in telephony and other telecommunication services.

World Trade Organization (WTO) The permanent trade policy and dispute resolution forum created in 1994 by the nations that had formed GATT.

X.25 An ITU-recognized protocol for user access to data networks.

X.75 An ITU-recognized protocol for interconnecting data networks.

X.400/X.500 A series of protocols governing electronic mail and directory functions.

X-band A portion of the frequency spectrum in the 7- to 8-GHz range used primarily for satellites with defense and intelligence gathering applications.

Zero+ dialing A calling arrangement involving verification of credit or payment through intervention by an operator or computer.

About the Author

Robert M. Frieden is an associate professor at Pennsylvania State University, where he teaches telecommunications management, law, and economics. He also provides legal, management, and market forecasting consultancy services in such diverse fields as personal and mobile communications, satellites, and international telecommunications business development.

Before accepting an academic appointment, Mr. Frieden served as deputy director of international relations for Motorola Satellite Communications, Inc. In this capacity, he provided a broad range of business development, strategic planning, policy analysis, and regulatory functions for the IRIDIUM mobile satellite venture.

Mr. Frieden has held senior policy-making positions in international telecommunications at the Federal Communication Commission and the National Telecommunications and Information Administration. In the private sector, he practiced law in Washington, D.C., and served as assistant general counsel at PTAT System, Inc., where he handled corporate, transactional, and regulatory issues for the nation's first private underseas fiber-optic cable.

Mr. Frieden holds a B.A. with distinction from the University of Pennsylvania (1977) and a J.D. from the University of Virginia (1980). His contributions have been recognized by *Who's Who in American Law* and *Who's Who in American Education*.

Index

Spectrum management, 239–55
  domestic, 246
  FCC and, 251–53
  international, 246–50
  NTIA and, 253–54
  scarcity role in, 242–46
  See also Spectrum(s)
Standards
  broadcast television, 96–97
  dueling, 89
  end-to-end, 83
  function of, 83
  global, failure, 87
  implementation benefits, 84
  incompatible, 93
  interface, 83
  international consensus, 93
  single, 89
Standard setting, 83–99
  complexity in, 85–87
  ETSI and, 94
  global interconnection and, 86
  international, 87, 91–93
  ISDN and, 126
  Japan and, 95
  market vs. national sovereignty and, 98–99
  models, 89–91
    new, 90–91
    traditional, 89
  in multimedia environment, 98
  in multinational environment, 98
  national industrial policies and, 85
  national strategies, 87–88
  network equipment design and, 85
  perspective, 84–85
  products, 91
  "pure private goods" and, 88
  regional bodies, 94–95
  See also Standards
Strategic alliances, 262–63, 269
  carrier franchises, 275
  failure of, 278–80
  global, 283–85, 359
    challenges, 285
    new ventures, 284–85
    outcomes, 285–86
    types of, 283–84
  government-created new market niches, 276
  incumbents and, 275
  multinational, 273–81
  multiple-carrier, 278

new carrier franchise opportunities, 276–78
new facilities, 275–76
outsourcing/one-stop shopping, 275
proliferation of, 365
structuring, 280–81
successful, 280
types of, 275–78
Strategic partners, 365
Submarine cables, 53–54, 101
Sweden, 288
Synchronous optical networking (SONET), 363

Technologies, 49–57
  compression, 353
  convergence of, 352–54
  developments, 50–54
  "enabling, 54–55
  future, 56–57
  innovations, 353
  ISDN, 125
  management, 7
  policies, 3
  satellite, 189–90
  spectrum compromises, 240
Telecommunications
  anticompetitive tactics in, 160–62
  as commodity subject to narrowing margins, 358–59
  deregulation case studies, 287–310
  equipment and services, 151–70
  equipment manufacturing industry, 93
  foreign investment in, 153
  information services and, 113–14
  infrastructure, 173
  integrated networks, 25
  in international development, 173–84
  investment payoff, 177–79
  liberalization of, 24
  network services, trade in, 153–55
  new world order, 22–25
  policy proposals, 165
  privatization in, 259–86
  system diffusion, 174
  traffic balance, 134
  value-added, 114
  See also International telecommunications
Telecommunications development, 173–84
  assistance, 176–79
  developing country solutions, 179
  failed strategies, 176–77
  investment payoff, 177–79
  ITU role in, 183–84

Recent Titles in the Artech House
Telecommunications Library

Vinton G. Cerf, Senior Series Editor

Understanding Networking Technology: Concepts, Terms, and Trends, Second Edition, Mark Norris

Understanding Token Ring: Protocols and Standards, James T. Carlo, Robert D. Love, Michael S. Siegel, and Kenneth T. Wilson

Videoconferencing and Videotelephony: Technology and Standards, Second Edition, Richard Schaphorst

Visual Telephony, Edward A. Daly and Kathleen J. Hansell

Wide-Area Data Network Performance Engineering, Robert G. Cole and Ravi Ramaswamy

Winning Telco Customers Using Marketing Databases, Rob Mattison

World-Class Telecommunications Service Development, Ellen P. Ward

For further information on these and other Artech House titles, including previously considered out-of-print books now available through our In-Print-Forever® (IPF®) program, contact:

Artech House	Artech House
685 Canton Street	46 Gillingham Street
Norwood, MA 02062	London SW1V 1AH UK
Phone: 781-769-9750	Phone: +44 (0)20 7596-8750
Fax: 781-769-6334	Fax: +44 (0)20 7630-0166
e-mail: artech@artechhouse.com	e-mail: artech-uk@artechhouse.com

Find us on the World Wide Web at:
www.artechhouse.com